THE LOGIC OF MICROSPACE

THE SPACE TECHNOLOGY LIBRARY

Published jointly by Microcosm Press and Kluwer Academic Publishers

An Introduction to Mission Design for Geostationary Satellites, J. J. Pocha
Space Mission Analysis and Design, 1st edition, James R. Wertz and Wiley J. Larson
**Space Mission Analysis and Design,* 2nd edition, Wiley J. Larson and James R. Wertz
**Space Mission Analysis and Design,* 3rd edition, James R. Wertz and Wiley J. Larson
**Space Mission Analysis and Design Workbook,* Wiley J. Larson and James R. Wertz
Handbook of Geostationary Orbits, E. M. Soop
**Spacecraft Structures and Mechanisms, From Concept to Launch*, Thomas P. Sarafin
Spaceflight Life Support and Biospherics, Peter Eckart
**Reducing Space Mission Cost,* James R. Wertz and Wiley J. Larson

*Also in the DoD/NASA Space Technology Series (Managing Editor Wiley J. Larson)

The Logic of Microspace

by

Rick Fleeter

AeroAstro Inc.

Space Technology Library

Published Jointly by

Microcosm Press

El Segundo, California

Kluwer Academic Publishers

Dordrecht / Boston / London

Library of Congress Cataloging-in-Publication Data

A C.I.P Catalogue record for this book is available from the Library of Congress

ISBN 1-881883-11-6 (pb) (acid-free paper)
ISBN 0-7923-6028-1 (hb) (acid-free paper)

Published jointly by
Microcosm Press
401 Coral Circle, El Segundo, CA 90245-4622 USA
and
Kluwer Academic Publishers,
P.O. Box 17, 3300 AA Dordrecht, The Netherlands.

Kluwer Academic Publishers incorporates
the publishing programmes of
D. Reidel, Martinus Nijhoff, Dr. W. Junk, and MTP Press

Sold and distributed in the USA and Canada
by Microcosm
401 Coral Circle, El Segundo, CA 90245-4622 USA
and Kluwer Academic Publishers,
101 Philip Drive, Norwell, MA 02061 USA

In all other countries, sold and distributed
by Kluwer Academic Publishers Group,
P.O. Box 322, 3300 AA Dordrecht, The Netherlands.

Printed on acid-free paper

Page layout and editing by T. L. and F. S. Ponick
"Microspacecraft" illustrations by Heidi Given

Printed in the United States of America

Table of Contents

Part II: Missions and Reliability
Section 2: Reliability 209

To the microspace missions none of us have thought of yet,
and the creative people who will.

Preface

Welcome to *The Logic of Microspace*

I'd like to say welcome to a brand new book, but in the interest of keeping my closet skeleton-free, I'll admit it's not entirely new. But neither are small satellites. While microspace practitioners might build a satellite in a year or even six months, it still takes a year or more of brainstorming to really get the bugs out of the mission concept, bureaucracies need years to get them under contract, and the launching process is a bit more than a matter of stepping up to the turnstile with the appropriate number of quarters to drop into the slot. Or maybe that's a pretty good model, since it would take a few years (four is my estimate) to drop ten million quarters into a slot. Books are not that different — writing them takes forever minus epsilon, and the production process is long. Plus, in another assault on "new," *The Logic of Microspace* includes, as one of its three major sections, an updated version of the wildly popular (ok, the popular) *Micro Space Craft*.

What one million dollars still buys

What noble motivations spurred me to write a second book on microspace, stripped of the excuse of naivete and ignorance I enjoyed when I wrote *Micro Space Craft*? None that I can think of, but plenty of baser ones. For instance, we sold all the existing copies of *Micro Space Craft*. The business side of my brain could not bear the thought of turning away a customer, even for a product with no profit margin (thank the humanistic brain hemisphere for that pricing decision). In almost four years, a few things had changed. Like Dr. Evil, the uncharred members of the publishing machine that brings out these great works were forced to break the news to me that a 64kbyte chip was no longer considered an example of breathtaking density and brash newness. Stuff like that. Unlike Dr. Evil, I'm pleased to say that nowadays in the microspace world, one million dollars is, even more so than in 1996, a lot of money.

Engineering is not alone — there is management

Speaking of hemispheres, I also realized that very little of my own time was being spent on deploying automobile antennas out of trunks, putting golf balls into orbit and building rocket powered Mustangs, some of the subjects treated in *Micro Space Craft* along that merry road to understanding how microspace works. Very much of my time is in fact consumed trying to make microspace programs happen, which is really the philosophy of microspace (the answer to the question "Why do we want to talk to Rick Fleeter" often posed by the leaders of AeroAstro's erstwhile clients). And having done that, the rest of my time is absorbed in trying to get microspacecraft to emerge from the end of a program, also known as management of microspace projects. A second book was my answer to those questions, and that second book, merged with the updated version of the first, is *The Logic of Microspace.* Almost.

Aesop did not sleep here

Too much logic can seriously erode readership. Know any best sellers on inductive reasoning or Boolean logic? And the ultimate reason we go to space is quite different from the ultimate reason we grow corn. Without corn, microwave ovens in offices lie fallow, and animals go hungry. Without corn to feed animals, the price of Egg McMuffins goes up, and people on the margin starve, while people not on the margin grouse about getting poorer, farmers' livelihoods are destroyed, inflation rises, and the President does not get reelected, and neither does his VP. Oh, that already happened too? You can see how tough it is to be Dr. Evil.

Sure, some spacecraft have enormous utility, but space is motivated by more than commercial imperative alone. We shape our future according to what we want to get from space. Do we want to discover extraterrestrial intelligence? Or visit comets and asteroids and think about the beginning, and possibly the end, of Earth? Or entertain ourselves with telerobots on the moon, or neat pictures from the surface of planets or of astronauts floating around inside a space station? Motivation is key — it's why trees get tall, birds fly, and people get out of bed, gridlock the Beltway, and eventually boot their office computers.

A Wrinkle In Microspace is the third part of this book. It is a novella about some motivations for doing small, low cost space. It is also the first adventure drama ever written about small satellites and the people who build and use them. And if, like me, the only reason you pick up the copy of the *New Yorker* in your dentist's office is to read the cartoons, you'll be pleased that my editors describe *Wrinkle* as "richly illustrated." Actually, if you spend most of your workday reading the Federal Acquisition Regulations or the Unified Tax Code, and possibly even if you don't, you'll find that all three parts of the book are richly illustrated, each in its own way.

Going the twofer one better

Three really fun books including lots of pictures for the price of one uninteresting one filled with small print and possibly depreciation schedules. You can't go wrong. Plus you get this free bulleted list of what's in here:

- All the technology you need to know to know why you can't just fire all the engineers on day one;
- Enough philosophy of small space to understand why it's the fastest growing sector of the space business and most important in creating the space applications of tomorrow, combined with some management insight on how to succeed in breeding this new species of space program; and
- A fable to provide a few new reasons to get out of frozen storage and design, manage, pay for, study about, or otherwise help make happen, lots of new microspace projects.

Introduction to MicroSpaceCraft: (The Original Edition)

The way we (like to think we) were

Sizzling in the hot Virginia summer sun and humidity, I try to imagine January. Let's see: snow piled up over the gas grill; icicles hanging from the gutters; driveway is a mixture of ice, snow and largely ineffective sand; the non FWD (TWD?) car cowers in the garage; ducks face windward standing on sheets of ice on the pond out back. All the images are there, but without the bite. The twain, even in imagination, don't fully meet.

Can we remember when space was mission, rather than budget, driven? In the late'80s Air Force generals pronounced small, low cost spacecraft "worthless" and a "distraction." Today, excepting some special cases, if a program can't lay a claim to smallness, low cost, and rapid schedule, it is either dying or already dead.

Aerospace's fall from grace with the sea of humanity

Why? Shrinking budgets is the obvious and partially true answer. Aerospace's springtime of raw attractiveness for its own sake is over. Society has been there, done that and won't spend large sums on an infatuation with the siren song of space. The dollars are smaller now, and relevance to more people means moving projects away from the exclusive purview of a few industrial giants. Relevance in part forces lower cost, more rapid schedule and a more diverse community of developers including students and young professionals - people who don't have decades in one job to nurse along stone by stone construction of a modern pyramid. Immediacy and relevance are attributes of low cost programs of more limited scope.

The world has changed in other ways than economic. The rate of change, in science and particularly in technology, is increasing. Classical aerospace programs require decades from concept to flight. In that time they are typically obsoleted by several generations and often uninteresting beyond an initial "turn it on and see if it works." It has become more important to fly a new technology immediately, learn and progress forward, than to wait long years to create and fly the most capable possible facility.

Savoir Faire — Can't leave home without it

These are societal imperatives. But in our technology-based society, change happens only when there is a confluence of our desires and our technical abilities. What created the mass markets for answering and fax machines in the '80s? The desire to stay in touch coupled with the technical ability to improve our connectedness. In the '90s the cellular phone is booming because our ability to provide mobile telephone service compliments our increasing mobility. Small satellite technology is now so important both because we want it (all those societal factors) and we can do it - a highly capable spacecraft which a few years ago weighed tons and was built by thousands can now be built, thanks to revolutionary progress in electronics and the ingenuity of small satellite designers, by 10 people on a tabletop.

No Fear, No Limits, No Grades

Except for occasional (ok, frequent) philosophical interludes this book is an introduction to the core technologies behind small satellites. It is not a textbook, it does not profess great rigor and it is far from exhaustive - where necessary I may have shortchanged Fourier in favor of Fun and taken a few scenic but definitely off road shortcuts to get from Anabru to Zurbranchburg without straining any attention spans. But it covers the entire field of low cost space activity in a way that is meant to be accessible - as are the satellites themselves - to a wide audience. This is satellite engineering for the rest of us - not just for people who've spent 5 or 10 years studying the subject in college and in engineering jobs.

We have an 800 number

Some of the topics covered were serialized in The ISSO (International Small Satellite Organization) and *New Space Newsletters*, though all of them have been updated and improved for this book. The book chapters stand alone, and the bitesize morsels of education they contain include no prerequisites and no grades — pick it up wherever you'd like and browse.

I wasn't born this way

Thanks to Professor Dan DeBra, whose classes at Stanford were a delight and an inspiration, and who encouraged the articles which lead to this book. The only thing Dan showed me I ever found incomprehensible is his stamina on a bicycle. Professor J. Kestin taught me the discipline to write and to continue to learn. I am grateful to have been allocated a small part of his time while he was alive among us, time which could have been spent on serious, talented students. Another inveterate biker, Richard Warner, encouraged and cofounded our company, AeroAstro. Richard suggested many of the book's topics, created several illustrations, and coauthored with me several sections. This plus his predictably unpredictable advise has been invaluable. I thank Josh Cohen who composes, out of the ASCII mess I create and email to him from long distance and noisy phone lines, readable, interesting and attractive articles. Professor Rudy Panholzer of the Naval Postgraduate School first suggested my writing this book — a suggestion I found easily dismissed at the time, but which ultimately seeded the entire effort. Fran and Terry Ponick have somehow managed to maintain their cheery demeanor and professionalism, even after years of working with me to take the messes I create and turn them into something quite amazingly resembling a book. Their skills and their continuing encouragement have been wonderful. My wife Nancy has learned to tolerate loss of a husband in exchange for an icon, sitting at the kitchen table, staring at a computer screen and tapping a keyboard. And my mother, who says my writing is great, even when I know it's lousy, who said my piano playing was great even when it was truly unbearable (the good things never change, right mom?), and who has kept AeroAstro together single handedly, by providing us an unending supply of the worlds greatest bread, aka Hannah's Bananas.

Part 1

MICROSPACECRAFT

Chapter 1

Why Are We Here?

Parents like to think of their children as more handsome, more successful, and happier than the parents themselves are, but nowadays especially the focus on offspring is not so much achieving Lake Wobegone "above average" status, but rather consummate genius. The misconception that satellite engineering is a fantastically complex undertaking has really been a boon to my own parents' wish fulfillment, and probably many others.

Filial piety being a concept reserved not solely for the Orient, it is easy to ignore how useful some rather simple, inert satellites actually are. The moon comes to mind in this regard. It has no digital electronic systems, no radios, no photovoltaics, no batteries. The moon sports not a single moving part ("active component" in today's jargon).

Despite having not a line of software to its credit, the moon is more useful than most of us acknowledge. It provides an efficient, even aesthetic, nightlight. It is a passive reflector used to relay radio signals between distant places on earth. Landing on it provided a major challenge to our own space technologies, not to mention returning from it. The moon's gravity has been used to assist interplanetary spacecraft on their trajectories and to test Einstein's gravitational theories. The moon also creates the earth's oceans' tides, themselves pretty handy not just for creating beaches, inlets, and breeding territories for animals, but also for producing electricity and for harvesting clams. The moon is incredibly reliable, with an MTBF (mean time between failures) measured in billions of years.

In the future we could be lucky enough to find ice on the moon, or maybe minerals containing useful substances like oxygen to use in establishing a lunar outpost. It can be argued that building a space station is redundant, because the moon provides a ready base for operating in space.

The moon is such a clearly good idea that it is interesting to wonder what the world would be like if the moon weren't there and NASA were to propose building it. The environmental impact studies alone would cost billions. It is

The moon: an ecologically friendly nightlight.

doubtful that building a moon would ever be approved. Fishing interests would decry the moon, claiming its tides would destroy ocean navigation as we know it. Astronomers would point out that a significant percentage of viewing opportunities would be totally spoiled. I suppose artists would decry the loss of night's velvet cover of darkness, unaware of the charm a large yellow moon can add, hanging huge over the horizon, to a warm summer evening, or the mystic chill it gives us when it is high overhead of a midwinter night's snowscape. The moon speaks to us without radios and in its own language, a language people, plants, and animals all understand.

Technology changes things, even low-tech stuff like earth's moon, and people don't like change. As a technologist, you have to accept that and work with it, realizing that nobody is going to thank you for delivering change, even change for the better. Know any cities or countries named after scientists and engineers? Einsteinville? Archimedesland? Gausstown? People prefer politicians, despite what you read in the press. Leningrad (gone but not forgotten), The Tom Bradley International Terminal at LAX, Cape Kennedy. Politicians are predictable, and they haven't made anything fundamentally better, or really changed anything about our lives, in hundreds, even thousands of years. At best they cook stuff up with the ingredients society provides them, but they are players on the existing field. They don't create new worlds, or even new fields.

If it's not too late, you might consider violin making. If you can figure out how to make violins exactly like they did 500 years ago, you'll be rich, famous, and popular beyond your wildest dreams. But if you invent a new piano that is played beautifully, mysteriously, hauntingly, without touch or training, simply by modulation of your own brain waves, the art world would assassinate you for alleged debasement of human culture. Have you ever walked into the office of a CEO of a major corporation, or an Ivy League college dean's office, or the office of a celebrated politician and found a bunch of aircraft radios, gyrocompasses, and air navigation charts festooning the place? Maybe a GPS receiver? Of course not, they're all new and crass. But if you find a yellowed old globe missing most of the detail of the New World, maybe a wooden ship's wheel, and a compass that Columbus could have used, complemented by a worn, useless brass sextant thrown in for good measure, that wouldn't surprise you at all. Change has very little constituency. You have been warned.

The moon has an important lesson for we who build artificial satellites. Complexity and usefulness are not two sides of the same coin. They are often traveling companions on our technological journeys, but a single drop of water says just as much as the powerful river and the massive ocean it flows into. Where complexity and usefulness part ways is often where you get the greatest overall value. Books, paper, bricks, arrowheads, soap. One of my many language instructors (I needed a lot of tutoring) gave me some good advice. She said, as I left the security of her living room about to cross two oceans and apply my faltering language "skills" on a real world full of native speakers:

"Only say what you know how to say, don't try to say more."

If you want to build a satellite, do it. Build what you can build. A junior high school class can build a satellite. That satellite can be observed and tracked in the night sky or heard on a radio for a few days as it orbits overhead. A single college class could build a satellite with a radio repeater, and a group of students working over several years can build a stabilized platform with a pretty capable computer, digital radios, and some scientific instrumentation.

All of these projects are just as much a valid application of satellite engineering as the most complex devices our societies have produced— satellites like Voyager that sent back images of Saturn and Uranus; Pioneer, which transmitted to us for decades from beyond the edge of our solar system; and TDRS, one of the most complex communications terminals ever built in daily operation from 40,000 km (24,000 miles) above earth.

People build things in a fundamentally different way from nature. We're always in a big hurry. To survive in a Darwinian world, our minds have evolved an intense focus on the individual, and the individual lifespan is short. Similarly, we are preprogrammed to be deterministic. So for us, it's not good enough to wait for a lot of trees to grow along a river, for some of them to die and to fall over in a bunch someplace across that river to create a bridge that our descendants can walk across 100 or 1,000 or 100,000 years from now, even though that certainly is a viable way to create a bridge. Plants work that way. People don't. We chop down a bunch of trees, build a truss structure, and in a few months we have a bridge. In the process, we also create the field of civil engineering and a whole technology grows up—rope bridges, truss bridges, suspension bridges. And we fit bridge technology into the other technologies we have synthesized. For example, we make wide concrete bridges that cars drive over at 65 mph.

Same human behavior at work in space. In 50 years of frantic effort we have surpassed the ancient, majestic moon in several respects. We have satellites that point at the earth, even at special spots on the earth, or at particular stars or planets. We have satellites with active radio receivers and transmitters for relaying television, radio, telephone, and data. Our satellites carry advanced optics for astronomy and reconnaissance, radio receivers for spying on people's telephone conversations, electric power supplies using solar energy, nuclear energy, and chemical energy that we use to power refrigerators, lights, heaters, even personal computers in orbit. Like civil engineering, a whole technology has grown up around how to do stuff in space. Whereas our medieval ancestors studied the moon, the planets and the stars, we study calculus, analytical geometry, dynamics of rigid bodies, orbital mechanics, automatic control, and thermodynamics.

But as another of my patient mentors (all my mentors were patient, which sort of says it all) once explained, starting from first principles is fine if you don't want to get past first principles. This is therefore not a set of chapters on engineering fundamentals underlying spacecraft engineering; in fact, you might wonder if this is actually a coherent book at all! It's more of a collection of short subjects—sort of a satellite fan's *Readers' Digest,* or an aerospace engineering equivalent of an aboriginal approach to fishing and hunting. A person who learned from some elders passes on to you the ancient ways of catching a salmon, avoiding crocodiles, or building an igloo. There's a lot of science behind salmon, crocs, and igloos, but neither you nor your mentor is concerned about it. There are pilots and there are aerospace engineers; there are musicians and there are cello builders; there are hungry people and there are gourmet chefs. This book, and particularly this section, is for people with an appetite for space.

Building, launching, and using spacefaring stuff has developed a pretty significant bag of tricks. The tricks are not magic but there is a great danger in treating them. You can die. Of boredom. OK, I have a short attention span, so that danger lurks ominous in my mind every morning. Treatments in this treatise err on the side of the quick read. Blame it on a spoiled child of the second half of the twentieth century known locally as the author. So my apologies if technical sensibilities are trampled. It's a lesser of many evils.

The bag of tricks includes what we know about orbits and getting into orbits; about radio communication; about keeping our orbiting creations at the right temperature; surviving in the space environment; figuring out in advance of launching something whether it stands an ice cube's chance at a Texas Bar-B-Que in August of surviving "up there"; stabilization and pointing things; space computers; data storage and software; remote sensing from space and other satellite applications; electric power production and storage; why we ever invented clean rooms (and why satellite customers always want to see them). A few geopolitical tidbits are also included like radio frequency allocation; why it is fundamentally impossible to get a small satellite program funded adequately even though it's one hundred times cheaper than large satellite programs; and why the real frontier of satellite technology is to figure out how to build satellites in less than 15 months even though it takes 8 years to get the bureaucratic machinery rolling to turn on the contract in the first place. See above disclaimer on technical purity; this is not the Church of Aerospace Technology. We're closer to the dusty ol' sarsaparilla bar back behind the train station to space. Plant your spurs up on the tabletop, pour yourself a cool one and enjoy the ride.

Chapter 2

Propulsion — Or, How to Get There?

An illusion of youth is that there are right answers. For instance, you know there are the right parents to have; you just haven't got 'em. There are the right cars to drive, but your parents don't drive 'em. There are the right places to be from, but they're always in other countries. At some point, not coincidentally about when you have to actually rent a place and finance a car, reality becomes a process of getting by with the achievable, striving for the desirable, and learning to enjoy the process.

Such was definitely the case with the automobile, and particularly the internal combustion engine and its need for a transmission. The people who first tried to replace steam engines with internal combustion (gasoline) engines in cars had a problem. These early devices operated efficiently only over a very narrow range of engine speeds. Some clever engineer whomped up a set of gears to allow the car to travel at different speeds and placed the gears between the engine and the wheels, along with a "clutch" to separate the engine momentarily so the gears could be switched. This temporary patch, meant only to fix up a small problem with primitive internal combustion engines, is still with us 100 years later in the car's evolution. We use the prettier term "transmission," leaving "gears" for bicycles and gearheads, but call it what you will, and even automate it if you want, it's the patch that became the jeans. Propulsion for spacecraft has made similar non-progress over its 10,000-year history. The fundamental problem in space is there's nothing to push against. Fish and swimmers push water around to move themselves. We walk by pushing back on the floor and pushing ourselves forward. We climb ladders by stepping down on their rungs. A bicycle or car tire pushes back on the street to push the car forward. It is not terribly inaccurate to say that aircraft push themselves forward by pushing back on the air, and their wings provide lift by pressing down on the air. In space, except for the occasional photon or errant hydrogen molecule, there is nothing to push on.

Early propulsion system bug fix.

Luckily for NASA, the Chinese solved this problem a few millennia ago by building a gadget that brings its own propellant along with it. The propellant is burned, and the hot gases it produces are directed backwards behind the vehicle, pushing the vehicle forward. We call this particular gadget a rocket, since the term "rocket scientist" offers significantly more ego-satisfying gazorch than "gadget scientist." This same performance is repeated every time a rocket is launched. Huge amounts of hot gas are thrown backwards behind the vehicle to push it forward, and we all gasp and reverently stare at this sublime invention of the human intellect.

It's worth noting, however, just how obtuse this "solution" really is. Imagine that instead of driving your car with an engine that turns the wheels that push on the street and move the car forward, you applied a rocketry-based technology. You would carry a mass of propellant, like water, in the car, along with an enormous pump to pressurize it and force it at high speed through a nozzle. With a fantastic effort you could get that water jet up to an exit velocity of 1 km per second (0.6 miles per second or about 2,000 miles per hour). To accelerate your car up to freeway speed, you'd have to chuck about 100 kg (220 pounds) of water overboard. To drive from New York to Boston (300 km) would require pumping a good-sized swimming pool full of water out the rear of the car, requiring frequent stops to refill even a rather humongous water tank, the weight of which itself would require a lot of water to get up to speed. Using an engine in the conventional way, we can achieve this trip with just 20 or 30 easily stored pounds of gasoline (about 5 gallons or 20 liters). Is this pretty silly? Put yourself in a rowboat. Maybe you weigh 70 kg and the boat weighs another 30. Instead of carrying your built-in muscles and a few kg worth of oars, you bring a slingshot and a pile of rocks, a big pile that weighs almost enough to sink the boat, maybe 1000 kg, a metric ton. Every time you take a 100 g pebble and shoot it out the back at about 100 m per second (200 miles per hour - a very impressive slingshot!) you accelerate the boat less than 0.01 meters per second (less than 0.02 miles per hour). Besides being a danger to maritime navigation shooting all those pebbles out the back, you'd have no room for passengers or freight.

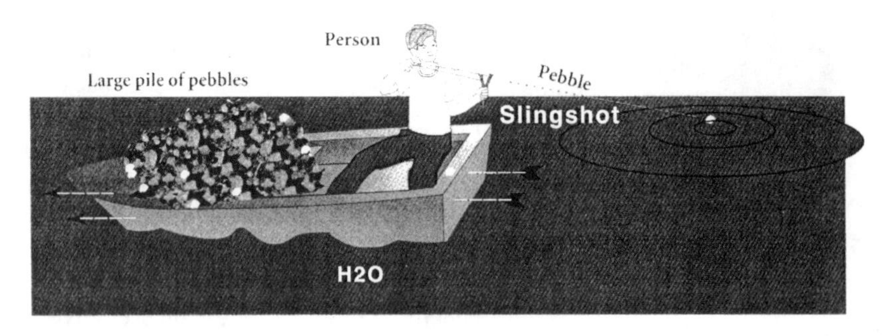

Primitive propulsion system.

What I'm saying is that if rocketry were the only means available to get from A to B, nobody would. Like other inelegant patches we've learned to not just live with but actually respect, like $900 automatic transmissions, telephones that make our voices sound like they are reproduced through string and a Dixie cup, portable computers and CD players with batteries that last 45 minutes, and the housing of our immortal souls in bodies that, with luck, last about 1 part per billion of the lifetime of our surroundings (but nonetheless invariably outlast our savings accounts), we are taught from birth that rockets are really cool devices, worthy of our unquestioning admiration. After all, it doesn't take a rocket scientist to appreciate a pillar of fire and smoke a mile high, even if all that fire and smoke contribute absolutely nothing to moving the rocket on its way except to impress the politicians who allocate money for the next launch.

Like marriage, death, taxes, and college graduation, we got ourselves into this mess with only the best of intentions. No less than Newton himself (the person, not the Personal Digital Assistant) formulated the postulate that to move forward, something has to move backward. Probably even Alfred E. Newman could repeat the immortal words "for every action there is an equal and opposite reaction." All of the examples cited above speak to the fact that you need a medium to push against. Earth, water, air are all very handy. If you have none of these, and space does, which is why we call it space, after all, you have to bring your own.

In fact, Space is not totally empty. It's full of stuff like photons (the quantum mechanical packet of light); a little bit of interstellar gas, but less than one trillionth the density of air at earth's surface; and that universally available ingredient each of us encounters in tremendous supply, wherever we go whether on earth or off it, stupidity. While stupidity drive is not even on the drawing boards yet, the concept of scooping up interstellar gas and throwing it out the back end of a rocket has already been thought of. It works, but. There is so little material out there, that you have to be going sort of WARP-6 (6 times the speed of light) before you bump into enough interstellar matter to make a sensible rocket, leaving us with the gritty little problem of getting from 0 to WARP-6, a speed that Einstein and others have already revealed to us is not attainable due to relativistic considerations. Reality, as they say, bites.

If there is one reason why we don't have 2001, Star Wars, or any other futuristic vision of space activity extant on the particular sheet of the Riemann surface we think of as reality, it is propulsion. Shooting a rocket into orbit is an inaccurate phrase. A rocket lugs itself into orbit as laboriously as we would in a rowboat with a pile of rocks and a slingshot and gives up 99% of its mass by throwing it overboard. Pieces of the rocket itself are even thrown overboard to lighten what remains. I think a precedent for this technology was in the movie "Airplane" where they tossed a bunch of baggage overboard to lighten up the stricken aircraft. If your car were a rocket, it would throw so much of itself away in the process of getting to orbit that only the radio and maybe the steering wheel would actually get there. The rest, including you, would either be converted into hot gas, like the rocks in the rowboat, or get dumped to lighten up the "payload" that the rocks were accelerating.

2.1 Rocket Travel Advice: Pack Light

Two and only two key factors are under our control in making a rocket as efficient as possible. One is what I have focused on so far, maximizing the fraction of the rocket that is propellant. The more mass you throw backwards, the faster you can move forwards. Rockets typically put less than 1% of their mass in orbit. For very distant missions like going to Mars, the so-called "payload fraction" or "mass fraction" can be less than 0.1%, which is one pound of payload for every 1000 pounds of rocket. An equivalent is building a big four-door Buick whose payload would be maxed out by one skinny Chihuahua or a six pack of Miller. Maybe you remember just how tiny the Apollo service module was perched atop the gigantic Saturn V.

Rocket vehicle designers focus on mass fraction. Their craft is to make as much of the total vehicle mass into propellant, in other words, to minimize the mass of structure, electronics systems, and other dead, meaning non-propulsive, weight. The fact that we actually throw pieces of the rocket away on the way up speaks to the level of desperation modern engineering must go to. We can't afford to take even an empty fuel tank with us. When it's empty, it gets dumped. Needless to say, the tremendous size of rockets relative to their payload and this proclivity for throwing pieces overboard makes rocket travel more costly than planes, trains, and automobiles, not to mention bicycles, surfboards, and elephant backs. It's also less safe if you live under the climbout path, which is why launch sites tend to be at the edges of oceans or deserts.

Lest anyone jump to obvious conclusions, throwing junk away is not the main reason rockets cost big bucks to deliver relatively small payloads to orbit. The main reason is the need to resort to rocketry as the propulsion approach in the first place. Once you break the surly bonds of earth and the atmosphere and get into carrying your own fuel and the oxidizer to burn it, and also the stuff that you want to push back against, you have bought into a very difficult problem. Tossing a few spare parts into the local ocean instead of bringing them back with you costs nothing compared to the cost of building a vehicle that is 90+% propellant and still maintains any payload capacity at all.

2.2 More Rocket Travel Advice: Carry a BIG Slingshot

The second parameter determining rocket performance is the one that rocket engine designers focus on. The point of lobbing mass backward is to push ourselves forward. The faster a given mass is lobbed backwards, the greater the acceleration we get from it. A significant fraction of ancient Egypt's labor was directed toward the development of the pyramids, and a significant fraction of the intellectual energy in the United States during the '50s, '60s and '70s, and even the '80s and '90s, has been focused on creating machines that throw mass backward at ever higher speed.

Unlike other modern-day public works projects, rocket performance has at least one easily quantifiable measure of success, which is how fast we are throwing stuff backwards. We could measure a rocket's efficiency by measuring speed like feet or meters per second. This number could pretty much tell you how efficiently you are using the limited fuel resource carried on board. In fact, engineers divide this number, in meters per second, by the gravitational constant, that is, by the acceleration earth's gravity gives to things when they fall. This number, which you might as well know since it is responsible for everything from your fight with the bathroom scale and other sags and bulges characteristic of the aging body, to that family heirloom of your husband's mother's that you unceremoniously knocked off the kitchen counter on your last visit to Cleveland, to the amount of work it takes to get things like rockets off the ground, is 9.8 meters per second per second or 32 feet per second per second. That is, in every second that something is falling, it accelerates another 9.8 meters per second. At the end of two seconds, it is moving 19.6 meters per second, or about 40 miles per hour, ignoring effects like atmospheric drag, which is the main reason that potato chips don't shatter when they hit the kitchen floor, but heirloom vases do.

If you divide an exit velocity measured in meters per second by an acceleration measured in meters per second per second, you end up with the units of seconds. Of course if you divide the exit velocity in feet per second by acceleration in feet per second per second, you also get units of seconds. That's a nice thing about this otherwise questionable exercise in division. For example, an exit velocity of 100 meters per second (m/s) divided by 9.8 meters per second per second (m/s^2) gives you a little over 10 seconds. Rocket people call this number the "specific impulse," a very key term in rocketry, because it measures how much propulsive force we squeeze out of every bit of precious propellant. The higher the specific impulse, written Isp, the faster the precious mass we are ejecting is moving, and hence the farther we can go given a certain fuel load. Isp is the miles per gallon of rocketry. Because rocket vehicles are critically limited by the fuel they can carry with them (Remember the prodigious amounts of water needed to operate a rocket automobile?), Isp is the most important single measure of the quality of a rocket.

What are some typical Isp numbers? Some rockets just have stores of pressurized gas, usually nitrogen or helium, that is expelled through nozzles. This is not too efficient, but it's very safe and can be metered highly precisely for close maneuvering. The propulsion packs integral to space walking astronauts' space suits use this approach, and the Isp achieved is typically in the range of 35 to 65 seconds, corresponding to effective exit velocities of around 500 m/s or 1000 miles per hour. That really doesn't sound so shabby, but it's the bottom rung of the rocketry ladder.

Chemical systems that burn various propellant combinations, including the big rockets you see on the space shuttle, achieve Isp ranging from 250 seconds to over 400 seconds with exit velocities over 8000 miles an hour. The highest performance rockets are electrically driven devices that accelerate charged particles (ions, hence the name ion propulsion or ion drive) in an electric field. They can deliver Isp well over 2000s, which is an exit velocity greater than 40,000 miles per hour! But ion propulsion is limited to very small mass flow rates. While the efficiency of propellant usage is very high, the thrust is very low, typically less than 1/1000th of a pound of thrust—less than the propulsive force of a house fly. Nonetheless, applied over many months in space, the minute forces generated by these highly efficient systems can propel spacecraft through the solar system and beyond.

2.3 The One and Only Equation

I try to write without equations. Nothing against equations, but people don't like to read them and I like people to read what I write. But there is one equation so basic to rocketry that it's called "the rocket equation." It is simple, yet it contains the Yin and the Yang of rocketry all in one line. The rocket equation describes the relationship between the need for mass fraction efficiency sought by vehicle designers and propellant usage efficiency, Isp, sought by engine designers. Let's just get the equation out there, then we'll talk about it:

$$\Delta V = g \, Isp \, \ln(M_i/M_f)$$

The Rocket Equation

Looks simple. What is it? V is the change of velocity. You start off stationary, light your rocket engine, burn all the propellant, now you're going 5000 miles per hour. Hence your ΔV (pronounced Delta Vee) is 5000 mph. The acceleration given by the earth is a constant known as g. It is 32.2 feet per sec per sec or 9.8 meters per sec per sec. Isp is the rocket engine designer's holy grail, measured in seconds. Multiplied by g it yields units of speed or velocity as meters per second or feet per second. So you can think of the product g Isp as the effective exit velocity of the rocket plume. The higher, the better.

Mi is the mass of the rocket vehicle with all its propellant, with its structure, and with its payload, ready to go. It's the initial mass. M_f is the mass of the rocket exactly like M_i, but after all the propellant has been burned or expelled. It is the final mass.

Which leaves us with ln, the natural logarithm. Rather than explain logarithms, I'll just mention a few facts about them. There is a button on a scientific calculator labeled ln. If you type in the number 1, the answer you'll get if you then hit ln is 0. This means that if the M_f and M_i are the same, you have by definition no propellant on board, and the ΔV is going to be zero. My dad is a busy guy, and he likes to see how much distance he can squeeze out of a tank of gas before stopping at a service station. He is not the type of guy to warm up to a bunch of equations with Greek letters and subscripted variables, but has tested this special case of the rocket equation many times by attempting to continue to drive with $M_i/M_f = 1$, i.e., no gas. Luckily, he has a cellular phone and a patient wife without a similar inquisitive proclivity.

2.4 Thrills, Chills, and Spills with the Rocket Equation

The ln of 2 is about 0.69. How, you might ask yourself, did you ever get as far as you have in life without realizing this? Tack a rocket engine on the back of an old Mustang with the blown engine already pulled. The Mustang, sans engine and all the other parts you gutted from it, like brakes, is left weighing about 1000 lb. Strap on a tank big enough to hold 1000 lb (about 150 gallons, the size of two big hot water heaters) of hydrogen peroxide, a rocket propellant whose exhaust is "just" torch-hot water vapor and hydrogen. Evil Kneivel almost crossed the Grand Canyon on a hydrogen peroxide rocket.

The ratio of M_i divided by M_f is 2. You start off with initial mass of 1000 pounds of Mustang shell and empty tank, plus 1000 pounds of propellant, for a total of 2000 lb. When all the peroxide is exhausted, you've got the 1000 pounds left. Peroxide decomposition is not terrifically energetic, though you may disagree if you spill even a tiny drop on your hands. Typical exhaust velocities are 1500 meters per sec, or about 3000 miles an hour. So the Isp is about 150 seconds.

That's all we need to know. Using the rocket equation, we multiply ln (2), which is about 0.69, times the Isp, 150 s and g, 9.8 m/s/s, which gives a V of 1019 meters per second, or about 2000 miles an hour.

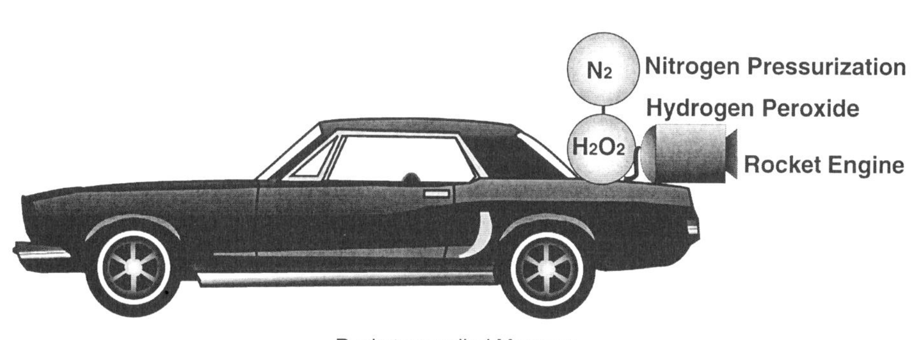

Rocket-propelled Mustang.

Saying that you should leave this trick to experienced drivers is a bit of an understatement. In fact, at around 250 mph the tires start to melt, the Mustang begins to lift off the track, and air drag gets pretty huge. A lot of your V gets eaten up evaporating rubber, lifting your body and that of your Mustang off the track, and heating the air. You'll be lucky to see 300 mph fleetingly before a messy crash. But you have combined two classic ingredients of drag racing—incredible speed and colossal crackups. If you live, you'll be a hero!

This might sound a bit insane. It is. But it's quite a popular stunt in drag racing, a sport, like many, that redefines the boundaries of sanity. It also illustrates something very important about rockets and the rocket equation. Remember our somewhat clumsy trip to Boston with the water rocket? With just 10 gallons of hydrogen peroxide you can get your '57 Chevy from 0 to 100 mph in less than the time it takes to say "Holy Whiplash, Batman," which is great for space travel. So going to the moon is a matter of applying the right V in the right direction and coasting for a long time. No air or road drag in space—once you impart the ΔV, you coast forever.

But with 10 gallons of gasoline you can shove that Chevy through all the air drag and bearing drag and tire drag and engine drag of a trip from Stanford to Tahoe where you can go up to the top of large, snow covered mountains, point your skis straight down the fall line, and personally redefine sanity. What's the difference? There's lots of differences. For one thing, gasoline engines burn air, so they cheat. Rockets carry all their own reactants. The Isp of a gasoline engine can be at least double that of a rocket, if you charge yourself only for the fuel and not the "free" oxidizer. For another thing, its very inefficient, energy-wise, to push yourself forward at 50 mph by throwing rocket propellant out the back at 3000 mph. None of that kinetic energy of the rocket jet, none of its heat and light and smoke producing energy does anything for pushing you frontwards! In a gasoline engine all of that dramatic stuff happens inside the cylinders, where it is captured ultimately as heat that pushes the pistons down and provides motive power. Of course, a properly tuned gasoline engine does not melt the pavement along with any tailgating cars in its wake. The same cannot be assured for even a modest hydrogen peroxide dragster.

2.5 Next: A bit more detail on Isp

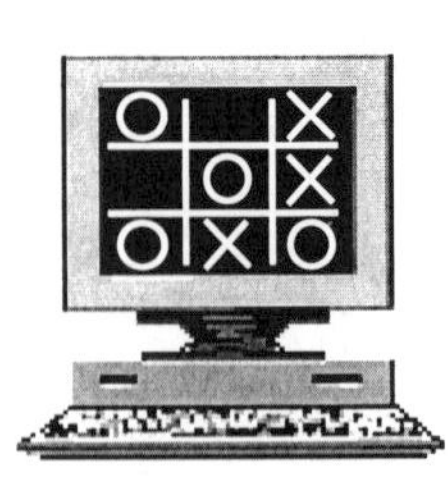

But first take this test: Put a 1 next to questions you answer TRUE. Put a zero next to questions you answer false:

1) I read the directions when I buy a new Walkman.

2) I actually know how to program phone numbers into my cordless telephone.

3) I've written software for a computer that did something, like played Tic Tac Toe or calculated biorhythms.

4) I sometimes buy small parts (wires, clip leads, etc.) at Radio Shack.

5) I agree with the statement "What's the big deal about programming a VCR?"

6) I have been to an auto junk yard at least once in my life.

7) I have sewn something myself: a sleeping bag, a shirt, blouse, a button.

8) I prefer to eat my own cooking over going to a restaurant.

9) I own more tools than just a pliers and a screwdriver.

10) I can identify the major parts under the hood of my car.

Add up your ones (and zeros if you'd like), then use the grading scale below to know everything necessary to shoehorn yourself into one or two of life's little cubbyholes:

2.6 Score: Your Lot in Life

8 – 10 — Face it, you're a hopeless techie at heart. Society is suspicious of you. Avoid politics. Read this section.

5 –7 — You are great to have along on a road trip, but a failure at cocktail parties. Skim this section.

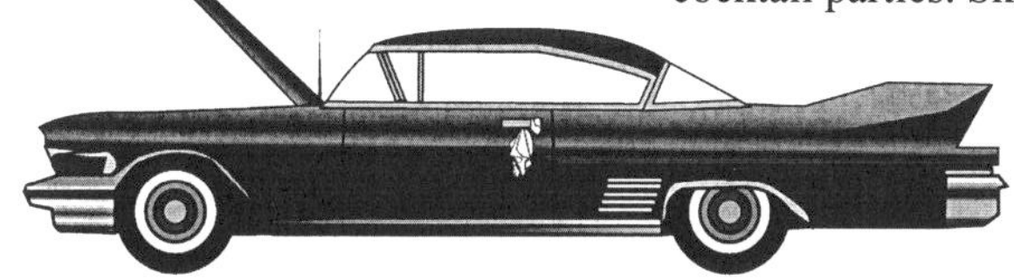

1 – 4 — You are or will be a highly successful lawyer, accountant, TV person-

ality or businessperson, since you don't get bogged in details. You are highly attractive to the opposite sex, and it's enough to know that this section has some details that engineers worry about. Skip to the next section, "Anatomy of a Rocket."

0 — You are destined to lead, possibly as President of the United States or a Vogue model. Definitely skip to the next section: Anatomy...

If life were simple (and if turkeys were in fact spherical), the measure Isp would be exactly equivalent to a measure of speed of a rocket's ejected material. But rockets do a little bit more than just accelerate hot material and eject it out the back end of a nozzle. Looking at the illustration below, the inside of a rocket engine is filled with high pressure gases and some other materials. Unlike the more normal pressure vessel shown next to the rocket, an engine has a hole on one side for the exhaust to come speeding out. A normal pressure vessel, like a can of Diet 7-Up or shaving cream, is sealed and static. The pressure forces are equal on all sides of the container, so they cause no net force. The asymmetry of the rocket motor means that the pressure forces are not balanced, and even if no material were to come out, there would be some propulsive force because of that imbalance.

In practice we measure just two things, the mass flow or consumption rate (how fast we are using up propellant) and the propulsive force. Nobody really cares what the magnitude of the pressure effect and the mass ejection effect are separately. In fact

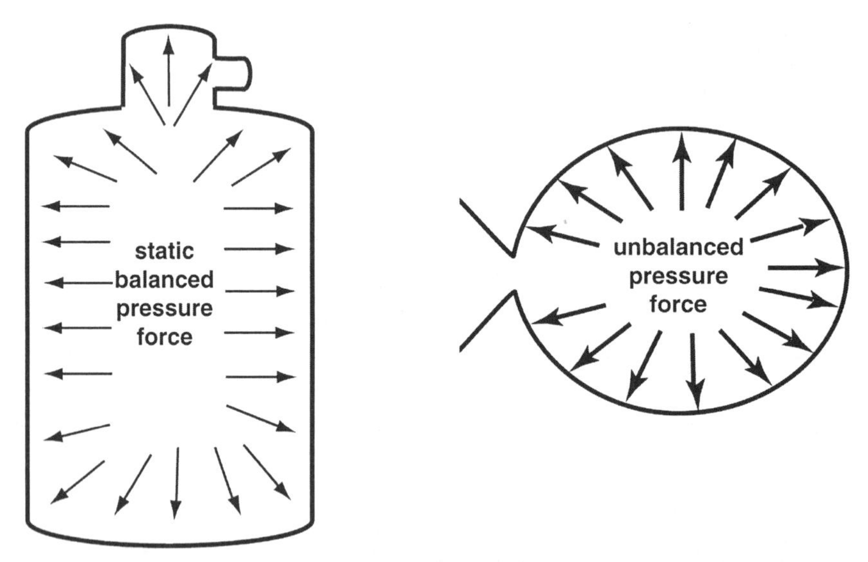

Shaving Cream Propulsion (L) vs. Rocket Propulsion (R). Constant combustion of propellants keeps rocket chamber pressure high despite an open hole on one side.

the mass ejection is by far the dominant contributor. Note too that it is impossible to have the asymmetrical pressure distribution without the hole (nozzle throat). Once you have a hole with a high pressure gas inside, mass is definitely going to come out of it, so the two effects are closely linked. We lump the extra oomph gotten from pressure in with the mass ejection derived thrust and treat it as one effect, and from that calculate an apparent exit velocity. The mass consumption divided by the ejection rate, multiplied by the lumped-together exit velocity gives us the propulsive force, and that is Newton's Law. Thus, the propulsive force, including the mass ejection and unbalanced pressure effects together, divided by the mass ejection rate gives an effective or apparent exit velocity. Dividing that exit velocity by the 9.8 m/s gravitational constant gives you the true specific impulse, Isp, of the engine.

2.7 Anatomy of a Rocket

The skeleton in the rocket scientist's closet is that rockets are in fact incredibly simple devices. They have only two or three basic parts. They don't have all kinds of spinning wheels and pistons going up and down like airplane gas turbines and automobile engines. They don't have better than one million active switches arranged in complex topographies like microprocessor chips. They don't have zillions of neurons all firing off like the brains of mice and men, or a soup of complex chemical agents like enzymes, proteins, lipoproteins, and complex sugars creating all kinds of growth and motive actions. They have, in essence, just one chamber where fuel and oxidizer are burned and an exit nozzle.

Looking at most rockets is like looking at the engine compartment of your old MG. It's a maze of tubes, valves, nuts, and bolts. The MG is, in fact, complex, albeit needlessly and hopelessly unreliably so, but the rocket mainly uses that stuff to cover up its basic simplicity, thereby maintaining the efficacy and clout of the term Rocket Scientist.

The nozzle has a converging, then diverging shape. The propellant gas accelerates to squeeze through the throat of the nozzle where it reaches sonic velocity, that is, the local speed of sound. Gas flowing faster than the speed of sound is further accelerated in the diverging section of the nozzle, sometimes called the bell of the nozzle. Usually the gas inside the engine is under very high pressure, from 10 to 1000 times normal sea level atmospheric pressure. As the gas spreads out in the diverging nozzle, its pressure steadily drops. Rocket engines that operate at sea level, which are the ones that get a vehicle started from the earth, have relatively small bells because they do not want the gas pressure to be lowered below the outside pressure of one atmosphere. Rocket engines that operate in space, where the outside pressure is virtually zero, use much larger bells relative to the size of the engine. By allowing the gas to flow through a larger pressure drop, they achieve about 20% higher exit velocity and higher Isp than their sea-level counterparts.

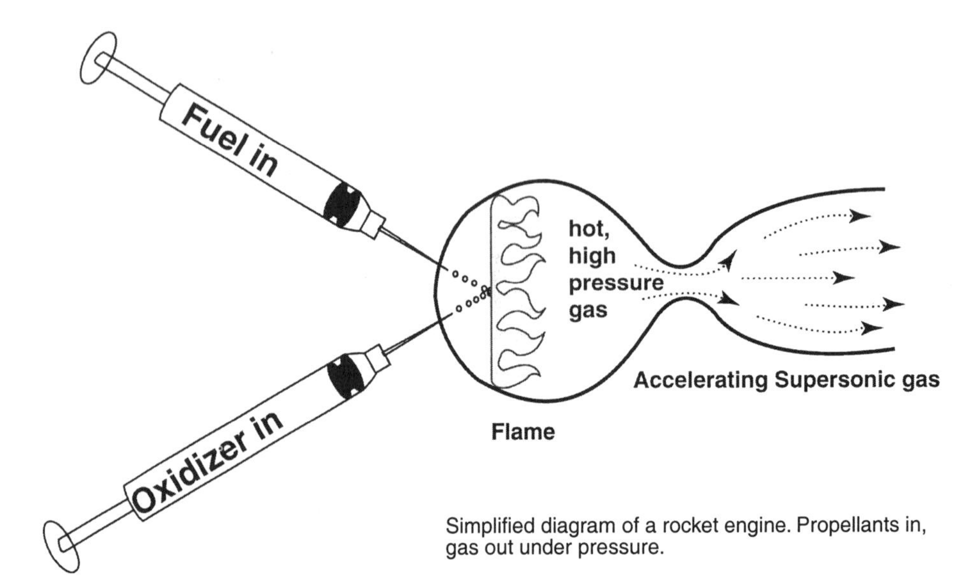

Simplified diagram of a rocket engine. Propellants in, gas out under pressure.

2.8 What to Feed a Rocket: Propellants

Once you get to know them, rockets are as individual as plants, houses, or people. Unlike plants, they don't invade your body with pollen or die when you forget to water them. Unlike houses, they don't have termites or need the grass cut or contain a bunch of relatives with whom you must learn to live in apparent harmony. Unlike people, they don't argue with you about religion or politics or call you during dinner time to sell tickets to the state policemen's dance. In a nutshell, that's why I got into this business. It has afforded me the income to buy a house, surround it with plants, and create the necessity of going to technical conferences to talk about rockets with a bunch of people.

Some plants stand around in your living room at Christmas time and look red and green. Some are farmed in the billions of kilograms to make transfatty acids for people to eat at McDonalds, and some stand around in ancient forests waiting for bulldozers. Similarly with rockets; there are myriad ways of designing and using them. Some rockets are those fiery monsters you see lifting the space shuttle off the ground. Others 1/1000th as big boost satellites from the shuttle to higher orbits. Some once again 1/1000th as powerful are used to gently correct a spacecraft's or an astronaut's position and attitude.

TABLE 1. Table of Rocket Propellant Categories and Compounds

Categories	Compounds
Inert Gas	• nitrogen (simple, safe, Isp around 45s) • helium (ditto, Isp around 60s) • ammonia (stores more densely but have to separate liquid and gas, Isp 50s)
Liquids	• hydrazine (monopropellant, highly toxic Isp 220 s) N_2O_4 / mmh (highly toxic and corrosive but stores permanently at room temperature, Isp low 300s range, expensive) • LOx kerosene (LOx must be loaded just before launch, cheap, clean, Isp in low 300s) • LOx hydrogen (expensive, hard to make and store, very high Isp — over 400s)
Solids	• Isp upper 200s to 300s
Electric	• pulsed plasma • teflon propellant (Isp 1000 to 2000s) • ion thrusters: Xenon propellant (ditto) arc jets, many propellants (Isp 500s to 1000s)
Hybrids	• LOx / rubber (safety comparable with LOx Kerosene, now in R&D)

Probably the most common categorization of different rockets is by the propellants they consume, sort of like dividing the dinosaurs into meat eaters, vegetarians, and the omnivorous ones that eat anything. A few of the basic types of rocket engines are mapped out in the table below. All rockets use either a non-reactive propellant like nitrogen gas or some combination of 1, 2, 3, or even more discrete chemicals. Electric propulsion usually uses a single, non-reactive propellant combined with a lot of electric power. Liquid rockets generally carry their propellants as liquids in separate fuel and oxidizer tanks. Solid rockets premix the fuel and oxidizer into a solid crystalline or rubbery material that is molded and bonded into the rocket motor casing itself.

It's a world of options, which tells you a few things right there. One is that we really haven't come up with The Solution. For instance, cars. Pretty much they all have four wheels. Yeh, there's vehicles with two wheels, three wheels, 16 wheels, but four wheels has gotten to be pretty dominant. In rockets nothing is truly dominant. For another thing, this chapter could get infinitely long if we're going to cover the field, so we're not and it won't. Trust me on that one. What with rockets being so ridiculously simple, we thermodynamicists had to figure a few ways to complicate them. You are just not going to get a very big house, afford very many plants, or go to very far away meetings with lots and lots of people at them, if the only knobs available for twiddling are pressure inside a chamber and nozzle geometry. Most of that was worked out in the 18th century, and people are bound to notice that before signing your nth paycheck.

2.9 Life Cycle of the Thermodynamicist

Talk about human ingenuity! Hey, we solved this problem years ago. What you do is: You go to public school for 13 years and survive the ire of your peers as you study hard and do well, earning yourself the various titles of nerd, jerk, gearhead, and geek. Then you go to college for four more very similar years of hard work and social rejection, except possibly for the camaraderie of a worthless bunch of nerds, jerks, gearheads, and geeks. Then on to four or five more years of graduate school, which is quite a shock after the previous 17 years of rejection for being too brainy. Suddenly everyone around you is telling you that you are probably just too dense to hack it.

After several years of insinuating your inadequacy, they give you the Ph.D. with a smug intimation that it was not so much for your brains or achievement as it was just the most efficient way to rid themselves of you and make room for others more worthy. Then there's maybe a year or two of post-doc-ing, mainly in order to allow you sufficient time to get over the previous five and possibly to discover things called the "out of doors" and something else called "sex." If you don't overdo it too muchly on those two goodies, you emerge into what remains of your youth, or at least young adulthood, handicapped with a world view totally out of synch with what passes for reality.

In society at large, otherwise known as where you are supposed to dwell happily and prosper, you are guaranteed, albeit by people who have never actually tested the hypothesis, that your whole huge bag of new, subtle little thermodynamical tricks will justify and facilitate your attainment of the house/plants/meetings/people stuff. All of these attainments eventually consume so much of your time that you really don't have much left for twiddling knobs on rocket engines. Such is life, reduced to one story, albeit short, sweet, and only a little sad, for most thermodynamicists. So before the phone rings, let's have a quick look at those knobs.

One is temperature, which is itself a measure of energy imparted to atoms and molecules. The higher the temperature, the more excited the atomic constituents of the propellant gas get. Hot gases accelerate more rapidly as they travel out the nozzle, converting some of their undirected thermal energy into increased velocity. This effect goes as the square root of the absolute temperature. Typical combustion products can be up to 9 times hotter than room temperature on an absolute scale, so a rocket that burns propellant is about three times more efficient (has 3 times higher Isp) than a rocket using "cold gas," that is, pressurized gas stored at ambient temperature.

Another trick is the design of the propellant molecules themselves. Let me sweep a decade of mind-rupturing study of classical and quantum mechanics, physical chemistry, thermodynamics, statistical mechanics, and mathematics away here and just say that the ideal propellant gas is made up of the lightest, simplest possible molecules (Why didn't they tell me that in the first place?). Why? Molecules have a few choices of what to do with the energy they get from being pressurized and heated. What we want them to do is convert that energy into the highest possible exit velocity. What they also tend to do is convert that energy into things like spinning themselves around, flexing, occasionally breaking their chemical bonds, and vibrating.

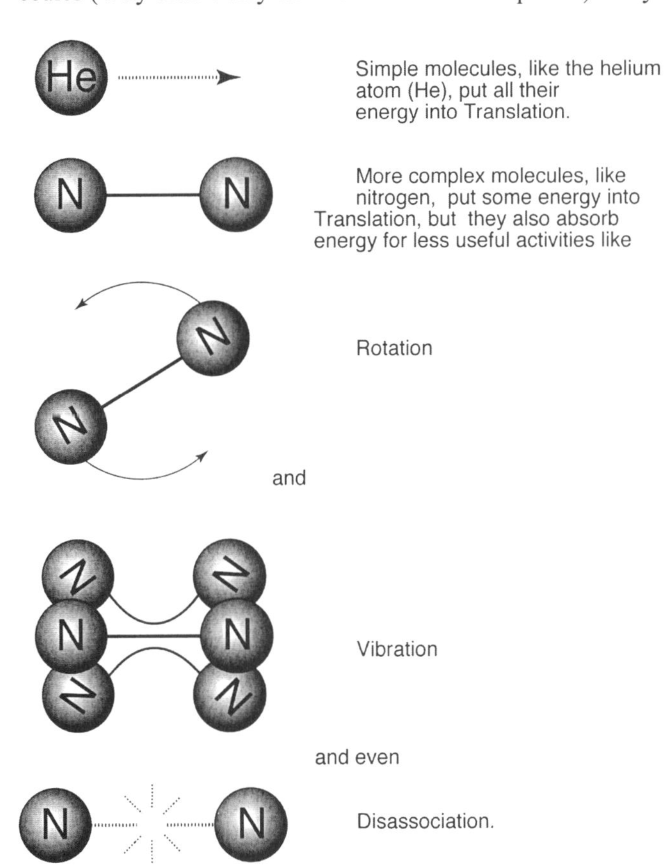

Something called equipartition of energy says that for every hour your kid spends doing her homework, she'll spend another hour arguing with you about doing it, an hour on the phone with friends, an hour playing with her food, and an hour in front of the television. Molecules also put about equal amounts of energy into all their alternatives, and why not? From their point of view, one is as useful as another. It is our Western, logical deductive mindset that tells us that self-rotation, vibration, and even destruction is not productive, but racing out the back of a nozzle at 20 times the speed of sound is highly productive. Scientists are the ultimate imperialists. These internal modes of molecular motion, somewhat akin to a proletariat value system with emphasis on God, family, and principle in action, sap energy that otherwise could have been used to accelerate them out the nozzle, grow more rice, increase stock dividends, or whatever.

Playboy claims that hiring only cute women for bunnies is not discrimination, but rather a step in delivering a quality product. Thermodynamicists hide behind the same thin veil of logic, ready to be pierced by some legal or programmatic stiletto. Given the same number of non-productive energy modes, a big fat molecule raised to the same temperature as a trim little molecule goes a lot slower. In fact, the speed is inversely proportional to the square root of the molecular weight at any given temperature. For that reason big molecules, even those with few or no rotation and vibration modes, are very undesirable in chemical propulsion systems. Which is to say, we don't use them except when forced to do so in deference to some external constraint or stimulation.

Nitrogen molecules have a molecular mass of about 28, while hydrogen, which has the same repertoire of vibration and rotation modes, has a molecular mass of 2. This means that hot hydrogen has almost 4 times the Isp of hot nitrogen! Not only that, hydrogen's light atoms and strong chemical bonding make the molecule stiffer and tougher than nitrogen, so it is harder to excite the vibration modes of hydrogen, and much harder to break its bonds, making hydrogen more than 4 times as efficient as nitrogen as a propellant from an Isp point of view. Unfortunately, hydrogen is hard to store. It is only a liquid at temperatures near absolute zero. The liquid is not dense, requiring large, massive, and heavily insulated (that is, even more massive) storage tanks. The gas is very non-dense, so if you don't bother to insulate it, the tanks get really big. Thus although it's an efficient propellant, hydrogen tends to make highly inefficient rocket vehicles. A trade off we'll get into a bit later on…

Another problem with big molecules—those with lots of atoms linked together —is that the bonds between the atoms tend to break. Wonderful, subtle, elegant and complex physical reasons explain this, but I'll restrain myself and describe only the two undesirable effects of this tendency of large molecules to come apart, particularly when hot, which they are if we use them in a rocket engine. One is that breaking bonds takes lots of energy, which then isn't available to achieve high exit velocities. The other is that often the new molecules formed when the old one is broken are messy.

For example, molecules we eat tend to be very big, as molecules go. A lot of cooking actually amounts to selective bond breaking. If you leave your toast in too long when the boss phones to complain about the shoddy work you did on the Smith account, you'll return to find that breakfast is a black hunk of inedible stuff. What happened was you overheated the bread and broke a bunch of chemical bonds. A lot of the atoms you wanted to eat escaped as gases, and a bunch of black carbon, bonded tightly by Van Der Waals forces to your best glassware, was left behind. The exact same thing can happen in rocket engines. The leftover carbon not only doesn't contribute to propulsion, but it can clog your engine (sounds like a gasoline ad, doesn't it?).

Unfortunately, chemicals like helium and hydrogen that are light, tightly bonded, and compact tend to be the least dense, hardest-to-package materials. They often are not highly reactive, so it's hard to raise their temperature.

What do you want to look for in shopping for propellant? Something dense with high molecular mass, ergo big molecules and a liquid, or possibly solid, at room temperature. Something stable that won't fall apart, explode things, or eat its container. Preferably something non-toxic that reacts rapidly, but not explosively, with something else dense, stable, and similarly non-toxic. The idea is to create one or more reaction products that are non-dense, meaning they have low molecular mass and a small number of atoms for each molecule; incredibly hot; stable, so they don't react with the materials the rocket itself is made of; and preferably also non-toxic.

Solid exhaust products like carbon and aluminum oxides are a minus. They erode rocket nozzles and decrease performance because they cannot accelerate and expand as gases do when they traverse the nozzle, and because people on the ground don't like them. Observe the remarkable parallel between the above list and the dating process. The list of preferred traits is like those ads you see in the personals. Do you think the blonde SWF who is fun loving, leggy, has a great smile, loves kids, the outdoors, and quiet evenings by the fire with someone special, is a professional and a non-smoker and just wants to enjoy life with a SWM of similar persuasion is telling the whole story? Frankly, we haven't found the solution to this problem yet, neither in mate selection nor propellant development, which is why we have such a profusion of candidate rocket propellants, SWFs and SWMs. They each have some attributes and some faults, and we pick for each application one with attributes we need and faults we can live with. You may not have been aware that Dr. Joyce Brothers is not an MD, but a thermodynamicist. You read it here first.

The table up there about fifteen hundred words ago mentions a few of the leading molecules rocket thermochemists selected over the aeons, or at least decades, of rocket research. In terms of pure elegance, my favorite propellant combination is hydrogen and oxygen. They burn with a very hot flame to make water (vapor) as the propellant that comes out the back end. There are absolutely no particles, and no toxic

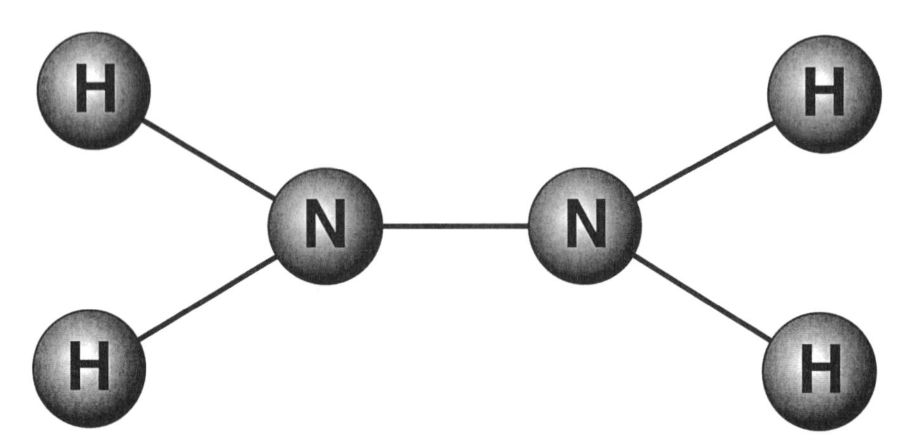

Hydrazine molecule—the classic monopropellant, Hydrazine decomposes to ammonia and nitrogen.

chemicals, either in the reactants or the exhaust. Water has excellent properties as a propulsion molecule. It is very tightly bound and stable and it is light, with a molecular mass of 18. The combination of high flame temperature and small molecular mass gives outstanding Isp—well over 400 seconds.

The major disadvantages of hydrogen and oxygen are that both propellants are cryogenic—not storable for long, and the flame is so hot that it's hard to build engines that stay cool and unmelted! Also, liquid hydrogen is not very dense, making the tankage very large and thus heavy, which costs you some of the great Isp advantage. Also, these propellants don't spontaneously ignite (a property rocket people call hypergolic), meaning they need to be ignited, which is an added complication. An Ariane rocket was lost once because the hydrogen/oxygen upper stage of the rocket failed to ignite. The Space Shuttle Main Engine (SSME) is hydrogen/oxygen, but its clear blue flame is almost invisible in the midst of all the fire and smoke of the solid rockets strapped on to the shuttle. Another classic propellant is hydrazine, a molecule designed and produced almost exclusively for its value as a rocket propellant. Hydrazine has two nitrogen atoms and four hydrogen atoms, which is the same composition as ammonia, only ammonia has one additional hydrogen atom per nitrogen atom. Hydrazine is a mono-propellant. Most other propellants, like hydrogen and oxygen, are bipropellants; you have to combine fuel and oxidizer to get a flame. Not hydrazine! It decomposes all on its own to form ammonia, nitrogen, and hydrogen, all of which are environmentally benign, and there is no solid component so it's clean burning and efficient. The mean (arithmetic) molecular mass of the exhaust is very low at 16, and the decomposition reaction gives off enough heat to get Isp around 220 seconds, which is pretty respectable without need for carrying two reactants. Even though one of the most common propellants in use today, hydrazine has some pretty serious problems. It is immediately poisonous, and, even if you get a dose too low to be lethal now, it is a powerful carcinogen. It is toxic below the concentration threshold to smell it, so if you can smell it, you could be poisoned already. Besides that, it doesn't take much to cause it to decompose, just a trace of ferrous metal-like steels and iron, or of carbon, and almost all organic materials have carbon in them. If it decomposes in a closed container like a fuel tank, there is a violent explosion. But if the container is 100% free of these contaminants, hydrazine stores forever. This long shelf life is one factor contributing to its use on long duration missions.

While hydrazine is a liquid at room temperature, it freezes at 34°F (1°C), so it needs to be kept warm. Because it does not require air or oxygen to burn, extinguishing a hydrazine flame is tricky. Mostly you hose it down with water until the concentration goes down so far that the decomposition stops. All that water is then contaminated and special provisions need to be made to treat the hydrazine before disposal. Because of all these factors, hydrazine handling is a specialty of its own, and the skilled people who do it wear fully protective gear. Setting up a facility for handling satellites with hydrazine takes time, money, and a lot of conversations with people like the EPA. When all that is said and done, it might be no more dangerous than oxygen and hydrogen, which are also easily ignited. At least the decomposition of hydrazine is virtually as clean as that of hydrogen/oxygen, so if you blow yourself up, the neighbors won't be endangered.

Hydrazine can also be combined with oxidizers, particularly nitrogen tetroxide (two nitrogen atoms, four oxygen atoms), as discussed earlier. This combination makes an entirely storable system with good performance (Isp greater than 300s), and it is hypergolic. You get instant ignition when you bring the two chemicals together, but you are handling two volatile, poisonous, corrosive, and potentially carcinogenic liquids, which drives up the cost of the system considerably. This combination of propellants has been used in military missiles, including ICBMs, and many rockets with military heritage, including the entire family of Chinese (Long March) rockets. The propellants are, however, quite expensive, costing over 25 times what oxygen and hydrogen cost for the same total amount of transportation.

Hydrogen peroxide, a water molecule with an extra oxygen atom, totalling two hydrogens, two oxygens, and the subject of our Mustang dragster, is more popular for garage work. The Isp is low (150 seconds or below) but the products, being oxygen and water, are safe. The problems with peroxide, as it's called, are that it too is unstable, particularly at high concentrations. It is normally used diluted in water; contact lens cleaner with 0.5% of peroxide in water is a very vigorous cleaner. For rockets, a minimum is 70% peroxide, and for orbital rockets numbers like 98% are kicked around. Like hydrazine, it decomposes spontaneously, particularly at high concentration. If that happens in a closed container, an explosion is the result. Evil Kneivel used it at 70%. Peroxide can also be used to burn kerosene as a component of an interesting

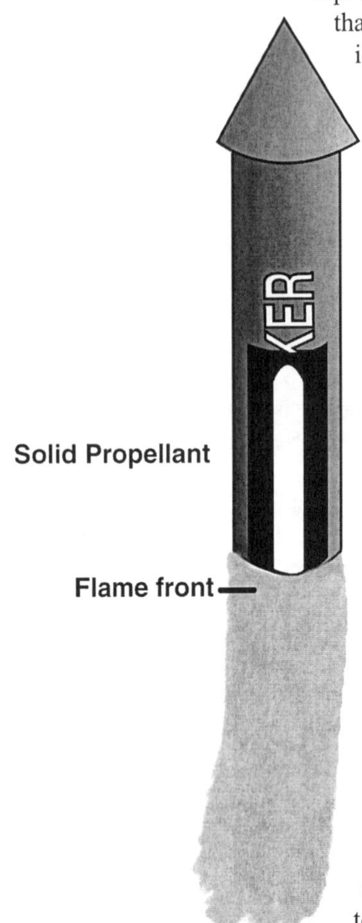

bipropellant. The British Black Arrow got into orbit on that combination, but because you can't store peroxide without suffering continuous decomposition, it is not used much any more.

The workhorse propellant combination is so-called LOx/kerosene, liquid oxygen and kerosene. Not as sexy as LOx/hydrogen, the reagents are cheap and the Isp performance, at around 300 seconds, is very good. The oxygen is not storable, so its primary application is in launch vehicles.

The simplest propellants are inert molecules like nitrogen, helium, and ammonia, which are released through a nozzle—aerosol can rockets. Helium has the highest Isp, but is hard to store because it's not dense, so the tank gets heavy. Ammonia, and also freon and propane, store as much denser liquids, but separating the liquid and gas under zero gravity can be tricky.

Solid propellant rocketry depends mainly on long-chain hydrocarbons as fuel (polymerized kerosene, if you like) and on so-called perchlorates as oxidizers. Often metal particles are added, such as aluminum, or metals are incorporated in the propellants, to raise flame temperature to get higher Isp. The fuel and oxidizer are cast together and need to be ignited, usually electrically, to begin operation. Both the propellants and the reaction products (exhaust) are highly toxic. Creating solids with more benign properties is a subject of current research.

These are just samples of the hundreds of propellants and propellant combinations, but these few constitute a big part of the total field today. As cost becomes more of a driver than military readiness, we are seeing more attention paid to the oxygen hydrogen and oxygen kerosene propellants, though they are not a solution for on-board propulsion for satellites because they aren't storable. Here the trend may be toward electric propulsion, using propellants that can be teflon rods or inert gases like xenon and argon.

2.10 Your Mission: Choosing a Rocket Technology

There are three major categories of propulsion applications: launch from earth to orbit, orbit transfer, and orbit/attitude modification. Each application has its own special considerations and optimizes with different propellants and propulsion technologies. Let's shop propulsion systems for these few missions.

Earth to Orbit

Launching into orbit from near earth requires large thrust, larger than the weight of the rocket in order to get it off the ground. That requirement can be a problem for liquid propellant systems, because all the lines carrying fuel and oxidizer must be big enough to handle huge flow rates. You also have to have big, fast pumps to move it and a fuel injector and chamber that can handle it all. All that hardware weighs plenty, making it all the more difficult to get off the ground, requiring yet bigger valves and pumps.

A solution is the solid rocket motor, shown schematically in the figure on the next page adjacent to a liquid rocket system diagram. Solid rocket propellant was used in those 10th century Chinese rockets, and is the stuff amateurs make cardboard rockets fly with (for example, Estes rockets), and drives the Pegasus and, before it, Scout rockets that have become so important to small satellites. Solid rockets have no pumps, valves or tanks. The oxidizer and the fuel are in the solid phase, and they are premixed into a nearly inert glassy or rubbery substance. Once ignited, however, the premixture burns vigorously and creates a hot exhaust gas that is expelled through a nozzle and produces thrust. Solids don't leak and don't require cold storage, so they are ideal for strategic weapons and other military applications. They can be held ready

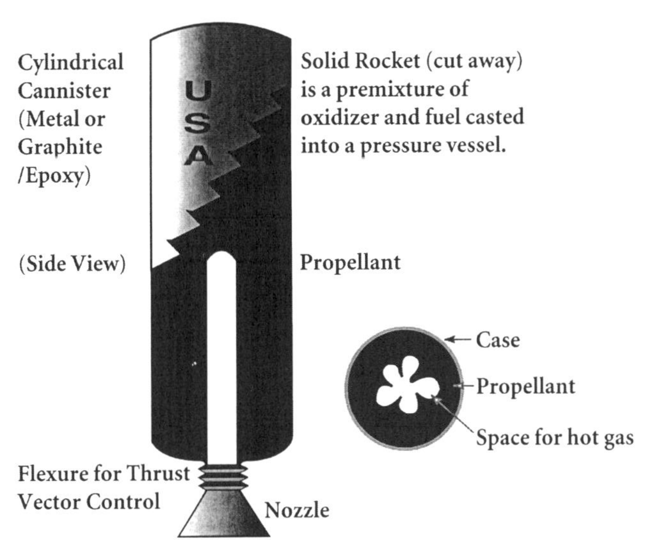

Cutaway diagram of a solid fuel rocket.

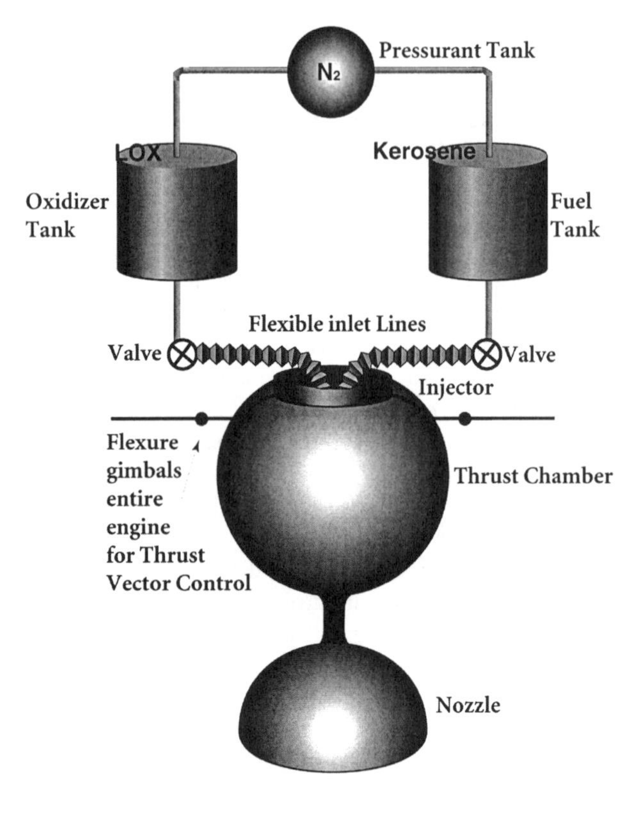

A liquid propulsion system. *Note:* In many rockets, the nitrogen pressurant's job is done by pressurization pumps.

for launch for years, even decades, and then fired immediately. They are also incapable of leaking, making them quite desirable on board ships and submarines.

For at least the last 40 years, proponents of solid propellant technology have argued these advantages, while liquid propellant rocket companies have squared off to do battle for the contracts to build modern rocketry. This argument is no less likely to be resolved than a final decision on whether Pepsi really outtastes Coke or All gets collars whiter than Tide. For some applications, solid technology really makes sense. Other times liquids are the only way to go, and there is a vast territory in which either one could and both have been used successfully.

When shopping for your next rocket, while you are being treated to dinner at Frére Jacques by the Solids boys, you might keep a few negatives in mind to balance all the dazzling positives. Solids tend to have lower Isp than liquids, so in general, rockets built from solids are larger and heavier. They must be handled with great care from the moment they are manufactured until they are used to ensure they are not inadvertently ignited. Because a solid rocket is already loaded with premixed fuel and oxidizer, it cannot be carried on many public roads in the US, nor shipped on aircraft or passenger carrying ships. Liquid rockets are shipped empty, weighing only 10% of their launch weight and completely safely since they carry no propellant. Solids are fueled when you buy them from the factory. Quite commonly visitors can't even enter a room storing solid propellant rockets until they have received an extensive safety briefing. Rockets using solid propellants must be built in specially equipped areas far from other people. Once they are started, solids cannot be shut off.

Because of the extraordinary care required in their manufacture and the special chemicals they use, solids tend to be expensive, though liquid rockets can also be quite expensive depending on the technological approach employed. The difference is that solid engines require highly capital intensive factories for fabricating, mixing, and bonding the solid propellant into the casing, and the propellant chemicals themselves are costly. Liquid costs tend to be in developing the engine, but recurring costs are typically, though not always, lower. Today's solids are not environmentally friendly, and their use may be curtailed in the future due to exhaust toxicity. Solids burn rapidly, leading to very high thrust levels, which is a big plus for launching, but it can also be a minus. A high thrust level subjects the payload inside the rocket to large acceleration loads. In launching from earth's surface, it is preferable to contour thrust to delay maximum acceleration until clear of the densest stratum of the atmosphere.

Got all that? If not, try the Creme Brulée. It's excellent. Make casual conversation. Here's a bit of rocket parlance you'll need. Solid rockets are called motors. Liquid rockets are called engines. Don't EVER mix those up. Like asking for La Coke instead of Le Coke, it's deadly. You might not get invited out again on another boondoggle!

Why the difference? Clearly you never studied French or you'd know the answer: Because. Because engineers, not users, dominate rocketry and they're careful people. Motor, according to engineer speak, is a prime mover that carries its own fuel/oxidizer (propellant) within it. Engine is just the device that converts propellants into motive force. Your car has an engine. Unfortunately, your grass cutter has a motor, according to common usage, which engineers would say is incorrect. Your grasscutter has an engine, too, and a gas tank. In my opinion, this definition would tell you that what makes your electric razor work is an electric engine, but most people call them motors. Best to fall back to that rationale for why it's Das Boot, Die Fernseher, and Der Spiegel, or why it's La Mano and El Tiempo, but La Braza. Because. Solid Rocket Motor. Liquid Rocket Engine. That's the mantra. Hey, it makes them happy, and around now the check is gonna show up. Go with what works.

What Really Happens When People Build Rockets

Philosophical arguments of the relative merits of solids, liquids, Vodka, and Old Crow can bog you down, and that's not what rocketry is all about. Maybe it's most instructive to just look at what people really choose for different applications. What do we see when looking at launch vehicles and their engines? All the largest boosters, Proton, Saturn V, Ariane IV and Ariane V, Long March IV, Delta, and Titan, rely primarily on liquids, and all crewed vehicles rely on liquids. The Shuttle uses both, but the fact that a solid rocket failure caused the loss of the Challenger raised the question of the suitability of solids for carrying people into space. Certainly, for launching astronauts solids are not usable without liquids because of the high acceleration.

By contrast, all the smallest rockets use solids—Pegasus, Taurus, Lockheed's LLV series and the venerable Scout. Except for possibly the LLV series, all of these vehicles were built expressly for or derived from military missions, so their use of solids could have something to do with that. Most of the small entrepreneurial launch vehicle companies (SELVs) plan to use liquids exclusively, arguing that they achieve the lowest recurring costs combined with safety and environmental compatibility.

Most commonly nowadays, we see a mixture of liquid technology used to get high Isp and controllability along with low cost, coupled with solid propellant engines "strapped on" to provide extra thrust for initial liftoff. Europe and the US have advanced solid technologies, while the Chinese and Russians have focused on liquids, and their launch vehicles reflect these technical factors. Sometimes the first stage of the rocket is a liquid "core" with solid strap-ons. The second stage might be solid, and the upper stages might be liquid, reflecting the need for controllability to achieve precise insertion of a payload into a particular orbit.

As if the decision between liquids and solids isn't confusing enough, a third option has been in the development of hybrid rockets. Amroc was founded to pursue this promising technology. A hybrid uses a solid, rubber-like fuel, but burns it with a liquid oxidizer, typically liquid oxygen. The hope for the hybrid is to combine the controllability and clean-burning advantages of liquids with the simplicity of solids. In fact, hybrids have proven difficult to perfect, and may bring the disadvantages of both systems into a single package. Unlike solids, they are not constantly launch-ready, and they lack the high Isp of liquids. The fluid handling equipment they require typically adds costs in ground support to liquids, while, like solids, the entire rocket must be pressurized, incurring large costs for the motor casing and for bonding the solid fuel to that casing. Liquid rocket engines are relatively small, though their fuel tanks might not be. The small engine can be tilted slightly, gim-

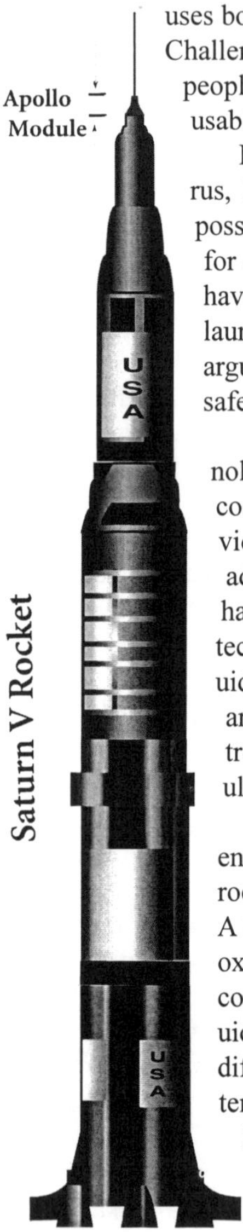

Apollo Module

U S A

Saturn V Rocket

It takes a lot of rocket to take a tiny payoad to the moon.

baled, to give thrust vector control. Solids, to be primary engines, must use flexible nozzles that are more expensive and less reliable. The hybrid would require a flexible nozzle since the long, narrow casing cannot be gimbaled.

Upper Stage and High Performance Rockets

We're in orbit. But now we want to get to an even higher orbit, like the geosynchronous orbits used by some communications and weather satellites up 40,000 km (24,000 miles) from earth. Or you want to get started on a trajectory to Mars or beyond. Whatever engine you choose should have excellent Isp performance because you had to lug all of its propellant up from the ground on your primary launch vehicle, a process that cost you at least $5000 for each pound lugged.

Interestingly, you'll still find both solids and liquids up here. Solids have lower Isp, but they don't have the overhead of tankage, pipes valves, pumps, and fill/drain fittings, so for many smaller boost engines, the highly elegant solid design is best. The most famous of these upper stages is the Star series by Thiokol. These are spherical titanium shells filled with solid propellant with a partially imbedded nozzle built in. They achieve mass fractions as high as 98%, meaning 98% of their mass is propellant and the other 2% is the titanium shell, the nozzle, and some attach points—an amazing number.

Larger upper stages tend to be liquids, because as the mass of propellant grows, the relative importance of the hardware shrinks. Liquid hydrogen/liquid oxygen is the favored propellant combination, with Isp numbers way over 400 seconds. Crank that into the rocket equation, tell your accountant that propellants cost $5k per lb to lug to orbit, and this stuff looks great. Both have to be kept cold, an application of the term "cryogenic" that has nothing to do with storage of cadaver wannabes, but you have to use them pretty soon after launch. If you are liquid, but not cryogenic, you are "storable." Lots of fuels are storable, kerosene, for example.

Few oxidizers are storable because, like oxygen, they tend to be gases. Nitrogen tetroxide has lots of oxygen in its molecules, and those oxygen atoms come off pretty easily. N_2O_4, also known as nitric acid, when mixed with water, which it does readily even with any odd water molecules wafting by in the air, or the moist, cozy lining of your nose, eyes, mouth, and throat, is nasty stuff. A few other molecules are added to it to keep it from devouring the tanks it's stored in. The smoke it creates in air while producing nitric acid is a kind of witch's brew red known in the biz as IRFNA, Inhibited Red Fuming Nitric Acid. You learn to respect chemicals by seeing an open bottle of IRFNA react with plain old air.

Tiny Rockets

Once you're coasting along at GEO, or en route to your rendezvous with a planet or two, your propulsion needs are over with, right? BUZZZZZZ. Guess again. Geosynchronous orbits are not stabile. You need ΔV to stay synchronized, to keep your orbit directly over a particular spot on earth's equator. You tend to drift not only east or west of the spot, but also to pick up a bit of north/south motion. The primary lifetime limitation on many GEO satellites is that they run out of propellant and leave their workplaces. A marvelous example of our creations imitating their creators, these orbital robots spend the rest of their lives ambulating pretty much uselessly somewhere in the vicinity of their former usefulness. Like many primitive peoples, we push these feeble elders into the deserted heath. In the case of satellites, the last of their propellant is expended to push them to an orbit higher than geosynchronous where they don't interfere with the productive labors of their younger counterparts, and they spend eternity orbiting the earth.

Just how much ΔV or propellant is typically carried on board a geosynchronous comsat? One answer is: as much as possible. For many missions the propellant tank(s) are filled at the launch site to bring the satellite mass up to the very maximum the launch vehicle can carry. This can constitute half the total satellite mass. Remember the Ford Mustang we took drag racing a while back? It was also 50% propellant, and hit about 1000 meters per second ΔV with an Isp of only 150 s. Modern comsats achieve Isp over 300 seconds, thus they might carry 2000 m/s of V capability to provide orbit correction and to control the spacecraft attitude.

Interplanetary spacecraft coast literally hundreds of millions of miles, aiming at targets at most a few thousands of miles across. As media coverage of these missions is wont to report, this is tantamount to kicking a soccer ball in New York through the goalie net of a soccer stadium in LA. That's not quite how it works. We kick the old ball in the general direction of south-of-west-ish, and then as we close in on California, we keep batting the ball around to a closer and closer approximation of the trajectory to the net and the final kicks are applied in the final days of a trip that lasted hundreds or sometimes thousands of days. Those kicks are applied by tiny rockets on board the interplanetary spacecraft.

In both the geosynchronous and interplanetary cases, the costs of lofting all that rocket propellant are huge. But the thrust needed is minuscule. We have plenty of time to make trajectory changes by applying gentle force over long periods, a stark contrast with launch vehicle propulsion, which doesn't need to be so efficient. For launching, you launch from the ground, and bringing propellant to a launch site is cheap. You need beaucoup de thrust because gravity is trying to reunite you with the ground and hence you want to get up and into orbit ASAP.

The propellants are different. The rockets are different. Even the people who build the rockets are different. Huge launch vehicle engines and motors highlight the engineering of giant structures and handling of large quantities of chemicals. I visited the plant outside of Rome where Ariane V's strap on solid rocket motors are cast. The cleaning vat used just to get the casing ready for the adhesive is 8 stories deep and 20 feet across, the world's largest washer/dryer.

But the engines that correct the trajectories of satellites are tiny. Some weigh less than 1/2 pound (220 grams). There are propellant valves whose total mass, including the seal, the actuator, and the electrical connector is less than 10% of that weight. What lengths we go to for higher Isp! Early tiny rockets used monopropellants, but now we have tiny injectors that precisely mix fuel and oxidizer into a combustion chamber the size of a thimble. Electric augmentation has been used to make the propellant warmer than just the temperature achieved by the energy of the chemical reaction. The latest trend is pure electric propulsion, which is heating propellant by passing it through electrical arcs in arcjet thrusters, some of which have double the Isp of chemical rockets. Coming soon are plasma and ion thrusters with double again the Isp of arcjets. Ion thrusters strip electrons off typically massive neutral molecules such as xenon. The ions thus formed are accelerated across an electric field to achieve exit velocities as high as 25,000 m/s.

2.11 The Final Frontier: Infinite Isp

Space is not empty, but the space near our sun is particularly filled with stuff. For one thing, it's loaded with photons—corpuscles of light—shining on us from old Sol. Our friendly star is also putting out a pretty steady flux of high energy particles known as the solar wind. The solar wind is like earth's trade winds, very dependable, very constant in both direction and strength, except for occasional gusts during solar storms, which would not have any significant effect on sailing conditions. Ultimately, its advantages could induce us to go solar and harness an infinite source of V.

Solar sailing has been proposed repeatedly for decades. Sailing on the solar wind requires no propellant, just as sailing earth's oceans requires only the equipment and the skill to master the wind and the sea. Unlike terrestrial winds, the solar wind, having traveled almost 100 million miles to get here, is pretty weak. To get even a millionth of a pound of thrust requires unfolding tens of square meters (hundreds of square feet) of sail, and then manipulating the sail to reflect solar wind particles and push the spacecraft by means of the recoil of their momentum transfer to the sail. Unlike navigation of the high seas, where sailing predated power, solar sails must not only solve their own technical problems, but also the philosophical biases of contemporary propulsion culture, which is to carry chemicals with you and not depend on somewhat ephemeral natural sources.

2.12 Propulsion for Small, Low Cost Missions

You've been having such a great time reading all about propulsion, thermochemistry, and rockets that you almost forgot that dead-end job of yours, clinging to a few threadbare contracts for small, low cost satellites. Here are a few things to tell the boss in order to free up some overhead money for that trip to the AIAA Joint Propulsion conference in Vail or someplace.

Arguing against your major league boondoggle, bossman says, "Small satellites [*Particularly the underbudget jobs you always get. —Ed.*] rarely have propulsion systems." He's got you there. It's fast thinking or face the ice and slush on the local ski bump. You beg for attention. Ahem, sir? He's on the phone. Subject your favorite Arbeitmeister (or Arbeitmesterin) to the following bullets, which you have committed to memory:

- One way to stay cheap is to take a piggyback ride. Those rides don't always take you where you want to go, and propulsion on board is cheaper than a dedicated ride into orbit.
- People are already building small geosynchronous satellites that orbit 40,000 km above the earth. At that distance over the equator, there is not enough magnetic field, nor enough variation in the field direction, to control the satellite attitude with magnetic torque coils. Small attitude control torques must be generated with small attitude control thrusters.
- Hey, people are even talking about interplanetary small spacecraft—the same attitude control problem *plus* you need course corrections en route to get to your destination.

If this doesn't get you to the deep powder, you'll have to pull out the big guns:

- All those LEO (low earth orbit) clusters of communications satellites, like Starsys, Orbcomm, Teledesic, Motorola, Ellipsat, all need propulsion to keep the satellites evenly spread out in their constellations.

2.13 See you in the hot tub for some Gluhweine!

Yes, in its innocent youth, small satellites didn't have propulsion on board. In 1983, I was the closest approximation to a propulsion engineer in all of AMSAT (Amateur Radio Satellite Organization), compared with maybe 100 active electrical engineers worrying about radios and on-board computers. But things are changing. Propulsion is becoming a key component of many small, low cost satellites. For many of these, the combination of short mission life and low cost has resulted in use of compressed gas, usually nitrogen, which is stored in a commercially available, DoT-certified, gas bottle. But hydrazine and bipropellant commercial systems have already been flown on small satellites, and the prospects are for further increases in the number of propulsion systems on small satellites.

Chapter 3

Orbit Mechanics — Or, What Keeps These Things Up, Anyway?

I apologize, in this and the next two chapters, for possibly interfering with the sensibilities of those who take the subject of orbit mechanics really seriously and attack it with computers and equations (hopefully not in that order). Also, thanks to Richard Warner, who got me started on this virtually interminable subject (stay out of dark parking lots, Bub), and also made all of the original drawings that I hope clarified the hopelessly murky text. Richard also did his best to insert technical accuracy, but failures are my own. Here's our guarantee: If you find yourself lost in space due to an errant rocket burn, write the publisher for a prompt refund.

Editor's note and disclaimer of liability: This article explains the basics of what a satellite orbit is and in general terms what different kinds of orbits are available for small satellites. If any of the following describe you:

1. Busy planning a mission to Pluto at JPL using three hypercubes and two Cray-XMPs;
2. Wrote your Ph.D. thesis at MIT on nth body perturbations to satellites orbiting planets of time varying mass and mass distribution;
3. Edited a textbook on orbit mechanics or satellite guidance and control;
4. Work as a tenured professor of Aerospace Engineering at a major university;

Then don't read any farther. Or go ahead and read for the marvelous wit and witticism, but don't write to me and complain that your brain got rusted. This article is for the rest of us who toil in the somewhat less sublime spheres or art, commerce, and left-handed thread design.

3.1 T-Off Time

Being younger than Orbital's president, while he was launching monkeys I was playing golf barefoot at Lost Creek Country Club. By the time I got to monkey launching age, I had given up on rockets in favor of girls and Fords. But like a lot of kids, I too was fascinated with putting things in space. Probing questions haunted my

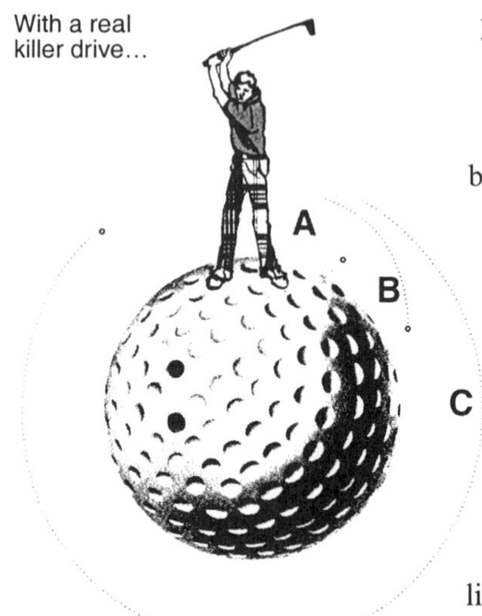

the ball "falls" around the world.

youth, like, "If I hit the ball hard enough, will it go into orbit?" (This serious, career-minded thinking pattern, developed at an early age, explains why I exist as a ward of the near-bankrupt aerospace state while my peers are all brain surgeons and Supreme Court justices.) My older brother opined that I would destroy the ball first. Unfortunately, at age 8 you don't know how to win an argument by defining terms and splitting hairs. And that, also unfortunately, is how older brothers get away with murder.

Leave that topic for my analyst, Ernst. It turns out that my 80-yard drive into the rough was actually in orbit, a very low orbit that collided with the earth after only a few seconds. A really long ball sails many more seconds, maybe passing over the next hill and out of sight before finally bopping someone teeing off for the next hole ("Fore!!!!!!"). This is the trajectory marked "A" in the figure. Now a Truly Awesome drive (we're talking Robocop, minimum, here) could go so far out that the curve of the earth becomes important. The ball still falls to earth but a lot farther away, like Trajectory B in the figure. While you and my brother might insist this isn't IN ORBIT, depending on just what you mean by the phrase, it is AN ORBIT. When Alan Shepard was the first person to ride a trajectory like this, high above earth's atmosphere on May 5, 1961, he made quite a splash (are there no lengths I won't go to for a joke?)

If somehow you get that golf ball going so fast that the earth falls away from the ball just as fast as the ball falls toward the earth (trajectory C in the picture), the thing is not going to come back to the ground. The golf ball is in stable orbit, that is, its trajectory never intercepts the surface of the earth, unlike any of my golf balls except the one implanted in the guy teeing off at the fourth hole. Now it so happens that Yuri Gagarian rode into space on this type of trajectory on April 12, 1961, thrcc weeks before Shepard. Sputnik did this in 1957, but the moon got into this type of trajectory around the earth quite a few years earlier than that.

3.2 Orbit Definition Good Enough for the Rest of Us

So what is this "orbit" concept? If you need to win arguments, go ahead and insist that the vase your mother-in-law gave your wife as a wedding present that had been in her family for 649 years was actually in orbit before the orbit's path collided with the tile kitchen floor after your big elbow bumped it off the counter. But at a party nobody likes a bore, and except for these special circumstances, Miss Manners would advise that you consider an "orbit" to be a more or less stable motion wherein a body continuously falls toward something big, like the earth or the sun, in such a way that its trajectory doesn't impact the big object. Examples of this highly unusual way of moving include the earth's path around the sun, the moon's path around the earth, the motions of most man-made satellites around the earth, and Marion Barry's performance as mayor of Washington, DC.

Having forgotten to also bar astrophysicists from reading this article, I will undoubtedly also be reminded that sometimes two objects go into orbit around each other, and that both orbiting bodies are of similar mass. Binary stars, they'll say. Fine. But to stay anywhere near the point of this chapter, we'll let binary stars escape my finely honed keyboard and stick with relatively itty bitty things (don't you love this technical jargon) going around big, round things (ignoring that the Kanji for big round things spells log). After all, small satellites are little bitty things relative to the big round earth around which they orbit.

3.3 How Mother Nature is Cruel to Eight-Year-Olds

There are a few problems with convincing your big brother that you can orbit a golf ball by means of a borrowed two-wood. Thanks to the pesky facts that the earth is pretty big, pretty massive, and surrounded by a viscous atmosphere (air), getting things to orbit is rough, particularly by swinging your mother's clubs choked 6" down the shaft. The slowest speed you must go to get far enough not to hit the earth as you fall is about 16,000 miles an hour (25,000 km per hr). At that speed you really don't want to be plowing through the air. The lawsuits from the sonic boom damage alone make it prohibitive, besides the fact that no known materials can withstand the aerodynamic heat created in the atmosphere at that speed. So to be in orbit and enjoy the comfort and convenience of not burning up, an altitude of at least 100 miles (160 km) is desirable.

Let's face it, eight-year-old barefoot kids wielding borrowed two-woods on Lost Creek's back 9 are not going to get a golf ball up to 16,000 mph and 100 miles. In that sense, my brother had a good point. Careers have been made on perfecting gadgets that accelerate and lift objects. People like Werner Von Braun and Robert Goddard spent their lives on the problem. Their toys are somewhat more complicated and larger than mine were. As shown below, they look like rockets, which is the generally accepted technology when you want to get to 100 miles altitude and 16,000 mph.

For those with a bent for numbers, about 10% of the work a rocket does lifts it to 100 miles altitude (h) and 90% of the work is accelerating up to 16,000 mph (V). So

More suitable toy than a 2-wood for pursuing a stable orbit by getting 100 miles up (h) and accelerating to 16,000 mph (V).

you can figure that getting on an airplane going 500 mph at 8 miles altitude buys you about a 0.9% edge in getting into orbit compared to launching from the earth's surface. Thus, the real gain of a Pegasus air launch configuration is not that it's a good way to get into orbit, rather that it's nice to be able to leave the ground from an airport instead of a launch pad.

At 100 miles, moving at orbital velocity, one orbit around the earth takes about 95 minutes. One idea for very fast transportation is to make an airplane that can achieve a near orbital trajectory. While not capable of reaching a stabile orbit altitude and velocity, it would, like trajectory B, "fall" almost half way around the world in under 50 minutes. This would mean London to Los Angeles in 35 minutes (with airline flights that fast, my writing would never get finished—but my laptop's battery-life would finally be near-adequate). Another major minus of suborbital travel is how painful it would be to spend an hour waiting for luggage at Dulles after coming in from Tokyo in 40 minutes.

Rockets that now provide this "suborbital" service are called sounding rockets, probably because they have been used over the years to take upper atmosphere soundings. Usually carrying an instrument payload to altitudes around 100 miles, they do not achieve full orbital velocity and hence their fall intersects the earth's surface. They can provide exposure to space vacuum and zero gravity conditions for about 10 minutes before reentry.

3.4 Orbit Altitude

I rode on the good ship Susquehanna recently, or at least its replica, which plies the harbor of Izu peninsula, Japan. ¥300 is all it takes to tour the harbor where Commodore Perry landed. And a hundred miles is all it takes to be in orbit. But who wants to tour the famous scenic harbor from inside the hull of the replica Susquehanna watching the paddle wheels go around? You want to be up top, breathing the air, taking pictures, and feigning throwing your friends overboard. To do that costs ¥250 extra, payable at the candy counter once the boat gets moving.

A hundred miles means you're in the crew's quarters of orbit. For one thing, there's still a little air up there. Not a lot, but at 16,000 mph you hit the few molecules there pretty hard and pretty often. The drag slows you down and pretty soon your velocity isn't high enough to keep falling past where the earth's edge is and, like that old vase, you are in for a landing with the earth, possibly on the kitchen tile floor. Luckily the atmosphere's density drops off exponentially. Thus at 100 miles altitude you're going to stay in orbit for a week or two, but at 175 miles, where the Space Shuttle usually hangs out, you'll last a few months. The Shuttle stays on orbit only about a week, not because of orbit mechanics, but mainly because the number of Disney videotapes that can be stored on board is limited and the crew gets cranky. At 300 miles, orbital lifetime (the jargon to use at cocktail parties once you've identified the particular creature you want to mesmerize with your command of aerospace engineering) increases to about 10 years. Just a little higher, like 500 miles, and your great-great-great-grandchildren are the ones to worry about where your primitive satellite smashes on its return to earth.

Getting up high has a big advantage in lifetime. Unlike riding the Japanese Susquehanna, the relative cost to go higher, once you've paid the basic ¥300 to get to 100 miles and 16,000 mph, is a very small additional amount of energy. So why not always go higher? Well, if you want to get every last ounce into orbit, you can lift more to a lower altitude. Also, the radiation dose people and electronic things receive goes up a little with altitude. Plus, if you want to make color glossy pictures of the shifting sands of the Sahara, your friends' ABM launchers and license plates in Moscow (Why do they always boast that some $1B satellite can do this? My mother-in-law did it on vacation in Russia last year with a cardboard camera she bought with the film built in for $17 at K-Mart). Well, anyway you can get better pictures of the earth from lower down. Some people like to experiment with the chemicals at the top of the atmosphere. And some people are just bad calculators or they ran out of money building a rocket that was supposed to go somewhere useful and just ended up at 100 miles.

Another advantage of being higher up in orbit is that you can see points on the earth that are farther away. Most spacecraft radios are "line-of-sight" like TV and FM radio signals, and by getting the satellite up higher the satellite can reach a larger region, shown in the figure.

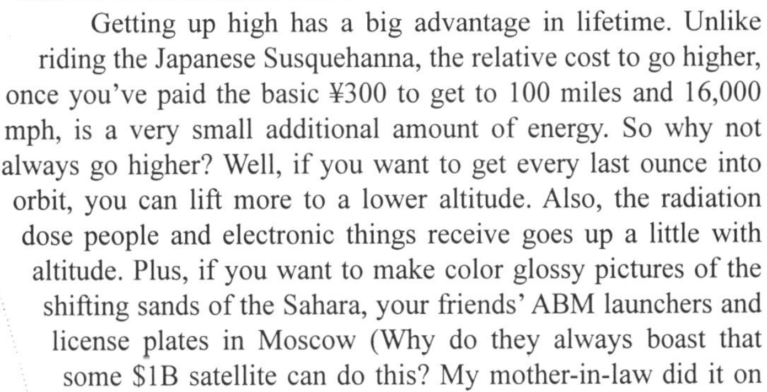

The higher you are, the more you can see.

A bit above 500 miles you enter a region of increased radiation intensity called the Van Allen belts, which extend to about 5,000 miles. You don't find too many satellites in this region. Matter of fact, the next popular orbit altitude is about 24,000 miles (40,000 km), a long distance, being about 2 earth diameters away (or 450 times farther than the distance from Toledo to Lima, Ohio). The ticket to put your satellite in 24,000 mile orbit is 3 to 5 times more expensive than a low earth orbit launch. Signals are 6,400 times weaker from 24,000 miles than from low earth orbit satellites at 300 miles. The exquisite radio antennas used to reach earth from that distance make that coat you bought your wife to get her over the loss of the crummy old vase look like pocket change, not to mention the tracking system you need to point those shiny antennas at the ground station.

3.5 So What's Everybody Doing Up Here?

Having already covered about 10,000 years of what humans have learned about orbital mechanics, it seemed like a good time for a pepperoni pizza (lots of small round shapes with big round shapes, according to Ernst). Pick up the phone, order one or two, and prepare yourself for geosynchronous orbits, Molniya and other eccentric orbits, sun synchronous orbits, what orbits look like from the earth, and some special tricks like earth escape and minimum sunlight orbit. Please, if the pizza does not supply enough heartburn, then while finishing it, think about the following homework problem:

The moon is smaller and less massive than the earth and has no atmosphere. Could you get a golf ball into orbit there? Assume Cher gets rid of her perm and Microsoft elects to build a sprawling complex within the lunar surface instead of the Lunar Towers IV Center now in preliminary design (but unlikely to gain lunar zoning approval), so very tall obstacles should not be an issue.

Chapter 4

Orbit Mechanics II: The Movie

They say the line between genius and insanity is a fine one. Having puzzled over the quiz provided at no extra cost as part of this nail-biting account of orbit mechanics, you're probably wondering what the call is. Well… what was the question? How hard is it to put a golf ball in orbit from the moon's surface?

The good news is that with no atmosphere and assuming a spherical moon, meaning no mountains or office towers poking up above the ground, you don't need to provide energy to gain altitude. How fast does the ball need to go to reach a stabile orbit scraping the surface of the moon? The moon's mass is about 12% of the earth's, and its radius is 27% as large. This means lunar orbit velocity is about 21% of earth's, which would make it 1,550 meters per second (about 3,465 mph). For comparison, a golf ball hit on earth with that huge velocity off the tee would travel, ignoring aerodynamic drag, about 79 miles. With drag, maybe 10 miles. So if you're the kind of guy who tees off at the Los Angeles Country Club and chips onto the green in Palos Verdes, you have a career ahead of you in golf ball launching on the moon.

The answer to the quiz is, sadly, no, not with a golf club. With a small rocket maybe. If you want to put things in orbit with a golf club, find a small asteroid where gravity is real weak.

Now that we're all relaxed and resigned to carry on with our mundane lives instead of putting golf balls into zero altitude orbit around the moon using 2 woods, we can attack some pretty substantial issues in satellite orbits like:

- Geosynchronous Orbits (GEO)
- Sun Synchronous Orbits
- Orbit Planes
- Ground Coverage.

4.1 Geosynchronous Orbits (GEO)

Probably the orbit most important to The Rest of Us (the people this article was written for) is geosynchronous orbit because satellite transmission of TV, radio, telephone, and telemetry (data) uses satellites in this orbit. This wasn't always true. Back when televisions used to come in something called black and white and phones used to have something called dials on them, and people used to cook food by heating the air around the food in big insulated boxes and listen to music made by the vibrations of small needles travelling in plastic grooves on rotating discs, back in the time before any of today's college students were born and I was playing the back nine of Lost Creek in peace, a few satellites were used for television relay from low earth orbit, similar to the orbits ridden by Yuri Gagarian and John Glenn. These orbits pass over any point on earth very quickly since they travel around the earth in an orbital period lasting about 95 minutes. This short time period is inconvenient because to transmit a TV broadcast via satellite between two places on earth, both places must be able to look up in the sky and "see" the satellite with their radio antennas. In a low earth orbit, this visibility period is short, typically five or six minutes. If this were all we could do with satellites, there would be no Bart Simpson. Let the gravity of that statement settle in for a while, eh? Well, there are two solutions. Geosynchronous orbit is one of them.

Wouldn't it be nice to put a satellite in orbit that doesn't circle the earth, but rather just "hangs" over a single spot on the earth's surface? Such a satellite would enable 24 hour a day TV transmission—Bart included. Gnarly, you might say.

Not to throw cold water on a good idea, but in general, a satellite that just hangs there is impossible. I distinctly remember mentioning that orbiting bodies can be thought of as continuously falling toward earth. The reason they don't simply fall to earth, making a large crater but not really doing what they're supposed to do, is that they move forward fast enough to fall over the arc of the earth's horizon. Two facts of adult life are now going to be exposed to the reader. If you are under 13 years old or if you are one to cling to faint hope, stop reading now.

1. There is no one named Santa Claus living at the North Pole. Nobody lives at the North Pole.
2. Satellites don't "hang" up there. (reverent silence)

Let's say we're in Low Earth Orbit (LEO) over the equator flying west to east (As the World Turns) at 350 miles (600 km) altitude. Every orbit takes 98 minutes during which time the earth turns a little, and it turns out that it takes 105 minutes to pass over the same point on the ground. What if we raise the orbit altitude a little? The orbit is now going around a bigger circle, and the earth's gravity is weaker because the distance to earth has been increased. The earth takes a whole year to get around the sun because we're so far from the sun that the path around it is a long way, about 560 million miles.

The combination of weakening the pull of earth's gravity and increasing orbit circumference with altitude means that orbit period increases with orbit altitude. At

24,000 miles (40,000 km) the orbit period is 24 hours, which is the same as the earth's rotation period. The figure on this page shows that a satellite in orbit over the earth's equator at 24,000 miles still goes around the earth, but at the same rate the earth turns, so the satellite appears to move in synchronized motion with the earth—hence the term geosynchronous (or, in the case of this particular illustration, Eiffel-synchronous).

There are a lot of nice things about geosynchronous satellites. They appear stationary relative to us earthlings and our backyard satellite dishes. Once you get the dish planted on the ground, it doesn't have to move to track the satellite's motion. Satellites in geosynchronous orbit are high enough to see half the entire earth at once, like the Apollo mission pictures of earth showing it as a big circular blue, green, and white disk. Seeing this much means that a satellite can relay TV, telephone, and other communications over long distances in a single hop.

SUN

The few disadvantages of geosynchronous satellites are due to the simple fact that they are far away. Traveling at the speed of light (or radio waves), it takes about 1/4 of a second for your electronically transmitted voice ("Mom, Sheila and I are getting married.") to reach from your condo in Culver City, LA to your parents' condo in Sarasota, FL. Add another 1/4 second or so for their voices to answer you (Mom: "My Baby!"; Dad: "Sheila who?"). This delay is doubled if the conversation is carried across two geosynchronous links, which can happen when calling, say, from LA to Tel Aviv. As difficult as this can make conversation for you, it's really hard on your computer's modem or your fax machine, because these devices rely on constantly hearing replies from the other end of the line. They are designed to wait for these replies in cases of distorted transmission.

A geosynchronous orbit maintains the same relative orientation relative to the earth, and the Tour Eiffel, throughout the day.

From a long distance, radio signals become pretty weak, mainly due to the beam spreading, just the way a light appears dimmer seen from a distance. Elaborate, expensive antennas are used both on the satellite and on the ground to reduce this spreading. The high power transmitters needed on the satellite despite these antennas are also expensive. All the fuss about direct broadcast satellites is based on using more powerful satellite transmitters and narrower beam satellite antennas so that you don't need a huge dish antenna in your yard to receive satellite broadcasts. I mentioned that there are two solutions to the problem of using satellites in low orbit for TV transmission. One is to raise the satellite orbit to geosynchronous orbit (GEO). The other is to fly many satellites at low orbit arranged so that one is always in sight of you and one is always in sight of the stations you're communicating with. Since they don't require such elaborate transmitters and antennas, these LEO satellites could be much more simple, and that is why small satellites are being proposed for LEO networks. The disadvantage of the approach is you've got a lot of satellites moving in their own independent orbits, and a lot of "cross linking" is required. Satellites at LEO see only a small part of the earth's surface, so to communicate over long distances, the signal is uplinked to one satellite of the system, crosslinked among satellites, and finally downlinked by a satellite in view of the receiving station. A further complication is that the constellation of small satellites moves relative to observers on earth, and hence the uplink, cross link, and down link satellites are constantly changing. (See Chapter 17 for more on how LEO satellite clusters work.)

But let's leave the world of telecommunications to the telecommunicators and get back into orbits. Orbits are described in part by their inclination, meaning the angle the orbit makes with the equator. Geosynchronous satellites are normally placed in equatorial orbit because they can only be synchronous with a point on the equator. This is because every orbit makes a complete circle around the earth. Further, this circle has to lie in a flat plane that intersects the center of the earth. This is because all objects fall toward the center of the earth (which is why people think "down" is such a simple concept.)

Thus, for that "geosynchronous" orbit altitude over the North Pole (yes, where Santa Claus doesn't live), the satellite would have to proceed south, pass the South Pole and reappear over the North Pole 24 hours (1 orbit period) later. Not a tremendously handy orbit, at first glance. But in low earth orbit, being at 90 degree or "polar" inclination has a lot of interesting advantages. A satellite in orbit inclined 90 degrees to the equator at 350 miles (650 km) "sees" every point on earth twice a day. This orbit is show in the figure.

The orbit path remains (to first order) stationary while the earth rotates.

Every 95 minutes the satellite makes a circle over the earth following roughly a line of longitude on the globe. But by the time it returns to the North Pole, the earth has rotated about 1/18 of a revolution. Thus, on the next orbit, it follows a new longitude line about 1000 miles west of the previous

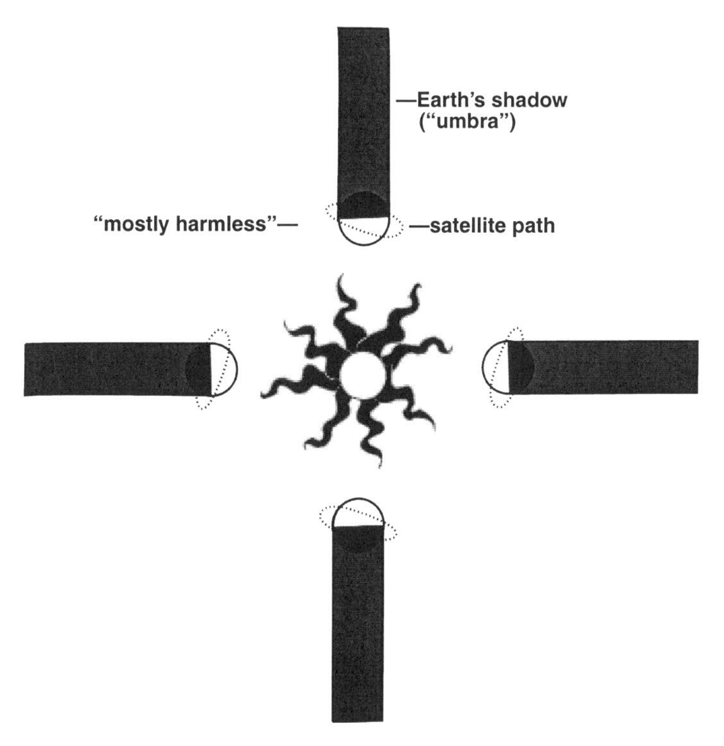

A sun synchronous orbit maintains the same orientation relative to the sun throughout the year.

orbit path. Imagine a satellite in polar orbit that passes over Cleveland at noon. An hour and a half later it passes over Denver. An hour and a half later, three hours after passing Cleveland, it passes over San Francisco, where it is local noon. By carefully selecting orbit altitude, it is possible to put a satellite in an orbit that makes exactly 1 or 2 or 3 or 4 or... 18 orbits in 24 hours. These orbits pass over the same spot on the earth at the same time every day. Except for one small problem. The satellite's orbit is in no way linked to the earth's rotation. It is pretty much stationary relative to a fixed reference frame, like the stars and the sun. So as the earth revolves around the sun, the synchronicity is lost (but a Police album is gained).

Fear not, those who want their own satellite to pass over their house every day at 5:00 a.m. and thereby to own the world's most expensive alarm clock. There is a special orbit called a sun synchronous orbit, which is shown above.

This orbit is inclined slightly away from pure polar orbit in such a way that the orbit plane (the imaginary disc whose edge is the satellite's path) rotates (precesses, if you keep that kind of company) once a year and the satellite stays sun synchronous indefinitely.

Since so few of these satellite alarm clocks have been sold, who uses sun synchronous orbits? Imaging satellites use them. They can compare traffic patterns or crop growth or nuclear submarine factory construction best by imaging targets at the same time of day every day, to get the same sun "color" and the same shadow patterns. A neat aspect of the sun synchronous orbit is that if you're synchronous at the right time of day, the satellite almost never goes into the shade of the earth (Can you say "umbra?" It sounds very chic.) Note in the figure that even though the satellite flies over shaded parts of the earth (areas where it is sunset or sunrise), it is high enough that the satellite itself can always see the sun. This means the thermal environment of the satellite can be very stable and it may need few or no batteries to power electronics since its solar panels are almost continuously illuminated (batteries might still be needed to handle peak loads).

Well, I don't know about you but all this satellite talk is really tiring and I'm ready to pop open a cool one and flake out in front of the tube with the Simpsons (try that one on your Japanese tutor). But that fat lady, whoever she is, is not singing until we get to talk about getting from one orbit to another.

Chapter 5

You Send Me: Orbit Mechanics III

Initial Orbit

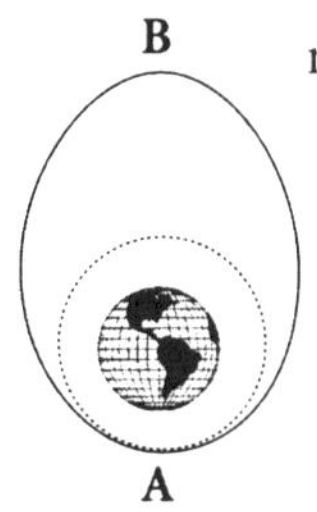

Transfer Orbit

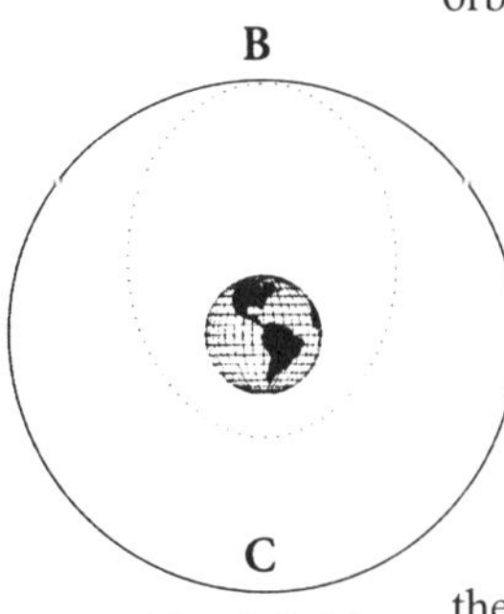

Final Orbit

A Hohmann transfer orbit is the most energy efficient way to go from one circular orbit to a different circular orbit.

My wife, Nancy, is always explaining to me that particles, satellites, cars, and other so-called inanimate objects do not have feelings, ideas, and aspirations. She calls them "it" or "that." I call them "he," "she," or "that guy." While most people find our points of view irreconcilable, I believe the only difference is that she never owned a 1969 MGC. Talk about personality (and usually a pretty conniving one at that). I promise not to depart on a musical interlude into how this car knew when it was being neglected and how it arranged for suitable revenge. Rather I mention this because in at least one sense she (Nancy) is right. Simple objects don't know much about where they've been or where they're going. Like horses, they live for the present, and are probably much happier for it.

This chapter, for those of you stepping into the middle, which means everyone since I started in the middle and intend to stay there right up to the end, is about modifying satellite orbits. Three such orbits are shown schematically on the left.

The innermost one is like a low earth (low, circular) orbit (LEO), the outermost one is like a geosynchronous (high circular) orbit (GEO), and the elliptical one is an elliptical orbit (breakthrough insight, that was). The elliptical one is more often thought of as a geosynchronous transfer orbit (GTO). The two circular orbits could also represent the orbits of two LEO satellites, one higher than the other, as for instance the Space Shuttle in low orbit and the Hubble Space Telescope higher up. What follows then is the story of how to get up and down.

Imagine you're one of two satellites, both now at point A on the diagram. One of you is in LEO and the other in GTO. How do you know which of you is which? At A, you are both pointing in the same direction, but the one in GTO is going faster than the one in LEO. This extra velocity makes your fall toward the earth less steep and you swing farther outward.

So now you're at point B, meeting up with another orbiting friend. Again, how do you tell who's in GTO and who's in GEO? As before, your friend in GEO is going faster than you, albeit in the same direction, at point B. Hence you fall a little more steeply toward the earth heading for your rendezvous with point A again.

This simple revelation doesn't tell you everything you'll ever need to know about maneuvering in orbit, unless you never planned to go there anyway, which is, after all, sort of the degenerate case. But it does tell you a lot. To get from LEO to GTO, all that's needed is to align a rocket motor with your local direction and accelerate to the velocity of a GTO orbiting body at its perigee (orbit low point). This is the first step in reaching GEO from, say, the Space Shuttle that drops things off in LEO. Note that wherever you burn your engine (point A), you return to over and over. Rocket burns at perigee raise the apogee (orbit high point), but only affect the velocity at perigee.

This maneuver is sometimes called Apogee Raising, but in reality you are Perigee Accelerating. To circularize into the higher, circular orbit, a second rocket engine burn, called Perigee Raising, is effected at the apogee.

These maneuvers can also be done in reverse by slowing the velocity. At some point in a circular orbit, the orbit becomes elliptical with a lower perigee. This is how people get themselves out of orbit and back to earth.

All of this maneuvering by changing the forward velocity to raise and lower apogees and perigees is generally grouped into the term "Hohmann Transfer" because Professor Hohmann first suggested it and calculated how much velocity change is required to effect a particular apogee or perigee change. There are other maneuvers, but this one is really the Big One. It has applications beyond just going from the Shuttle to GEO. For instance, say you're a satellite at GEO, and you're watching the weather over the West Coast of the U.S. and the Pacific. And your friend, a satellite whose job is to do the same thing over the U.S. East Coast and Atlantic, dies tragically of a burned out light bulb (don't laugh; that part actually happened). To make matters worse, it's late summer, boring weather time over the Pacific, but hurricane season in the East. So you want to hustle your buns on over to the East and fill in for your departed amigo. How to do it?

Remember, the periods of orbits are shorter as the orbit altitude (radius) decreases. For instance, it takes the earth a whole year to get around the sun, but it only takes the moon a month to get around the earth (it's a lot closer than the sun), and it only takes a satellite at 750 km (400 nautical miles) about 100 minutes to get around the earth. Geosynchronous satellites, by comparison, take 24 hours to get around the earth, which is why they are called geosynchronous. (For that matter, the fact that the moon takes a month to get around the earth is why the time period is

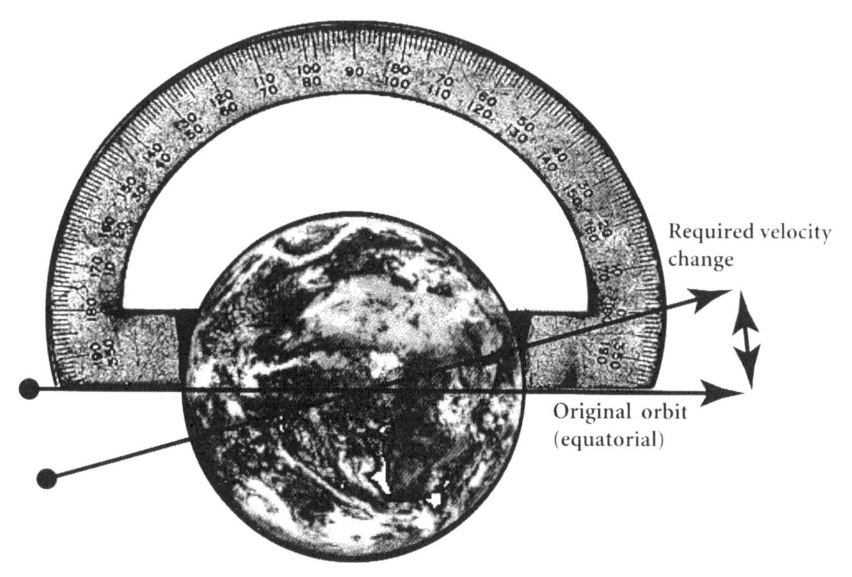

Small change in inclination requires small velocity change.

called a month, or why the moon is called the moon, depending on your point of view.)

But I digress. To move from synchronous orbit over the equator below the U.S. West Coast to the equatorial position south of, say, Bermuda, you need to speed up relative to the earth; that is, get into an orbit with a period of less than one day.

No Problem! says Prof. Hohmann. Just do a perigee lowering maneuver (burn your rocket to slow the satellite's velocity). You'll drop into an elliptical orbit with the apogee unchanged, but a lower perigee, and that new orbit has a shorter period, meaning that it goes around the earth in slightly less than one day. So relative to an observer on earth, you'll slowly move West to East across the sky. When you get to the eastern location you want, you again burn at apogee, this time in the prograde (as opposed to the retrograde) direction, raise the perigee back up to equal the apogee, and you are ready to watch for hurricanes.

This use of partial mini-Hohmann maneuvers is the way geosynchronous satellites keep themselves exactly over their desired locations. It is one of the two major reasons geosynchronous satellites have to carry a significant amount of propellant, and exhausting that propellant limits the useful life of the satellite.

The other reason these satellites need propellant is that geosynchronous only means something over the equator. Remember Santa's geosynchronous polar orbit. You'd pass over the North Pole once every 24 hours, over the equator on the way south once every 24 hours, and so on. But you would not appear stationary in the sky.

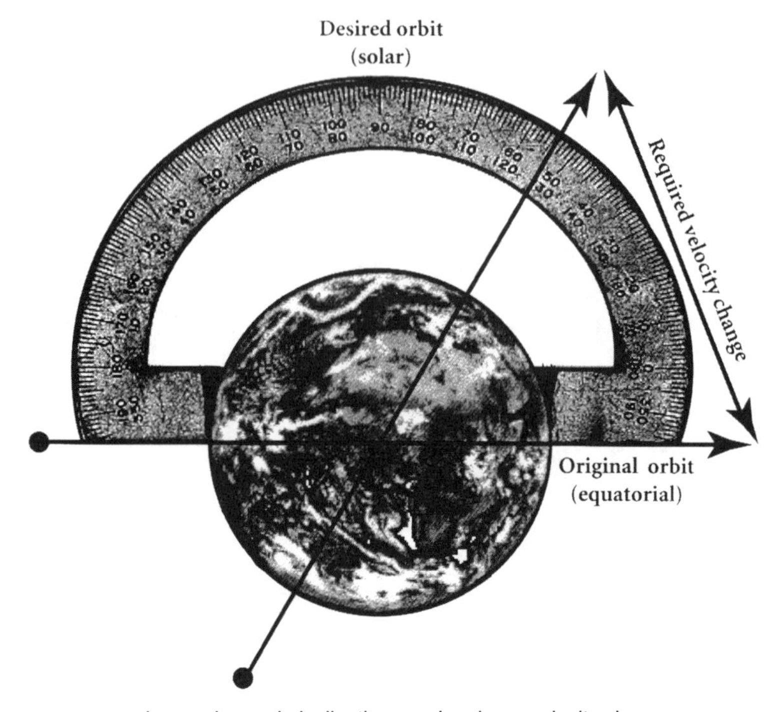

Large change in inclination requires large velocity change.

What do polar geosynchronous satellites and jolly elves with beards living at the North Pole have in common? They would both be very useful, were they to exist, which, Virginia, they do not. Geosynchronous satellites, besides needing to trim their orbit altitude, need to trim their orbit inclination to stay truly geosynchronous.

Imagine the erstwhile equatorial geosynch orbit somehow gets inclined by one degree, oscillating above and below the equator once a day. All the ground antennas would need to move around to track it. Since geosynchronous satellite dishes are not outfitted to be moved easily, this is a nuisance.

The last gasp of this prolonged treatise into satellite motion, then, is how to change planes, like go from one degree inclination to equatorial (0 degree) inclination, or go from equatorial to polar. The problem is the same. Going back to the line of thinking that satellites are rather naive when it comes to remembering and planning, how does a polar-orbiting satellite in 24-hour orbit (high up) know it's any different, as it crosses the equator, from another satellite in geosynchronous orbit at the same place? Its velocity is different. Period.

So as the geosynchronous-altitude-but-polar-orbiting satellite crosses the equator, he (or she) needs to use its rocket to provide enough velocity increment to cancel the north-to-south motion and add an equal amount of east-to-west motion.

Radically changing planes requires a truly huge amount of velocity increment (thousands of meters per second or miles per hour), so nobody does it because they can't afford the propellant. But to go from 1 degree to 0 degree inclination requires only about 2% as much velocity change as making the radical 90 degree to 0 degree change. But even a single degree is still a significant cost to the propellant reserves of most satellites.

The maneuver is: wait until you are crossing the equator (say north to south), and propel yourself northeastward to both cancel the southward component of your velocity and to replace it with an eastward component.

Which brings us to the conclusion of this odyssey in orbit mechanics. Ernst (my in-house Jungian phenomenologist) has warned me that a deep depression might set in once I don't have this project to occupy those lonely hours on airplanes and in airports and at meetings on totally unrelated topics where my incessant typing on the old Cambridge Sinclair Z88 is thought to be careful note taking, but is in fact my communion with orbits and readers. In fact, the earth's ecology will improve slightly as my AA battery consumption will go down. There's little compensation for you, but I'll be fine, occupying myself doing what other great men before me have done—leering at young girls, drinking to excess, and taking drugs in hotel rooms.

Chapter 6

Magnetic Attractions

—with Richard Warner

6.1 Introduction to Alchemy, Magnetism, and Cold Fusion

Gold is a malleable, nonmagnetic material. Unlike other metals such as iron, chromium, and aluminum, it is pretty much useless for building things—too soft. Nonetheless, it has been highly valued for thousands of years because it is beautiful, rare, and does not tarnish. Relatively recently we have noticed that gold is among the best conductors of heat and electricity, making it the material of choice for the fine wires of many integrated circuits and for coating electrical contacts. In the West we associate alchemy—a mix of chemistry and magic—with the production of gold from base materials like lead. But the ancient Chinese also knew gold well enough to describe the magical field in three simple characters that mean the art of tempering (hardening) gold, a process about as plausible as gold's production from lead.

Today's engineers, of course, would resent labeling their work as alchemy. We prefer to say we understand an analytic basis for the way things work. This whole book implicitly rests on that questionable hypothesis—that we can explain what we observe, and we understand the principles of what's possible and what's not.

What about cold fusion? First it was impossible. Then Congress was considering putting $50M or so into Utah to pursue it, and Nobel Prize winning physicists advanced new theories on how it could be. Today fusion in a teacup continues to occupy a niche in the scientific underground, suspended somewhere between understanding (which equates to acceptance) and rejection. Its disciples carry on their research, books are written on its remarkable promise for earth's inhabitants, and others are written on its impossibility. So do we understand the possible and the impossible? Is cold fusion the bumblebee of the '80s and '90s—the flying insect we can prove can't fly?

When in doubt, change channels, I say, and with 500 of 'em out there, why not? Most of what I know about nuclear physics I learned in Isaac Asimov books, and my considered opinion on Cold Fusion is that if other people understand that stuff so well, how come they can't settle their arguments about it? I also know pretty little about changing lead to gold, or the fact that gold can't be tempered, but I'll accept the word of the experts that it's not too practical a thing to do.

Where do magnets come from? If cold fusion and making gold out of recycled bottles were all the alchemy we had to deal with, I'd dismiss it. But then there's magnetism. Frankly, few engineers really understand it. Light and radio we understand; they're photons and wave functions. But the typical explanation of magnetism goes something like this. Why is a bar magnet magnetic? Because iron atoms each have little magnetic dipoles (think of them as tiny bar magnets) in them. Normally these atomic magnets are randomly oriented and cancel each other out. In a magnet, many of the iron atoms that make up the bar have their magnetic dipoles lined up so that they don't cancel out. And voilà, a magnet.

I got this rather unsatisfying message at Wiley Junior High School in 1967, the same year I was sent home for dress code violation despite my conservative button down shirt, pressed navy pants, dark socks, and penny loafers (with pennies, btw), when my personal article-left-somewhere-in-the-gym-locker-room du jour was my narrow leather belt. Empty belt loops were not acceptable in public school. Period. Sending a little kid out for a three-mile walk home and back in subfreezing weather to correct the deficiency was totally PC. I like to tell that well worn tale to the odd 12 year old to permanently gel our generation gap. They like that.

Anyway, it was in '67 that I got a feeling I wasn't going to get along with magnetism (or nice clothes) at all. Frankly, both have been trouble to me ever since. Dressing for success I just can't handle, and what's the alternative? Dress for failure? The Dress for Success paradigm doesn't leave any room for dressing for comfort, variety, speed, economy, warmth, or even utility. It's success. Or failure. Tough choice. Magnetism-wise (in the '50s and part of the '60s people used the -wise suffix a lot, along with -nik as in beat-nik, kibbutz-nik), I had this little proto-conversation (the kind you don't ever actually have, you just argue with yourself and get totally frustrated with the other person over, when in fact you haven't said a single word to them) with my well-meaning science teacher, who really knew something about biology, but unfortunately not much about physics: "Excuse me, Mrs. Wolf (we said Miss and Mrs. back then and didn't think anything about it. Today you just say excuse me with a subtle hard edge, and end your phrase as if it's a question, as in "Excuse me, I don't think so(?)" which translates as "You are an idiot".) Magnets are explained because of tiny magnets? What about the tiny magnets? Where does their magnetism come from? Oh, I got an answer. About 7 years later in college physics. I remember distinctly that the blackboard was filled with vectors and tensors, remarkably similar to the vectors and tensors filling the chapter in the textbook on Magnetism. There are all these quantum variables which, if you accept them as God's truth, and if you then invest a few years in learning the math to manipulate them, convince you that there is a fundamental principle lurking behind magnetism. The tiny magnets were just a bit tinier in college than they were at Wiley. At least the collegiate experience allowed me to lose the belt without consequence.

6.2 What Rocks Tell Us

Everyone should launch a satellite at some time in their lives that doesn't work. Known in the business as "a rock," you learn a lot from an information machine in orbit that cannot transmit any of its information—the tree falling in the forest of the info age. You learn empathy for Woody Allen, particularly that short flick where his overbearing Jewish mom dies, but ends up as a giant face in heaven staring down at him 24 hours a day. On dates, at work, at the grocery store, his Mom's huge face in the sky constantly mocks him. Dear, your glasses are smudged. Dear, you'll never get anywhere with that attitude. Dear, you were so cute as a little baby when I changed your diapers on the kitchen table. Etc. Your rock, orbiting over your head in its total, multimillion dollar uselessness, teaches humility. You may think you know how to build things, and how they work, but you are just a little bit wrong.

Geologists don't launch rocks, but they live with them. One of the most significant rocks they live with is the earth, which happens to have a huge magnetic field. Why? It is a testimony to our lack of understanding of magnetism that nobody knows why the earth has a magnetic field. Oh, it has something to do with the earth's molten iron core. Molten iron, however, has been available to us humans for at least 10,000 years, and maybe 100,000, and nobody has ever made a magnet out of it. In fact, one sure way to un-align all those tiny magnetic dipoles and demagnetize iron is to heat it, particularly to melt it. Magnetism depends on structured order of atoms, something liquids ain't got. Maybe it's the molten core being spun around and circulating in all kinds of nifty ways due to Correolus accelerations, gravity, thermal gradients, and ferrous boundary conditions from the solid phase crust? Sounds like "Return of the Really Tiny Magnets" to me.

Now, if you accept those bizarre quantum variables, which turn out to not even explain all of what physicists already know about matter and energy, but do explain some other things that physicists know about matter and energy, and you learn the math, and then you accept the molten iron, Correolus, special boundary conditions and scale effects stuff, you are going to be totally comfortable with the earth's magnetic field. I just prefer discomfort.

6.3 Magnetic Many Uses Game

Maybe it's just as well magnetism is a bit magical. This would explain our contemporary usage of it—he has personal magnetism, for instance. We don't know what personal magnetism is, but we know it when we sense it. So let's agree that magnetism is like music. We really have no clue how it works, or why we like it, but we do, and we find all kinds of neat uses for it—taking dates to rock concerts, masking awkward silences in elevators, stimulating us to buy more in grocery stores, etc. In the case of magnetism, particularly aboard satellites, there are really just two uses: figuring out something about your position and/or attitude being the more familiar one, and doing something about your attitude being the other one.

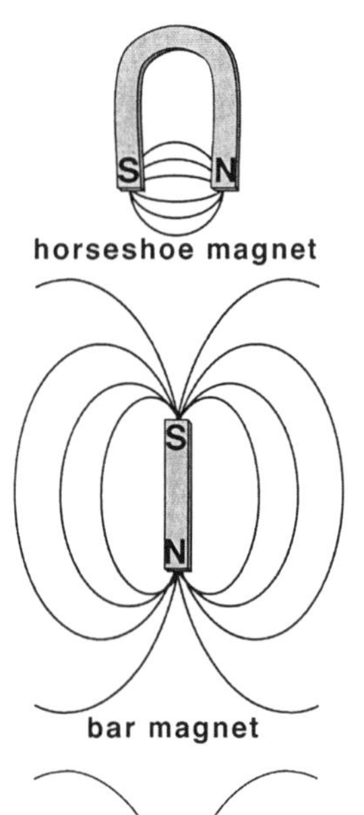

Magnetic fields I have known.

Both functions are quite practical to do at the low orbit altitudes most small satellites use. The picture shows that you are just as conveniently serviced by earth's magnetic field at LEO (low earth orbit) as you are on earth's surface. And since LEO satellites move around the earth quickly (an orbit about every 100 minutes), they see a lot of variation in the field direction. We'll soon see that this variability of the perceived field is particularly convenient for attitude control.

Satellites in very high orbits, like GEO (geosynchronous, 24,000 miles up), are too far away to see much of a field at all. And because they orbit very slowly, and in synchronicity with earth's own rotation, they see almost no field variations. Thus, navigating and steering via the earth's magnetic field is impractical for them. The situation is even worse magnetism-wise for interplanetary spacecraft, which see virtually no external fields except if they happen to be passing by Jupiter, or if they have very sensitive instruments to detect the minute magnetic influence of the sun.

This simplest model of the earth's magnetic properties (shown above) is a large bar magnet with its south pole in Canada and its north pole near Antarctica. It's a confusing convention, because the north magnetic pole of the Earth roughly corresponds to the south geographic pole. However, it makes all other magnets consistent; the north pole of a magnet wants to point geographically north.

6.4 Assume a Can Opener

It is handy to think in idealizations. Easterners like to think, for instance, that California is a Shangri La of beautiful weather. This image gives them something to aspire to throughout their lives, that they might eventually retire to some sleepy seaside community, soak up the sun, and relax. They skip over the crowding, obscene housing prices, earthquakes, floods, fog, rocky beaches, cold water, air pollution, skin cancer, and Mediterranean fruit flies. Or we want to think of our government as a benevolent and intelligent organization that ensures our personal safety, educates our kids, and protects us from poverty and disease in case of hardship. See how nice those idealizations are?

Here's another. Heat transfer is extremely dependent upon geometric factors. The only heat transfer problems people can actually solve when they learn heat transfer, involve the temperature profiles of objects with geometric simplicity like infinite flat plates and perfect spheres, though not too many of either are dotting my landscape. The perfect sphere is used to idealize the shape of a Thanksgiving turkey, and every budding engineer goes on to calculate the cooking time of a 14 lb bird. Whether you assume a mythical man-month, a spherical turkey, a rational, benevolent government, or perfect markets in equilibrium with access to perfect information, idealizations are abstractions that help us, but can also delude us.

I hope the material above will attenuate any great stress placed upon the reader who now learns the shocking truth that the earth is far from a perfect magnetic dipole (i.e., idealized bar magnet). The field has many fluctuations around it that render it quite unsymmetric. In fact, it varies from year to year, too. Mapping the earth's magnetic field is an ongoing activity that creates some job security for a few small satellite builders who need to put a satellite up there, know where it is as a function of time and which way it's pointing, and send to the ground the readings from an on-board magnetic field sensor, a magnetometer.

6.5 Satellite Compass: the Magnetometer

The most common magnetometer is called a flux gate magnetometer. Owning one of these used to brand you as a real techie since they were complex and pretty expensive. Nowadays, low cost flux gate magnetometers can be purchased as an option on your minivan, or inside of a digital watch, right alongside an integrating altimeter. If Columbus had one of those watches, Manhattan might have ended up in the Indian Ocean and Arthur C. Clark would be e-mailing his best sellers to a publisher in Calcutta from somewhere in the wilds of Long Island.

The advantage of a flux gate magnetometer over a scout's compass is that it tells you the field in 3 dimensions quantitatively. Or put another way, it describes the local magnetic field line to you in terms of four numbers, the first three being the relative strengths of the field in the X, Y, and Z axes of the instrument, and the 4th being the absolute strength of the local field.

Put yet a third way, it tells you what % of the magnetic field is along your line of site, what % is left to right, and what % is up or down. It also tells you what the absolute field strength is in some obtuse units like milligaus or microtesla. Note that the units of magnetic field strength are unrivaled in obscurity. Few engineers really have a "feel" for what a nanotesla is. All of us have a feel for what a pound of force is like, or a degree of temperature (Fahrenheit or Celsius). Scientists now believe that some migrating birds have a magnetometric organ in their brain. Possibly these birds feel that a few thousand nanotesla is a pretty nice thing, sort of like a 75°F day for us. Lacking such an organ, we humans must bear the burden of listening to a description of how a flux gate magnetometer works. Using an internally generated magnetic field created by flowing current through a coil of wire, it attempts to magnetize a piece of soft iron inside it along one axis. Then it immediately attempts to magnetize that piece of iron again, along that same axis, but flipping the north and south poles. It does this constantly, back and forth, back and forth. The magnetometer is calibrated to know how long it should take to flip the magnetization back and forth. Any variation in the rate that the iron sample actually gets magnetized, compared with the calibrated rate, is the result of an external magnetic field, which is exactly what it is trying to measure, reinforcing or partly cancelling the locally applied field. That is, if a component of the external field is helping to magnetize the sample in one direction, it is co-magnetized faster in that way and slower in the opposite way.

To measure the magnetic field strength along all three axes (X, Y, and Z or alternately front/back, left/right, and up/down), a magnetometer requires three flux gates, one on each axis. The magnetometers in minivans, for example, have only two gates, since minivan drivers tend to not care how much of the magnetic field is vertical. They want to know just the horizontal parts, from which you can determine, knowing that in fact the "interfering" magnetic field points north, what direction on the compass your minivan is pointed.

As complicated as this might sound, the end device is simple and easy to use. As an example, the now-extinct Schonstedt Instruments SAM-73C 3 axis magnetometer is a small box, about 6" x 2" x 2" with a single 9-pin connector on one end. You attach a single voltage supply on one pin, and three other pins output a signal directly proportional to the magnetic field measured along the three-instrument axis. Hook these three pins up to the nearest analog to digital converter and your spacecraft computer now knows the direction and magnitude of its local field. That is all there is to it, except for all the important little details that make building satellites fun. Some companies are beginning to produce highly miniaturized magnetometers that may prove ideal for small satellites. Ithaco makes a lovely little two-axis magnetometer that features very low power consumption; unfortunately, its price tag might discourage small satellite builders on a limited budget. Three of these would make a fully redundant three-axis system.

6.6 Navigating by Magnetic Compass

Creating a mathematical model of the earth's magnetic field is the *raison d'etre* for satellites that go out and dutifully measure that field every day. If we know where the satellite is in its orbit, and the mathematical model tells us which way the field, at that point, is oriented, we can determine which way the satellite itself is pointing. What we don't know is the attitude about the field line. A terrestrial example: Knowing which way is up tells you nothing about which way is east.

Conversely, if we know what the satellite's attitude is, perhaps by using a sun sensor and an earth sensor, we can determine where the satellite is in its orbit by knowing what the magnetic field locally feels like. If you aced Kalman filters in grad school, you might come up with a way to figure BOTH things out, by relying on the ever-changing nature of the field to gain more information over time. But if you aced Kalman filters in grad school, you probably are only reading this article to find little inaccuracies so you can "gotcha" the author. A mind, you know, is a terrible thing to waste.

The basic principle is evident in every ancient ship's compass, even though at this point we're beyond what Columbus, Magellan, and Lewis & Clark knew about magnetic navigation. They had just needle compasses, little bar magnets that line up with the earth's field. And they didn't have to worry about the third (vertical) axis, glued, as they were, to the surface of the earth. But LEO satellites' second use of the earth's magnetic field is unique to satellites.

A magnet, subjected to a local magnetic field, feels a twisting force or "torque" applied to it if the local field and the magnet are not aligned. Sounds complicated, but most of us have seen this principle at work. Find a small bar magnet, or magnetize a pin. Float it on a piece of rubber or cork in a bowl of water. The magnet turns until it is aligned with the earth's field. This torque caused by the skew between a magnet's field and the field in which it is immersed is what forces the needles of scouts' compasses to point north.

In a satellite, we can make an electromagnet by installing a coil of wire and passing a current through the wire, we have an electromagnet. If this electromagnet's field is not aligned with the earth's field, a torque is applied on the coil. Since the coil is attached to the satellite, the torque is applied to the whole satellite, trying to turn the coil into alignment with the earth's field.

Now we have a way to cause the satellite to change its attitude, essentially to steer it, without using rocket engines and propellants—just magnetism and electricity.

But with only one coil acting in one direction, we can only make one torque. We can play four more games to create a bigger variety of directions in to turn the satellite. First, just reverse the current flow in the coil, which flips its magnetic field. The direction it tries to turn to align itself with earth's field is also reversed. Second, give it three coils instead of just one, pointing left/right, front/back, and up/down—the three so-called principle axes. Third, add two directions of magnetization to each of the three coils, which gives six different torques that the satellite can create. You end up with a plus and a minus torque (or you can think of it as clockwise and counterclockwise torque) on each of three axes.

Trouble is, if the earth's field happens to be lined up with a coil, it's not going to create any torque, because torque results from the misalignment of the earth's and the coil's magnetic fields. No misalignment = no torque. Time for Trick #4, which is to wait. As the LEO satellite moves in its orbit, the earth's field appears to vary its direction. Near the poles it is more vertical, near the equator more horizontal, and the variations away from the perfect dipole model also create many opportunities to find fluctuations in the earth's field's orientation. Thus, if the satellite cannot create the torque it wants, but can wait typically 10 or 15 minutes, that torque becomes available.

6.7 Can You Say "Torque Coils"?

Rather than go to all the trouble to wind coils, build the circuitry to provide them with current, and control switching them on and off, you could create a torque with just a simple bar magnet. The trouble is, you can't shut it off! Some satellites have been flown with a single bar magnet as their attitude control system. The satellites point constantly with the bar magnet aligned with earth's field, causing the satellite to flip every time it goes over the north or south pole where the field goes vertical. Still, for most of the orbit, the satellite's magnet is about parallel to the earth's surface, since that is the orientation of the field when you are far from the poles.

For most applications, an electromagnet made of a coil of wire is the way to get torque. By varying the strength and/or duration of current flow in the wire, we can precisely meter out the amount of torque produced. These coils are, not surprisingly, known to satellite cognoscenti, as "torque coils."

Torque coils controlling ALEXIS satellite's attitude and spin rate.

There are two kinds of torque coils. The simplest is just a coil of wire wrapped around an insulating spindle. The budget-limited satellite builder can use a simple coil wound on a lathe, which makes a perfectly effective torque coil. You want to use as much wire and wind as big a coil as your mass and volume budgets allow, since a big coil with lots of windings gives the most magnetic gazorch for the minimum electric power. Be sure to choose a wire gauge that gives you the right resistance for the supply voltage you want to work with. Six coils like the one shown in the figure above have been steering the ALEXIS satellite for almost 2 years on orbit.

A coil with a slug of iron in it produces more magnet moment for a given amount of electric current than one without the slug. These iron core coils are standard for bigger LEO satellites, and are also found on small ones if space constraints don't allow packaging the larger air core coils.

Ithaco is the major supplier of iron core magnetic coils for attitude control. These are very linear, efficient devices, but a bit expensive. An air core coil uses about three times more power than an iron core for the same mass and characteristic dimension. But for small satellites, the power draw of the coils is seldom an issue. An overriding issue with very small, light satellites is residual magnetism when the coil is shut off, since very little torque is still enough to disturb the attitude of a very light satellite. Iron cores become slightly magnetized in use, and care must be taken to ensure that magnetization is fully canceled. This is not an issue with air core coils. They have no residual torque because they have no core, and hence no core to become magnetized

6.8 Magnets Chasing Their Own Tails

There is an interesting strategy to actually using any of these coils. To determine which coil or blend of coils to use and in which polarity to get the torque you want, you need to know the direction of the local magnetic field. A magnetometer is carried on board to measure that field. But turning on the coil produces a new magnetic field that is locally much stronger than the earth's field, and the magnetometer senses this locally generated field instead of the external field it is working against to produce the torque. If things work out right, you can get a nice unstable attitude control system out of this that is sure to provide hours of amusement for you and your customer.

Several strategies are available for turning on the coil. One is to simply switch it directly to the supply voltage rails, so that there are three possible states: ON+, ON- and OFF. This strategy is the easiest, but tends to be less than power efficient, because the power to a coil is proportional to the square of the resultant moment. It also limits the directional control of the moment. Another possibility is to control the coils linearly, which allows complete control of the resultant moment's direction and magnitude. While some interesting schemes can be used to provide a linear drive current very efficiently, AeroAstro provides people with clever little coil drivers so that they can just think about designing a satellite and not worry about how to drive inductors.

6.9 The FORCE: It Comes in Colors

Besides controlling your satellite according your own field instead of the earth's, you need to remember a few important caveats to protect you from magnetism's darker side. Magnetism doesn't discriminate. A permanent magnet or a field produced by torque coils both create torques when interacting with earth's field. But those forces are produced not just from the permanent or electromagnet you put on board for that purpose. If you have any ferromagnetic materials in your satellite (maybe those big Kovar power transistor packages) that become magnetized, they will constantly interact with earth's field, steering your satellite around for you, and probably not the way you wanted it to go. Any loop of wire carrying current is an electromagnet. So if you run the plus wire out to a big power consumer, like your satellite's payload, and then bring the ground or minus wire back via a different route, the current loop you produce creates a field that will cause undesired torquing of the satellite.

Fortunately, these "noise" torques are directly correctable by the control coils. But they still must be considered and minimized because canceling them wastes electric power. More importantly, your satellite is constantly being seesawed in the tug-o-

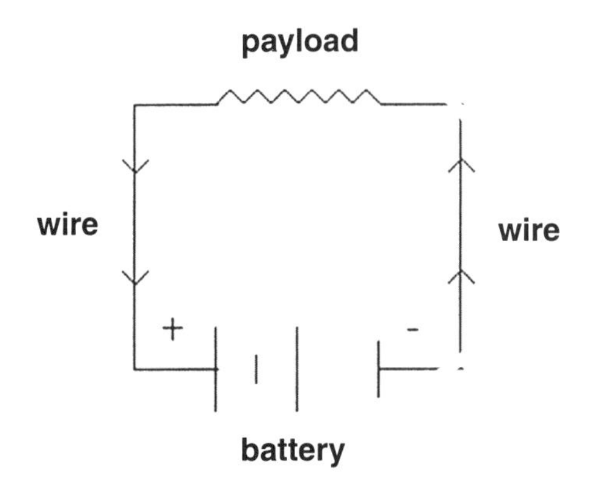

Current loops create magnetic fields.

war between the noise torques and the coils snapping on and off to correct them.

The tendency of electrically charged particles to travel along magnetic field lines is a great thing if you want to build a Tokomac fusion reactor or a magnetic bottle to hold antiprotons. You can file that with your other hints from Heloise for when special guests stop by. When you fly your satellite in LEO and enjoy that nice strong field to navigate and steer by, you are like a swimmer in the Gulf Stream enjoying its nice warm water. You have, in the latter case, at no additional cost, the opportunity to enjoy the company of Portuguese Men o' War (so far as I know these have not yet been renamed Persons o' War), jellyfish, stingrays, and the odd shark seeking refuge from cruising the chilling North Atlantic waters in search of a seal or downed naval officer. In the case of the earth's magnetic "stream," your companions are all kinds of charged particles that are wafted our way from the sun in what is known as the solar wind.

Atomic nuclei, electrons, and atoms missing a few of their electrons, causing them to be positively charged, really like our magnetic field. These tiny particles are "energetic," meaning they are moving real fast. They pass through electronics enclosures and either wedge themselves in your integrated circuit substrates, or zip right through them. In either case, they can deposit a bit of charge as they go, which has a nasty tendency to make 0s out of your 1s and 1s out of your 0s, which is disturbing, and can be fatal, to your on-board computing systems. These charged-up speed demons can damage materials just by passing through them—sort of like putting a tie tac through nearly the same spot on your favorite tie, the one that everyone in the office is totally sick of, every day. Eventually the office will get its relief!

Charged particles tend to be concentrated in the three places where the field lines themselves are concentrated: the magnetic North Pole, the magnetic South Pole, and one place off the coast of Brazil, aptly and ominously named the South Atlantic Anomaly. Satellite electronics designers spend a lot of time on this issue. We look for components that resist damage from these particles, and integrated circuit materials that do not tend to absorb charge from passing ions. These exist, but Murphy's Law prevents the component you really want to use, the one that would simplify the whole design and make your life at work really worth living, from being available in those materials. Instead, you end up designing around some power-hogging artifact of a previous decade just because it's pretty impervious to atomic guerrilla warfare. Alternatively, circuits can be buried inside of metal enclosures, thick ones, usually made of dense stuff like lead and tantalum. But at $1000 per pound launch costs, this solution tends to not create harmony with your program manager and customer.

6.10 For More Info:

Following are a few references on some space applications of magnetics. The basics of magnetism are covered in virtually all collegiate introductory physics textbooks.

Spacecraft Attitude Determination and Control, James Wertz, Ed., Kluwer Academic Publishers div. D. Reidel Publishing Co. Dordrecht, The Netherlands, 1978.

Psiaki, M., F. Martel, and P. Pal, "Three-Axis Attitude Determination via Kalman Filtering of Magnetometer Data," *Journal of Guidance, Control and Dynamics,* Vol. 13, No. 3, May-June 1990, pp. 506-14.

Psiaki, M. and F. Martel, "Autonomous Magnetic Navigation for Earth Orbiting Spacecraft," Proceedings, Third Annual AIAA/USU Conference on Small Satellites, Logan, Utah, Sept. 1989.

Chapter 7

Everything You Always Wanted to Know About Radio, Part I: Shatter the Myth of the Digital Miracle?

Let's! Does it irk you that your electric service account number has enough digits to specify the date of birth, sex, hair color, height, and weight of every human on earth? Does anyone really believe that before software all R&D programs came in on time and on budget? A 1 μs dropout of the AC power mains means a million people are late for work, and 100,000 person hours are lost resetting clocks flashing 12:00 AM. Ever lose 100 pages of your handwritten work because the phone rang at a critical moment? How much leisure time has AUTOEXEC.BAT given back to you?

Digitalization has subtler costs. Anxious to join the herd plunging off the edge of the digital cliff, engineers have deserted everything mechanical and analog. This is obvious just looking around our post-digital but otherwise Neanderthal world. You have a digital thermometer to tell you precisely how poorly your office temperature is regulated. Your all-digital car, which still has a spare tire in the trunk the size of a 1985 PC Portable, is still made the way they made 'em in 1937—sheet steel, reciprocating 6 cylinders, glass windshields and pneumatic tires. The digital "information center" is always at the ready to tell you that "Fuel Is Low" so that you don't abruptly stop burning one of earth's most precious, and noxious—when mined, shipped, refined, and burned— resources. The 737 is the loudest, least comfortable passenger jet airplane ever made, but its flight displays are gorgeous! Radio engineering, once the leading edge of electrical engineering, has fallen into disfavor with the digital fiber optic crowd. Radios—eeeuw, aren't they analog? (Horrors!) Anyway, everybody knows how radios work. We're practically born with radios and televisions glued to our ears and fixed in our line of sight. Most of us spoiled children of the second half of the twentieth century were conceived with either a radio or a television on somewhere in the room. As infants lying in cribs our breathing is transmitted by little Fisher-Price radios all over the house, a legacy of potentially tremendous value to aliens listening to earth's strange radiations! We don't have air engineers to study breathing, or water engineers to study drinking. What's the big hassle with radio engineering? Insert 12 D cells, heft that blaster onto the left shoulder, and boogie. Don't bug me about radios right now. I'm on the cellular phone. Radios have become so embedded that, like your car's engine, unless you own an MG or a '65 Ford Fairlane with 289 V-8, we really don't notice them or need to understand them any more.

Satellites have a few pesky qualities about them that make their dependence on radio rather significant. For one thing, they are far away. It is very hard to make wires 400 or 40,000 km long, hang them from space, and not get them all tangled up. Plus, the visual pollution is unacceptable to EPA. They move fast (satellites, not EPA)—7 km/s (15,000 miles an hour), which makes reeling out cable a bit clumsy anyway. Yet they are pretty useless if we can't exchange information with them. All satellites, whether they are relaying TV, radio and telephone signals, measuring the height of the ocean, tracking a volcanic plume, or watching for Gamma Ray bursts, are information machines. Radios move that information down to us.

I loved smoke signals as a kid. I was your normal eight-year-old pyrotechnic. Smoke fascinated me, mainly because where there's smoke, there's the possibility of burning down your parents' house, including the arrival of neat fire engines, water spraying all over the place and the promise of a dramatic rescue of the dog like the Dad on TV did. That's how little boys think, which kind of motivates one to rethink this whole latchkey child thing.

But smoke signals are radios. Native Americans didn't send the actual puffs of smoke to neighboring tribes to signal a shortage of water or whatever. The smoke modulated the transmission of radiation, sunlight. The receiver, a friend at the next settlement with eyes to detect sunlight, picked up this modulation, which consisted of dark patches of smoke interposed with clear areas. Modulated radiation = radio. Not a perfect definition, but it'll get us where we need to go.

Morse Code is a way to represent all the letters of the Roman alphabet, plus numbers and punctuation marks, with dots and dashes. "A" is dot dash (• –), "B" is dash dot dot dot (– • • •) and "Hey, I gotta get the heck outta the radio room or I'll be late for the kids' soccer practice. See you later, dude, say hey to Sarah and the kids" is

– – • • • • • • – –

Abbreviation was elevated to high art by the early radio operators.

Sending the long and short pulses of Morse Code by radio is not so different from sending smoke signals. You have a transmitter that you can turn on and off, and a receiver sensitive to what you are transmitting. The simplest transmitter may be a light with an aperture you can open and close. This was briefly popularized by Paul Revere—not the '60s rock band or the copper saucepans, the 18th century revolutionary who earned his living when not warning of Redcoats by making copperware. In fact, much of 19th century Europe was linked by optical telegraphy stations operating in a giant international network. It is still used today by ships at sea. SOS, aka Save Our Ship, is just

• • • – – – • • •

a relatively easy pattern to remember, even while some of the North Atlantic flows over the aft deck. Flashlights are good Morse Code senders.

But light isn't radio. It's light, right? Yes, it's light, but light is just one type of electromagnetic radiation, a type we sense with our eyes. Microwaves reheating last night's tuna casserole are another form of that radiation, as are x-rays that form images on photographic film used to look at our bones. Infrared radiation, the way you can "feel" from a distance that a turned off stove top burner is still warm, is also electromagnetic radiation.

When electricity flows in a wire, it creates a field around the wire called an electromagnetic field. This idea sounds very scientific and mysterious, and I still think it actually is, but it is easy to visualize nonetheless. A current flowing through a wire creates alignments in the atomic nuclei and electrons that form the wire material, allowing their electrical and magnetic properties, which are normally all random and cancel each other out, to align themselves and cooperate. One result is that a compass brought close to a current-carrying wire will point in the same direction as the wire. Electromagnets are just coils of wire carrying lots of current to make a very strong alignment of their magnetic dipoles. Very handy for picking up black Lincoln Continentals containing bad guys in James Bond films and dropping them unceremoniously into crushers—another fantasy common in 8-year-old boys.

If the current is steady, or varies only slowly, the fields are steady and about all we can do is make magnets and heaters. The "heads" of a cassette tape recorder or a floppy disc drive basically work this way. These wires with current flowing in them make magnetic fields that "write" on the magnetic materials inside the tape or disc.

But if we make the electric field fluctuate, that is, flow first one way, then the other, the magnetic field propagates, or travels. Why? Because some energy is required to reverse the flow, and some of this energy escapes to propagate itself away from the source. At very low frequencies very few reversals of flow occur, so this transmission is not too efficient. The current that flows in our electrical wiring at home fluctuates, that is, reverses direction and then reverses again to the original direction, only 60 times a second, which is considered VERY slow indeed. If you drive under a high tension (high voltage, but also high current, which is the key to strong radio energy transmission) cable with your car's AM radio on, you can hear the 60 Hz (Hertz = cycles per second) hum. Originally popularized in horror movies of the '40s when they were getting ready to electrocute some ghoul (yet another fascination of eight-year-olds), the 60 Hz hum has become, in our society, synonymous with electric power.

For some reason, frequency waves as low as 100 or 200 Hz propagate well through water. Used for transmitting radio waves to submarines, they are otherwise pretty useless. Aside from requiring huge antennas to resonate efficiently with their very long wavelengths, they are bad information carriers. The human voice can fluctuate at up to 8000 Hz, but a 100 Hz radio wave cannot vary any faster than 100 Hz, and practically cannot be modulated faster than 50 Hz. Carrying a human voice is practically out of the question, unless we want to transmit the voice very slowly and then play it back later at the right speed. In which case, a 10-second burst of speech could take 15 minutes to send. Thus, 100 Hz class radio waves, with very limited application, are called very low frequency (VLF) for good reason.

Radio waves really don't get interesting 'till you get to high frequency (HF). HF happens around 100,000 Hz, or 100 kHz, a value known as "medium wave" that you can find on AM radios in Europe and Asia. Our American AM radios start at 550 kHz. These frequencies have two significant advantages. The waves can be efficiently transmitted and received using more compact gear, and they vary fast enough that we can modulate them with human voice and other sounds that our ears can hear. The highest frequency, or rate of variation, that we can hear is about 20,000 Hz. A radio wave at 100,000 Hz varies five times faster, so it has plenty of ability to vary fast enough to carry, or be modulated by, all the sounds we can hear.

Voilà! Radio. For a long time, people were pretty excited to radiate at 100 kHz or so. After all, it isn't easy to make something vary at 100 kHz. The fastest your car engine can spin is maybe 50 or 100 Hz (0.1 kHz), so these radio transmitters, were they mechanical devices, would be really whizzing! But after a while...

The most common way to make electricity cycle is with a piezoelectric crystal. These quartz crystals are not unlike the quartz you find in nature as a clearish rock, or as a component of granite, which we all are told in elementary school is quartz and feldspar, even though we have no idea, nor could we care less, what quartz or feldspar are. God's little gift to radio is that when you squeeze a carefully cut sliver of quartz, a bit of electric current comes out, and when you do the opposite, put a current into it, the crystal contracts or expands a little. Really a very neat thing. Modern micromachines use this property of quartz to affect very small motions, like for aligning precise optics or soldering wires.into miniature electronic circuits.

In radio applications, we apply a brief spike of current to the crystal, and it expands and then relaxes and vibrates in that mode for a while, like a bell. As it vibrates, we get a vibrating electric current. These vibrations can go up to 100 million vibrations per second for very fine slivers of crystal.

In the early radio days, making such fine crystals wasn't possible. The pioneers used bigger hunks of crystals, and got resonances in the range of 100 kHz to 2000 kHz, two million Hertz, or two Megahertz (MHz). The standard AM radio band that extends from 550 kHz to 1.6 MHz was allocated because it was the highest frequency range of radio signals that could be practically produced and detected by the modern radios of the 1920s. Today we more typically use television and radio frequencies of 80 MHz to 160 MHz. The pioneers thought they were really sizzling to get 2 MHz, so they subsequently called the range 3 MHz to 30 MHz "High Frequency." We call frequencies of 30 to 300 MHz "Very High Frequency" or "VHF."

The press to get to higher and higher frequencies was motivated by several factors. One was space. When all the frequency space is used in a particular band, in other words, when transmitters broadcast on all the spaces between 550 kHz and 1.6 MHZ, your only choice is to go a little higher. By the end of WWII, practical radios had been built with frequencies over 1 GHz. That's one billion cycles per second! Today, without going to any technological extremes, we can produce radio energy with frequencies as high as 100 GHz, or 100 billion cycles per second. The range from about 300 MHz to about 3000 MHz (3 GHz) is called Ultra High Frequency (UHF). If you have ever shopped for olives, you know where we're headed. The smallest olives you can buy are Large. Probably 75 years ago olive growers were pretty impressed when they grew a few of those Large guys. But agriculture, like radios, moved on, bringing to our grocery store shelves the Giants, which are still easily mistaken for a bloated raisin. Grape growers were no way going to be outdone by the olive mavens. Then you get into Mammoth, which are nowadays the size you find in the lonely little salad bars cast into the corners of rural grocery stores as a gesture that says, "We know city people eat this stuff, but Lord knows why." Big olives, the kind you want to serve when the boss comes over for dinner, are known as Colossal and Super Colossal.

Why should electromagnetic waves be any different than olives? Today the rage is Super High Frequency (SHF). Somehow I always expect these waves to come flying through the window wearing blue tights and a red cape.

Enough radio talk. You probably have a few disks that need to be backed up, mail to read off the Internet, or maybe your answering machine's digital outgoing message needs refreshing and you need to get to the store and pick up some new software for the kids. No problem at all! Take a couple months' break. Radios have been around 100 years. They'll wait. No rush to find out what S, L, and X have in common besides the word "Suzuki"; why anybody really needs to know the speed of light; and whether a crack team of government proposal reviewers would ever put grant money into radio.

Chapter 8

Everything You Always Wanted to Know About Radio, Part II: Faster than a Speeding Bullet

No primer on radios is complete without the graphic on the following page, which shows the progression of radio waves from VLF power transmission, through the HF, VHF, UHF spectra, on up into microwaves, up through SHF, then into infrared and visible light, ultra violet, X-rays and Gamma rays. Wherever you roam in the spectrum, the physics is the same—alternating electromagnetic fields.

All this electromagnetic radiation propagates at about the same speed. What junior scientist didn't learn that light travels 186,000 miles (300,000 km) EACH SECOND. Fast! Bullets get up to maybe 1 mile a second, and sound only travels 1/5 of a mile in a second in air. 747s and other jets travel more like 1/7 of a mile per second, less than one millionth as fast as light. Even our satellites whizzing around at 4 miles a second are less than 1/40,000th the speed of radio and light waves.

Given that all electromagnetic radiation travels at about 300,000,000 meters a second, we can define the wavelength as the distance the radio wave travels in 1 cycle. For a signal on your AM radio at 600 kHz, the wavelength is given by:

$$\text{wavelength, } \lambda = 300{,}000{,}000 \text{ m / s} \div 600{,}000 \text{ cycles / s} = 500 \text{ meters (about 1650 feet).}$$

Note that fancy Greek lambda, λ, which is the symbol engineers use for wavelength. We also use another classy letter, Ω, omega, for frequency. So a wavelength of a signal in the AM radio band is somewhere around 500 meters long, maybe only 187.5 meters (615 feet) at the top of the band around 1600 kHz.

It's hard to carry an antenna 200 meters long around with you when jogging. If you look inside a portable AM radio, you'll see a wire almost that long tightly coiled up. That's your antenna. FM radio (88 MHz to 108 MHz) wavelength is 3.4 to 2.8 meters. Your car's antenna and the headphone cord from your Walkman to your ear are both about half that length, making them pretty good antennas. Cellular phones work at 850 MHz, where the wavelength is 0.35 meters or about 1 foot. A "half wave" antenna is the size of those little black curlicue things you see sticking up from

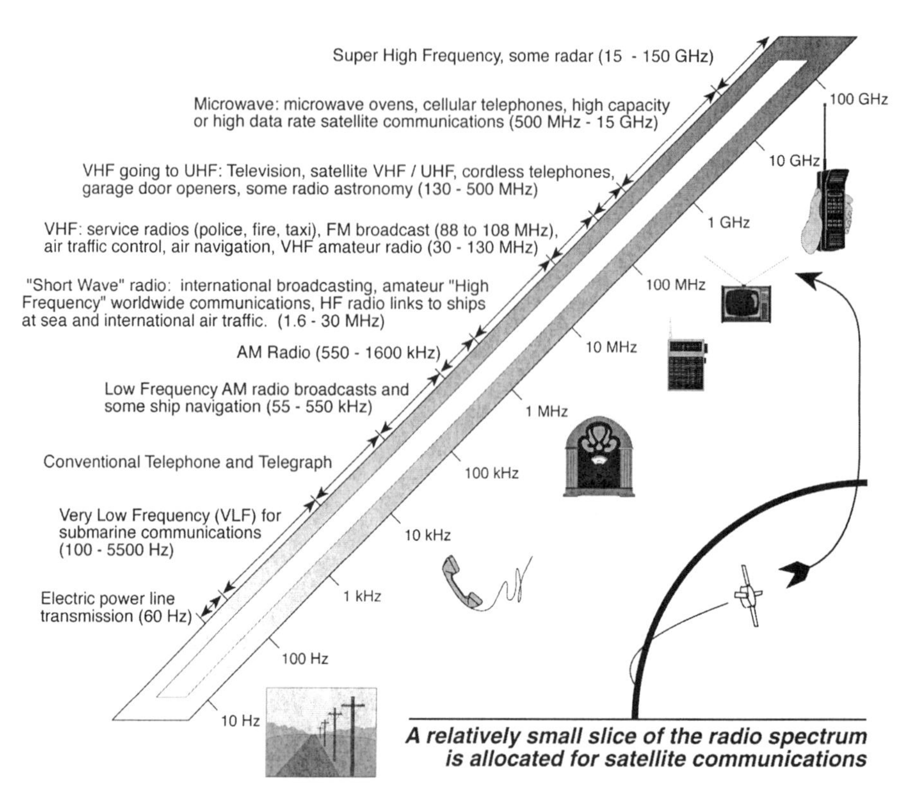

A relatively small slice of the radio spectrum is allocated for satellite communications

the rear window of almost every car in the world with a sticker price over $25,000. Efficient antennas for cellular telephones.

One little quirk of satellite radio nomenclature: most radio people outside the satellite world talk in terms of bands—ranges of frequencies or more often of wavelength. The latter is preferred since it's the wavelength we have to deal with most, like what length antenna, what size features on the circuit board, and so on. Not satellite jocks. They talk about L band, S band, X band. Lucky you, I'm too young to tell you all about how they made the first microwave radios.

Note here "microwave" refers to frequencies ranging from the upper UHF into SHF. In fact microwave ovens run around 2.5 GHz, 2,500 MHz, or a wavelength of about 0.1 meter or 4 inches. That happens to be a resonant frequency for water molecules, which are efficient antennas for that wavelength, and that's how microwave radiation cooks food. Unfortunately, cats are made of mostly water, which has generated a lot of misdirected speculation, epicentered in the eight-year-old-boy segment of our population, on the potential for inserting cats into microwave ovens. I like to think nobody has ever actually zapped a cat, and that it's just so much macho playground talk.

With all the excitement in recent years about the possible health effects of radio waves, you might be wondering about whether the disappearance of radio engineers is caused by other reasons than disinterest and fascination with bits and bytes. Certainly, the radiation of a microwave oven is dangerous above a certain level. Above that level, the cells in your body are heated and could be killed. Not good, and no controversy. This explains why, for instance, microwave ovens are carefully built so that the door has to be closed before they turn on. Despite initial controversy, I would suppose microwave ovens are a lot safer than the stoves and conventional ovens we all take for granted—no grease fires, no burned fingers, no gas leaks, just to name a few hazards that can accompany conventional cooking. I doubt a microwave oven has ever caused a house to burn down! The arterial effects of buttered microwave popcorn are, however, another possible consideration. But as you move away from the wavelengths that resonate with biological materials, which are mostly water and hence mostly 2500 GHz, and thus eliminate heating effects, the health picture is much less menacing. Standards have been set for maximum exposure to radio energy, and engineers who work around radios, particularly microwave radios, where the standards dictate the most stringent controls, need to be conscious of this limitation. But cordless and cellular phones, electric power transmission lines, desktop computers, and induction stoves radiate at wavelengths very far from those needed to couple to the molecules in our bodies, and their radiation is weak. Handheld cellular phones have 0.8 watts of transmit power, about the same as one of those little key chain flashlights. No clear evidence indicates that radio waves from these sources have any health effects. One study did show increased incidence of disease in people living near power lines, but it turned out that the power company used vegetation-killing chemicals under the lines to keep trees and bushes from shorting to them, and these chemicals were dangerous to human health.

It is virtually certain that safe, normal usage of radios has no effect on your life expectancy or health, but people will continue to research radio energy's biological effects for many years. While no clear danger inherent to our current handling and application of radios exists, possibly some of our uses of radio and power transmission may have statistically significant effects on health over large human populations. Current research on radio safety focuses on this subject.

Back to the radios themselves and the arcane nomenclature of the radio bands used to talk with satellites. Because resonant microwave cavities were created with various oddly shaped devices, early microwave engineers named the wavelength bands after their resonators. For instance, L band is around 1000 MHz (1 GHz). S Band is 2 GHz. X Band is 8 to 10 GHz.

You might already guess three reasons satellites like these very high frequencies. First, you need a high frequency carrier to send a lot of information fast, like a stream of images from a telescope, or to relay television. Second, weight and space are at a premium on satellites. Microwaves are short, so their antennas are much more compact. (Otherwise we'd call them macrowaves.) Finally, by the time people invented satellites, we'd found other terrestrial uses for all the frequency space up to several hundred MHz. To a technical person, this might sound a bit political. But I doubt the world was ready to trash all of its brand new television sets to make spectrum space for the first satellites. The real world is like that.

Just to keep us guessing that someone might be looking out for us after all, it turns out very handy that people were so attached to the sub-100 MHz spectrum. The highest levels of earth's atmosphere are busy shielding us from nasty ultraviolet, gamma, and X-radiation. In doing so, these layers become reflective to radio waves up to about 50 MHz. In the days before satellites, this feature of the ionosphere was the only means we had for transoceanic communication by radio. High Frequency (HF) radios transmitted East from the coast of Virginia, eventually bounced off the ionosphere, and were received in England. Lots of services still use this means of communication, popularly known as shortwave radio. Many of the world's countries broadcast their cultural and political images daily on the wavelength bands between about 6 MHz (50 meters) and 21 MHz (15 meters). Ham radio operators using low-wattage transmitters equivalent to the brightness of a flashlight bulb communicate all over the world bouncing signals off the charged ionosphere.

So there would be a couple of problems with using HF radio and ionospheric reflection for what satellites do—relay telephone and television and bring data down from space. For one thing, the frequency is low, so the data rate you can get is limited. A few television channels would fill up the entire usable wavelength band. Forget 200 channels of satellite TV. Ditto telephone calls, or plan on long queues to get a turn to use the available throughput capacity.

But what's maybe even more troublesome is that propagation of HF radio signals by means of ionospheric reflection varies randomly from minute to minute, and it varies strongly as the reflecting area moves from sunlight to darkness, also with the seasons, and even with the sunspot cycle. Thus, HF links are undependable and often noisy. HF radio is also unsuitable for satellite communications because ionospheric reflection strongly attenuates the signal along its path to the satellite.

For all these reasons, satellite communication has focused on the VHF, UHF, microwave, and SHF bands, that is, everything above 30 MHz. Terrestrial long distance communications rely on the HF bands from about 3 MHz to about 30 MHz. In fact, most satellite operations begin above 100 MHz. What's left between 30 MHz and 100 MHz? Police radios, aircraft radio, remote controls, cordless telephones, FM radio, and a bit of television—stuff that we do NOT want to propagate over long distances. Even so, many services are abandoning the lower frequencies for higher ones because their electronics become smaller and their antennas more compact.

What's been described so far is like drawing the geopolitical globe of the world. We've established:

- the locations of the oceans and the continents, that is, the use of VLF for submarine communications and power transmission,
- "regular" frequencies for AM radio for historical reasons,
- HF for international shortwave radio using ionospheric reflection, and
- VHF, UHF, microwave, and SHF for satellites, for line-of-sight terrestrial communications including cellular telephones, FM radio, television, and cooking dinner.

Like countries vying for the finite real estate on their land masses, various radio services compete and divide up the spectrum. Telephone and TV transmission, police radio, scientific satellites, garage door openers, microwave ovens, even hydrogen molecules, whose vibrational modes at microwave frequencies form the basis for radio astronomy—all of these have little patches of spectrum allocated to them. The continuing distribution and redistribution of this resource helps the editors of *Space News* keep their pages filled, employs hard working lawyers all over the world, and provides the focus of international conferences where the agreements on frequency allocations worldwide are hammered out, or not. I'll leave the political commentary for Art Buchwald and the Maison Blanc set, and we'll stick to our photons, shall we?

Tune in next chapter for the exciting conclusion to the Satellite Radio mini-series. A torrid discussion of radio power and amplification is followed by passionate confessions about ground station links. Issues of Doppler and modulation throw a bit of spice into the story. Don't miss the surprise ending!

Chapter 9

Everything You Always Wanted to Know About Radio, Part III: What's Up, Doc?

What's a watt? Let's not get too technical. A typical light bulb puts out 10 or 20 watts of light. The other 80 or 90 watts in a typical 100 W incandescent light bulb are dissipated as heat. A flashlight puts out maybe two watts of light. The standard S-band (2 GHz) spacecraft transmitter, used on many low earth orbit satellites including science satellites and Landsat, the NOAA polar orbiters downlinking weather data from low earth orbit, is a two-watt transmitter, also about as much brightness as a flashlight. Why so dim? Just like a flashlight, to make two watts of radio energy at 2 GHz requires much more power, like about 15 to 20 watts, which is a significant power draw for a small or medium size satellite.

A satellite coming into view as it rises over the horizon in a 750 km (400 mile) orbit is about 2000 km (1400 miles) away. We routinely communicate across these distances with power so low that it is the equivalent of expecting a flashlight flicked on in Boston to be seen in Miami. Yet we can detect it flashing on and off thousands, even millions of times a second.

The lingo for the connection of a receiver and a transmitter is "link." Many factors go into establishing the link between a satellite and the ground. The transmitter's power spreads as it propagates from the satellite. Assuming a nondirectional, or isotropic, antenna, which is often used on a small satellite so that it doesn't have to precisely point at the ground station, the transmitted power spreads into ever increasing spherical shells. By the time the transmitted radio power has traveled 2000 km, it is 4 trillion (4,000 billion) times weaker than it was 1 meter from the satellite. Starting out at about 1 watt of radio power for each square meter, the power density on the ground is less than one millionth of a millionth of a watt per square meter.

Yes, there is a lot of technology behind receiving that infinitesimal signal. But we should not forget that it is pretty amazing. It's a miracle we take for granted every day. If radio didn't exist and you proposed to DARPA or NASA to put a two-watt transmitter thousands of miles away and receive it perfectly with a relatively simple set of equipment, your proposal would certainly be dismissed along with the perpetual motion machines and momentum drive rockets. Luckily the radio pioneers, like their aircraft-inventing counterparts, came along before the government got into funding its development!

Besides Divine Providence, we do rely on a some ingenuity to intercept these radio signals. The most visible feature of satellite ground stations, their dish antennas, create a big area for intercepting satellite-transmitted radio energy and focus the signals being broadcast to the satellite. A six-foot (about two-meter) diameter dish has a beam about 6° wide—an amplification factor of about 1000 compared with a non-directional antenna. Ignoring higher order effects, it's interesting that this focusing effect is frequency independent. A six-foot dish could just as well receive light energy from a two-watt light bulb as a 2 GHz RF signal. We call such an "optical" dish by another name, a six-foot telescope. Given the fine surface required to focus the shorter wavelength light, a six-foot telescope is just a bit more expensive than a typical small dish for a satellite ground station.

Having collected all the signal it can, the receiver's efforts are focused on amplifying, or increasing the signal strength by a factor of way over a million times. The S-band receiver flown on the ALEXIS satellite, which is not atypically sensitive, had an amplification factor of one trillion, meaning that it multiplied the radio signal coming into it by a factor of a million times a million. The fundamental problem with extreme amplification is that both noise and signal are amplified. Thus, the receiver must be selective as well as sensitive enough to reject signals near to the signal of interest. Even with the lowest noise amplifiers and the most selective filters, the signal remains below the power level of the amplified noise. Special detection schemes are employed that "vote" on whether perturbations in the noise indicate a signal or just random noise fluctuations. Encoding schemes increase the amount of raw information sent so that when an error in decoding the received signal occurs, it can be detected and fixed, or the satellite can be requested to retransmit a part of its message.

While extremely low signal level detection is virtually unique to satellite links, even the simplest AM/FM radio employs many of these same techniques, including highly selective tuners and very high gain amplifiers to multiply signal strength typically by a factor of a million or more so than you can tune in NPR or MTV. The large distance and high data rate coupled with the power handicap of transmitting from a satellite compound a problem that has always been a technological challenge.

Unique to Low Earth Orbit satellite communications is correction for Doppler, which is the apparent frequency shift caused by relative motion of the transmitter and the receiver. As the satellite first approaches and then recedes from the ground station during its brief overhead pass, the satellite transmitting frequency, as detected on the ground, appears to rise and then fall, just as an approaching car's horn, a train's whistle or a jet airplane's roar seems higher pitched then lower pitched as the moving object approaches, passes, and moves away. Those examples are such clichés that you might think they're the only times you experience Doppler. They're not. How many waves does a body surfer cross each minute? Several, when the surfer is swimming out from the beach, versus basically zero as the surfer coasts in with the waves. That's

Doppler. The "red shift" that astronomers argue is evidence of the big bang is the shift of wavelength of light coming from astronomical sources receding from earth. Doppler.

A ground station transmitter appears to shift upward and then downward in frequency as its satellite passes overhead. The simplest way to maintain the link during this rapid variation in frequency is to have an automatic frequency controller on the ground and flight receivers constantly adapting to the changing carrier frequency. But with very weak signals, this control is difficult to implement reliably. Rather than build this complexity into the satellite, some systems include compensation for Doppler shift in the ground station design.

Ground-based Doppler is not painless either, but aside from where you put the complexity, there is a physical reason to compensate for Doppler, and that is reducing the time wasted tuning around randomly across the range of frequencies Doppler could give you until you find the signal you're looking for.

Assuming your link is not way overpowered, the time it takes to acquire a signal is inversely proportional to its bandwidth. At 9600 bps, it's not a big deal. At 300 bps, sometimes used for low-power beacons from satellites, with the magnitude of Doppler you get at S-band; many minutes or even hours can be required to acquire a signal using a single channel receiver. In the time that a LEO satellite is typically in view of the ground station, about 5 to 10 minutes, no information can be transmitted, if all the time is spent trying to compensate for the Doppler shift!

Thus, LEO satellite ground station systems are often designed to calculate, by knowing the orbit of the satellite, second by second, what the amount of the Doppler shift is, then change their transmit and receive frequencies steadily so that the satellite senses an unchanging uplink frequency despite the Doppler variation.

Another tool to aid detection is the type of modulation used. Probably the simplest modulation is the on/off system used in smoke signals and in the early telegraphs. But on/off is actually quite difficult to detect because the receiver cannot lock on to a transmitter that is off, and it takes time to re-lock each time the transmitter comes back on. Television and FM radio use frequency modulation, hence the term FM. The transmitter is on continuously, but rapidly varies its frequency very slightly according to the modulating signal, whether that is voice or the 0/1 bits of a computer data stream. FM has similar lock problems. The frequency is constantly varying and the receiver, seeing the signal through a mask of noise, easily loses lock to the satellite. Phase Shift Keying (PSK) is the most effective, practical modulation scheme. Neither the frequency nor the amplitude changes, ever, allowing the ground receiver to establish and maintain lock very reliably even in low signal conditions. What changes is the phase of the transmitted wave. As shown below, the radio wave is advanced or delayed slightly with each bit of modulation, and this phase shift is detected. PSK coupled with encoding, which involves sending the same information

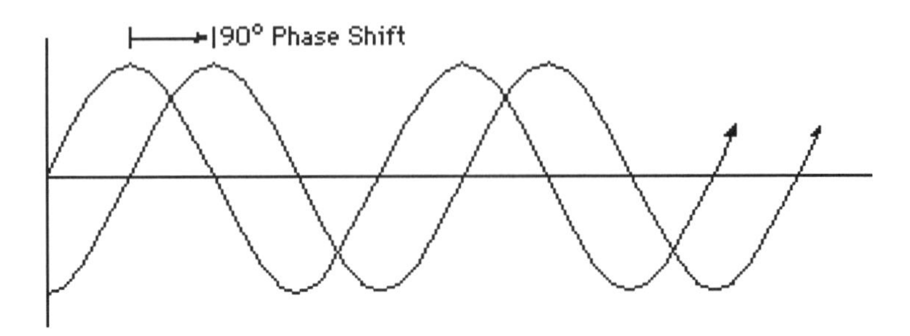

A sine wave phase shifted.

multiple times to provide a cross-check and a means to detect errors, provides a factor of 10 to 100 in increased sensitivity to very weak signals.

Phase shift modulation is by nature digital. The phase of the signal is either shifted or it isn't. Of course, you can shift it by varying amounts, but most PSK (Phase Shift Keying) systems use either 180° (BPSK for Bi-Phase) or 90° (QPSK for quadrature phase) shifts because differentiating between small phase shifts is difficult. Phase Shift receivers are typically designed to detect the existence or non-existence of a particular phase shift angle. In BPSK, this equates directly to detection of a 1 or a 0. In QPSK, the phase shift corresponds to two bits of data instead of one. For instance, the detection code might be:

Back when digital wasn't everything (remember LP records and clocks with round faces that were called "clocks" instead of "analog clocks"?), FM was popular and a lot of work was done sending tones. Touch tone phones are a good example of this scheme. Each tone or combination of tones equates to a "message." A filter in the receiver discriminates among the tones. The system is simple and reliable, but the data rate is painfully slow, which means that sending gobs of telemetered data, what people routinely do nowadays, is impossible.

Phase Shift	Information
0°	0,0
90°	0,1
180°	1,1
270°	1,0
360°	0,0 (same as 0°)

Many satellites were flown that could only detect about 10 tones. With each different tone, the satellite undertook a specific action, like turning a radio transmitter on and off. That's fine as far as it goes, but if you want to send a few color images in the form of 100 MB of data, and each tone gives you only 2 or 3 bits, you are basically left to dial 250,000,000 tones, which takes a while! Tone modulation is still used for simple systems. One of the most basic is command destruct systems on board launch vehicles, which don't have too much data to move around. Basically, it's do you command it to destroy itself or not. In this application, the simplicity of the tone system, and the fact that the rocket gets continuous reception of the ground station signal, nominally keying the "don't blow yourself up" tone, adds safety to the system.

Radio has become a very political technology. Only a finite amount of frequency bandwidth exists—the one-dimensional landscape of the electromagnetic spectrum. That landscape is constantly being divided and reallocated, a process that it seems like only lawyers have the patience to graciously endure. Ten years is not a long time to create an allocation. In fact, we technical types have partly created our own congestion. We like clear channels—slots reserved just for our use. In fact, we don't use them very often. A solution to this waste of geography is spread spectrum, where a signal is spread out over a large area. Another person using a different spreading pattern won't even notice you, and many hundreds or thousands of users can share spectrum without interference. The battle between the older territorial division and the newer Code Delay Multiple Access (CDMA) technologies is going on right now. The lines are drawn as you might expect them to be. The existing services that "own" frequency space like the older approach, and all the new guys are pushing CDMA.

Orbital Slots. What are they? It's easy to get the impression that there's only so much room for satellites up there. The circumference of the geosynchronous orbit track around the earth, 315,000 km long, is populated by only about 100 or so satellites. For every satellite, there is 3150 km of space, about 300,000 times bigger than the satellite itself. Each satellite appears to hang above a particular point on the earth because at geosynchronous altitude, the satellite's motion around the earth is synchronized with the earth's own rotation. (See Chapters 3 and 4.) They both go around once a day, the satellite moving in formation with the earth's surface below it.

What is tightly packed is the antenna patterns. These satellites all use the same frequency spectrum, and the ground stations uplinking to them have beams that become fairly wide way up at 40,000 km from the earth's surface. Similarly, the satellites use the same downlink frequencies. The downlink beams spread a lot too, usually more than the uplink because satellite antennas are generally smaller than ground station antennas, so their focus isn't as narrow. Satellites must be spaced wider than their beam width. It is the potential overlap of these beams that determines how many "orbital slots" are available.

Low earth orbiting satellites have a much shorter period of rotation around the earth. Typically, they complete a circuit in 100 minutes or fewer. Thus, they do not appear synchronous, and hence there are no orbital slots. By the nature of these low orbits, all satellites overlap each other's coverage areas, which is a potential interference problem. Some of the solutions to this problem include licensing the satellites in discrete areas around the world, allocating each LEO satellite system with slightly different frequency spectrum, or implementing various spectrum sharing plans where they mutually agree to accept some interference.

One potential solution to provide more spectrum space is communications by laser. The laser light is modulated just like a radio wave, either by amplitude variation (including rapidly turning the beam on and off) or potentially by frequency or phase modulation. Simple on/off coding is the primary candidate, because it is easier to turn lasers on and off very quickly than to modulate their frequency or maintain tight phase coherence. In fact, laser communication is desirable because of its huge information-carrying capacity, as much as a million simultaneous voice channels on a single beam. Lasers spread much less than radio because they intrinsically create highly focused light, the result of both the laser's light generation physics and the much shorter wavelength of light than radio. Thus, the orbital slot problem basically disappears and a very large number of satellites can be accommodated without interference.

Unfortunately, the same short wavelength that makes focusing the laser light possible also makes it almost impossible to penetrate clouds, which is a feature of radio most people never think about. After all, you expect your radio or TV to work even in cloudy weather! For laser communications to work, several ground stations are required to ensure at least one of them has a clear sky between it and the satellite.

Today's laser technologies are another issue. They require copious electricity to make a very modestly powered laser. Electricity on orbit is a precious commodity. While new technologies are being developed, laser communications in space remains only a research topic. Most of the effort is directed at satellite-to-satellite links which, since they do not face attenuation in the atmosphere, use very low power lasers.

Have you ever re-carpeted? I never paid much attention to carpet until I had to pick out a houseful of it. Colors? Sure! 10,000 colors and a semi-infinite number of combinations. There are also weaves, materials, stain protection, long and short pile, wall to wall, and borders. Even carpet tiles abound—little 12" squares of carpet you glue to the floor—instant carpet, plus you can make huge checkerboards and produce other digital art forms, albeit at low resolution.

If you go the tile route, several other doors present themselves for you to open, like vinyl tile, rubber tile, ceramic tile, even natural wood tile. Oak, cherry, bleached wood. Wood comes in some very interesting tiles, like long, thin ones people call "boards" and "planks." Your floor can say Manhattan or Mt. Vernon, or Miami, Milano, Maccau or Manchester or just lie there and take the abuse of daily life. Pretty soon your previously wall-to-wall suburban enclave looks like the inside of the Holden Arboretum with a few Pirelli rubber areas where people are likely to spill water or drip mud. You have graduated from carpeting to "flooring." Suddenly you see floors in all their myriad relationships and possibilities! To think my wife still considers a hardware superstore boring. It is my holy city and I am its disciple.

Some time ago we started out talking about radio. We all thought we knew what we wanted. Something of a grown-up, high-speed, digital pair of Dixie cups with an invisible string of radio waves linking them together, right? But light waves are radio waves, and the string can be a beam of light. The light doesn't have to travel freely in space. We can guide it through a fiber just as television comes to many homes in cables instead of through air waves. (Though not mine anymore. I decided to go for nostalgia and bought an antenna.)

Sunlight is a jumble of radio waves. People use sunlight to communicate, with flags, with smoke, with a mirror to flash to rescue aircraft. My wife got to saying, "It's a world of carpeting." But it's really a world of communications. Car horns, clothes, cash register beeps, conversations. Getting bits down from a satellite is pretty constraining, just as is carpeting a house whose inhabitants include people under the age of 8. Wires, cables, and fibers are definitely out. Forget sound—no air to carry it. But there are still lots of possibilities, like conventional radio, spread spectrum radio, and lasers, all of which can operate over an enormous range of frequency, power, and bandwidth. So far we have confined ourselves to frequencies from 0.1 to 10 GHz, which adds up to about 1% of the spectrum of emissions that we know how to transmit and receive.

Satellites, particularly small, LEO satellites, are a key element of the communications infrastructure that will radically alter the way we live and the resources of our earth that we will consume. Our ability to overcome the communications system obstacles we have lived with until now and to produce more throughput for more users with a broader range of applications will determine how big a role our small spacecraft technologies will play in this revolution.

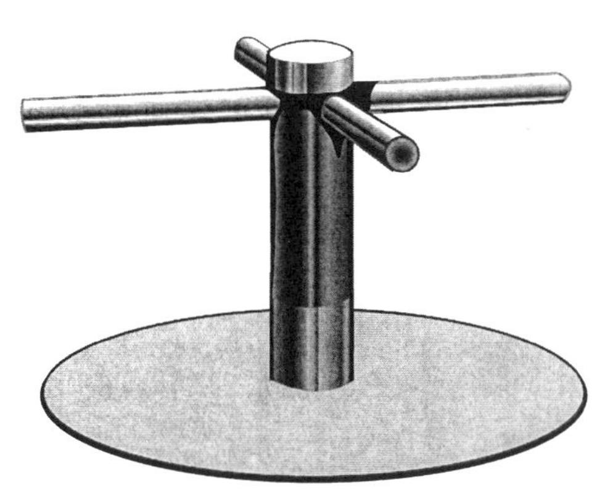

Satellite antenna.

Chapter 10

Thermal Dynamics: Tough Talk About Temperature

A short, virtually painless, and occasionally philosophical look at spacecraft thermostatics and thermodynamics

Everybody in college wanted to be a thermodynamicist. Don't ask me why. It's probably because of the big bucks you pull down once you get out, and of course it's very sexy, which is important in school. Naturally, society holds thermodynamicists in very high regard. During WW II, thermodynamicists got special rations of gasoline and tires for their cars so that their vital work wouldn't be seriously impaired by the war effort. And the prestige is a factor you can't measure, but it's certainly there. I have those TD (thermodynamicist) license plates, which gets me a lot better treatment with the Spago Valet Parking. Need a reservation at Four Seasons on a Saturday at 8:30 p.m. for two right by the window? Just tell the Maître de Hôte your special guest is a Thermodynamicist. Enough said.

The competition in Thermo classes made my college life Hell. Freshperson classes were chock full, and everybody elbowed for position. Grade grubbing was the rule. Before the six- and seven-year Thermodynamicist programs popped up to get you right into the pipeline out of high school, lots of students spent the whole summer before senior year cramming for the GRE to make sure they got into a Thermo grad school. Thermodynamicist. It's a ticket to country clubs, Lexus dealers, and fancy resorts. Unfortunately, that's what motivated most of the competition.

But I made it, and I can say it was all worth it, even those first two sleepless years of grad school: the 60-hour shifts in front of a micro Kelvin thermostated oil bath and a rack full of HP counters; the internship at an urban combustion tunnel with no budget and a single slit interferometer. Granted, my wife thinks I'm a nerd. Instead of the country club, I swim at the community center with the old folks' low-impact aerobics classes and the toddlers getting their heads in the water for the first time. No Lexus, either. I'm driving a VW, and I don't have time or money for the fancy resorts. I sup-

Optimum locale for studying thermodynamics.

pose I should've gone into private practice? All my college buds are shopping for wallpaper in Djakarta and throwing Bar Mitzvah parties in Jerusalem. Or they're spending Christmas at the Zugspitze. In bad years, at Aspen. But I do get to design satellites to run at the right temperature. Hey, it doesn't get a whole lot better than that. Who cares about that chalet on Lago Maggiore, a cellar full of vintage reds, and sending the kids to Brown and Stanford anyway? The real question is: what is the right temperature for a satellite, and how do you get it there?

No single temperature, of course, is right for everything on a satellite. At one extreme, some instruments might want to be as cold as possible, infrared detectors and very sensitive radio frequency (RF) preamplifiers, for instance. Most batteries want to be in the range 0°C (32°F) to 5°C (41°F). Most electronics want to be around that temperature or a little warmer (10°C, 50°F). Some things want to be warm, like ovenized oscillators and monopropellant thruster assemblies, which are small rockets used for attitude control and orbit adjustment.

Maybe you watched 2001 at an impressionable age, and you think of space as a very cold place. Maybe you think the space sun is very harsh because it isn't filtered by earth's atmosphere, hence the thick helmets on the Apollo astronauts. You are right on both counts, but only sort of.

Assume that a body is isolated. It doesn't gain or lose mass and has no internal source of heat. There are only three ways to change its temperature: conduction of heat, convection, and radiation. If it is "insulated" from these three effects, i.e., no heat can be transferred to or from it, we call its container "adiabatic." Conduction is how you burn the roof of your mouth on cheese pizza (I don't know about you but I hate it when I do that!). The hot cheese gets pressed against the roof of your mouth, conducting heat to your skin, overheating the cells near the surface, damaging, often killing them. That's what we mean by "burning" your finger or your tongue—overheating the surface cells. Our bodies also conduct heat to and from the surrounding air, which is why when it's hot outside, you get hot. Metal is a much better conductor than air, so it cools you off faster, and heats you up faster, than air. This is why a hunk of aluminum at room temperature feels cool to the touch. Room temperature is below skin temperature. Your body is gently having heat conducted away from it by the ambient air, but metal conducts the heat away much faster, so that we perceive it as cool in temperature.

Convection is sort of dynamic conduction. A windy 0°C (32°F) day is a lot colder than a still one. This is because as you conduct, or lose heat, to the air, you warm the air surrounding you. Without convection the warm air near your body actually helps keep you warm. Jackets work because they hold the air around you that you've warmed, keeping you warm. If the air is moving, you constantly lose your warm boundary layer and have to heat up a new one, which subsequently is convected away from you. Isn't convection interesting? No? No problem, since it's pretty meaningless to the heat balance of a satellite anyway, because (brilliant revelation ahead), satellites are not in contact with a fluid medium (i.e., air) to conduct or convect heat to.

A satellite is in physical contact with nothing. Think about that. Nothing touches it. An island in the truest sense. Simon and Garfunkel made millions off this concept! This solitude leaves only one heat transfer mechanism available to a satellite: radiation. The air is cold, I'm told, at St. Moritz in February, but the Alpine sun is warm. Its radiation warms you. Toasters radiate heat to bagels, and a whole industry has grown up around that phenomenon. One very civilized aspect of the "Tube," the London Underground, is the radiant heaters above the platforms, which warm you even though the air in the tube in winter time is cold and damp.

On a clear night, the earth's surface looks up to black space and loses heat by radiation, which makes freezing to death in Sacramento in May a real possibility, even if daytime temperatures are over 30°C (86°F). Desert areas are particularly cold at night because no cloud cover reflects radiated heat back to earth. The air is dry and "light," meaning no water vapor helps prevent radiation of heat to space.

With no air to conduct to, and nothing to blanket it from the nearly absolute cold of black space, a satellite lives in the ultimate radiatively dominated condition. How much heat can you or your satellite lose radiating to black space? Where is HAL when we need him!

Convection action, Chicago-style.

A thermodynamicist named Stephan noticed that the rate of heat loss from a hot object to its cold surroundings by means of radiation is proportional to the difference of their respective temperatures, each taken to the fourth power, or

$$Q = s\,e\,A\,(T_1^4 - T_2^4)$$

where Q is the heat flux, s, e and A are constants, and T_1 and T_2 are the temperatures of the body and its surroundings measured in absolute degrees, meaning degrees above absolute zero.

In the Fahrenheit system, room temperature is about 560 degrees above absolute zero. (Absolute degrees in this system are known as degrees Rankin.) In the Celsius system, where absolute temperature is referred to as Kelvin (Note: Not degrees Kelvin. This strange quirk of calling the temperature units "degrees" finally got axed in the Systeme International (SI) units system.) 295 Kelvin is comfy though not cozy.

The constant "s" is known as the Stephan-Boltzmann constant. One of those fundamental constants like the speed of light in a vacuum, the charge on the electron, or the amount of time it takes after a traffic light turns green for the teen racer in the black, winged, magged, and sound-systemed 5 liter Mustang behind you to honk his horn. "A" is the area of the object that is radiating, and "e" is the emissivity, a measure of the radiation efficiency. Objects with efficiency near one are efficient radiators. Near zero they are poor radiators, which makes them good radiation insulators.

To give you some feel for the numbers, a one-meter square (about 10 square feet) object at room temperature (25°C or about 78°F) that's radiating to black space loses heat at a rate of about 450 watts. That's about the heat flux of a clothes dryer in the low setting, one side of a two slot toaster, or what the sun radiates onto you if you are in LA, fashionably dressed in nihilistic black clothes, and walk over to your therapist's office at noon.

Enjoying radiant heat

In a gross sense, satellite temperature control amounts to control of the radiation flux. Of course, in a gross sense, all that's required to make an atom bomb is to bring together a supercritical mass of fissionable material. In both cases, and in a surprisingly large set of similar technical undertakings, a lot of work has gone into actually doing that which we so glibly conceptualize.

Two not so small complications to figuring out a satellite's temperature are commonly referred to as the Sun and the Earth. If you expended a significant fraction of your life wondering if the coincidence that the solar flux (heat flow per square foot or square meter) onto our bodies on a sunny day is about equal to the radiative loss from our bodies back to space on a cloudless night amounts to a proof of the existence of God, let me spare you further idle speculation. This rough equality is true, but not surprising. Our bodies cannot need to keep themselves very far in temperature from the temperature of our earth. Otherwise, we'd be incapable of survival and die of hypo or hyperthermia.

Note the significant corollary that our bodies cannot want to be at exactly the same temperature as our surroundings or the heating, ventilating, and air conditioning (HVAC) industry would collapse. Thus, while I may destroy an argument for a great cosmic plan, the same argument asserts that a direct link exists between Creation and HVAC, which may be theologically unique.

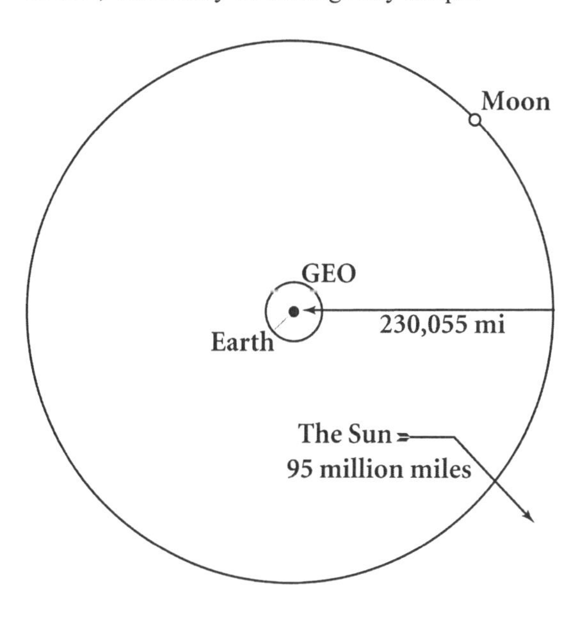

GEO orbit, moon, and the sun about 50 feet off the page, to scale.

The earth's temperature is itself driven by the heat it receives from the sun and the heat it loses to space. Thus, the earth's temperature is that at which the daytime heat input is about equal to the night time heat loss, our bodies' temperature is close to that of mother earth. Voilà! Your radiative heat losses and gains, like the earth's, should be pretty close to each other. An interesting aspect of both our own comfort and that of our satellites is that small temperature shifts affect heat loss significantly, because the heat loss is proportional to the temperature taken to the fourth power. A 5°C shift, only about 1.6% of the absolute temperature, results in a 7% shift in

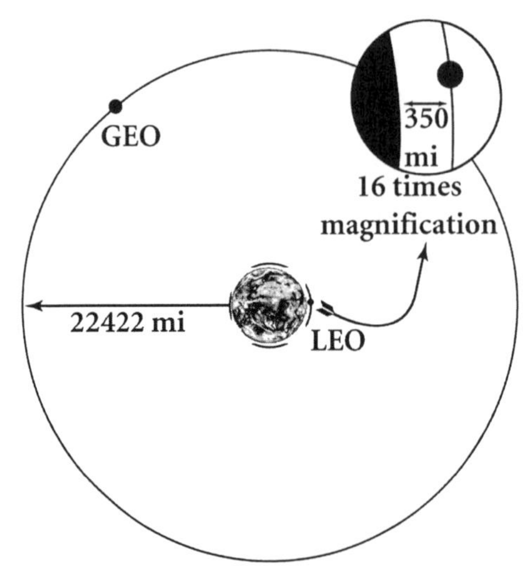

Earth, GEO, and LEO to scale.

the heat flux. Thus, if for some reason a satellite gets 5°C warmer, it loses 1.07 times more heat at night. But it also gains a little less heat during the day, both of which tend to cool it off.

The earth's effect on a satellite's temperature is a bit more subtle. At very high orbits, like geosynchronous orbits, the earth forms a small part of what the satellite sees, or radiates to. Its temperature is determined mostly by the flux from the sun, about 1,300 watts for each square meter, and its radiative flux to space, about 450 watts for each square meter.

Just like the earth, the satellite absorbs sunlight only over the part of its surface exposed to the sun, but it radiates to space from its entire surface. This ratio is (assuming a spherical satellite, which for some reason is a really funny punch line in a lot of obscure thermodynamic jokes) exactly 4. Thus the disparity between heat loss and heat gain is actually about 1300 W in compared to about 1800 W out. The satellite will be cold. The bright metallic, usually gold-colored coatings so commonly seen on geosynchronous satellites are thermal trim materials. These metallic surfaces are very efficient absorbers of solar radiation, but radiate weakly at the temperature of the satellite body. Thus they tend to warm the satellite. The exact amount and type of this surface coating, its precise placement, and its degradation in the space environment is a significant element of any geosynchronous satellite's design.

At low earth orbit (LEO) the job is a little easier, because the earth is close by, relatively speaking. Close this book and hold it at your maximum arm's length. Really reach way out there. It is about as large from your eyeball's point of view as the earth is from a geosynchronous satellite's point of view. Hold the book as if you were reading it normally (even though you're not at this point). Then bring it very close to your eyes so that you actually touch it to the tip of your nose, as if you were extremely near sighted. It is about as large to your eye as the earth is from a LEO satellite's viewpoint. Close! So what, right? Because the earth occupies so much of its view, the satellite is strongly radiatively coupled to the earth. Remember the area term in the radiation equation. About half of the satellite's surface area sees the earth. If the satellite tries to get warmer than the earth, it radiates heat to the earth and cools. If it tries to significantly cool itself, the radiation flux from the earth warms it.

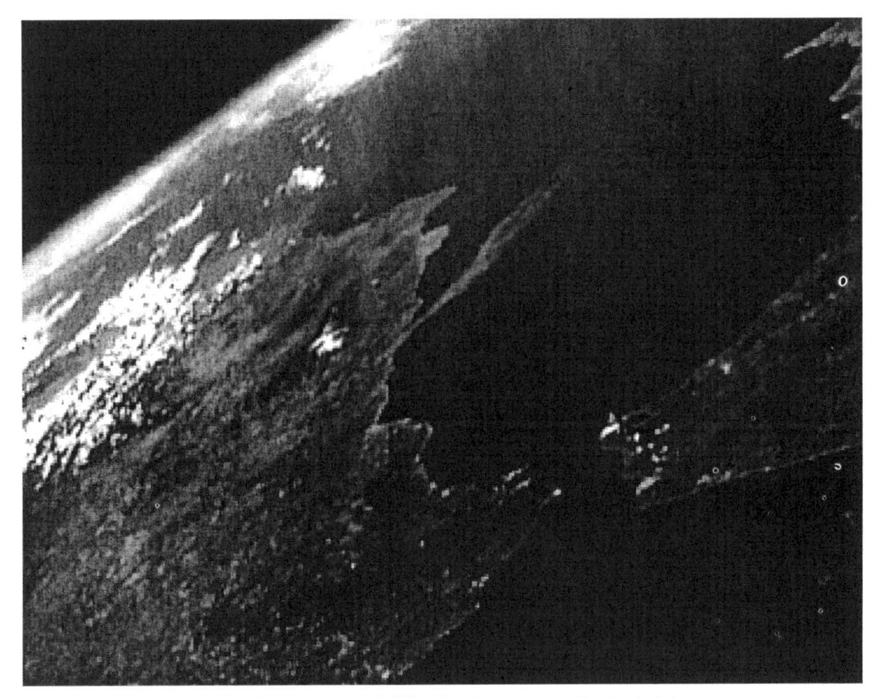

The Earth, seen from LEO, dominates your field of vision.

Just as it is no major coincidence that our bodies are near in temperature to that of our surroundings, which in turn are in equilibrium with the radiation in from the sun and the radiation loss to space, LEO satellites are hard pressed to reach a significantly different temperature than the earth. On average, the earth is a little above freezing, about 10°C or 50°F. And that, not so surprisingly, is the temperature of most LEO satellites.

"For that my genius son went to school for 22 years?" I can hear my father saying. To find out that no matter what you do you end up at 10°C? Hey, some kids turn out, some don't. My family doctor went to school for 23 years and every one of his patients eventually dies. But a lot of interesting stuff happens in the span between birth and death. Similarly, the variations around those 10°C averages are pretty important.

For example, what if we have a component we'd like to maintain at lower temperature than the rest of the satellite? What you do is first point the satellite at the sun, so that the object can fly in the satellite's shadow. Then thermally isolate, i.e., insulate, the object so that it can't easily conduct heat to or from the rest of the satellite. Create some radiation baffles, or barriers, to stop the warm satellite from radiatively coupling to the hopefully cold object. These barriers can be just curtains or walls coated with a low e (emmissivity) material. While you're at it, don't forget to similarly isolate the cold object from "seeing" the earth. Finally, given that it can't see earth or sun, expose it to what's left, namely black space. Depending on the care with which the insulation and radiation barriers are constructed, objects can be cooled to as much as 100°C (180°F) below the earth equilibrium temperature.

Another small detail that provides job security for us thermodynamicists is that while a 20°C (36°F) fluctuation is small in absolute temperature terms—less than 10%—it makes a big difference to the satellite's components. For instance, Nicad batteries are very happy at 10°C (50°F), but they are a bit cranky at 30°C (86°F). It could be dangerous to charge them rapidly at such a warm temperature. Plus, satellites move a lot of heat around. For instance, turning on a transmitter causes local heating around the transmitter as it dissipates power. So a lot of work can go into analyzing the precise temperature different parts of a satellite will reach.

Small, low cost satellite programs avoid some of these complications. The most basic approach is to thermally short all parts of the satellite together. Not letting any one part be insulated from the others prevents it from getting too hot or too cold. Since it is easy to transfer heat over short distances, thermally shorting the pieces of a small satellite amounts to just bolting them together, assuming everything is made of a good heat conductor like aluminum. The 50 kg ASAP payloads, like the AMSAT cubes (shown below) use this approach. They mostly do not get colder than 0°C (32°F) nor warmer than about 15°C (60°F).

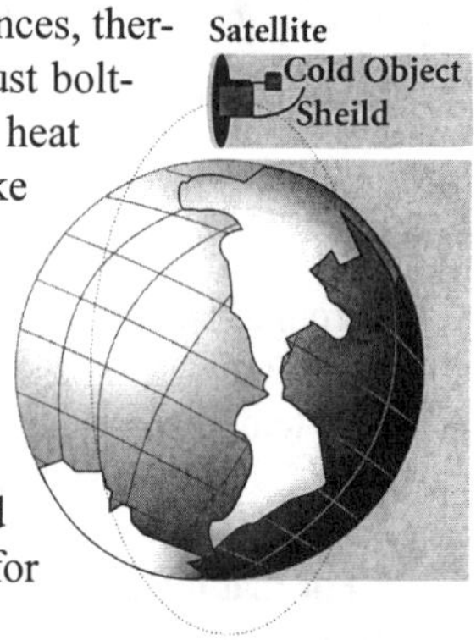

Satellite
Cold Object
Sheild

Sun-pointing satellite with shielding.

Which is not to say that the solution is as trivial as a few bolts. Small satellites have been flown with thermal blanketing, usually to reduce radiative heat loss on shaded areas of the satellite. They have louvers to compensate for variations in solar heating during interplanetary trajectories away from or toward the sun, with heaters to ensure that critical components are never too cold. Current designs even exploit heat pipes to aid in transferring heat away from highly stressed components.

Some satellites must turn their broad sides away from the sun during highly illuminated orbital seasons to reduce spacecraft heating, and local heating plays a key role in some satellite components. Gravity gradient booms can create oscillations in the satellite pointing attitude if their illuminated surfaces warm and stretch while their shaded surfaces cool and shrink. At the component level, without air to conduct heat away from microprocessors and other high performance components, they can get very hot. Many flight circuit boards have a special copper layer inside them to sink heat away from these components.

Interesting, isn't it? A satellite is its own microcosm, isolated from the universe except for the radiative exchange of energy. When you think about it, the solar collectors are radiation receiver/absorbers, the radios are highly tuned radiators, and the satellite absorbs and emits radiant heat. Even remote sensing instruments like telescopes or sensitive radio receivers are just other means to absorb incident radiant energy.

On earth we take our rich environment for granted. Heat is conducted to our surroundings. When something gets real hot or real cold, radiation and convection kick in to help smooth things out. Air brings us oxygen and aroma, a sensation of speed and some momentum, at least to blow our hair around. It's a medium for the pressure fluctuations that transmit sound to our ears. Air provides us a stable thermal environment to exchange heat with. The satellite's environment is much more spartan. With no molecular transfer, it lives submerged in a gossamer sea of photons. Radiation is

AMSAT cube, 17 cm. on a side.

the only contact the satellite has with the outside world. Thermal design regulates photon absorption and emission, and the distribution of the heat gained from those photons through the satellite's body.

Chapter 11

You got an Attitude, Buddy? A Primer on Small Satellite Stability and Control

It's a fact of life. Most things have an attitude. That's why I retired my beloved cat, Providence, to a horse farm in the Midwest. Small satellites (you knew I'd work around to them) have attitudes, too. Attitude determination and control are key features of satellite design and they're a special problem for small satellites. Antennas like to point at the ground for uplink and downlink. Solar panels like to point at the sun. An astronomical telescope may want to point at a star cluster or along a set of celestial coordinates. For sky survey astronomy missions, detectors like to spin slowly and scan the whole sky. A camera or directional antenna may want to point at Lima, Ohio, if that's where the news is.

Believe it or not, on the ground, in the Washington winter fog on New Year's Eve, at 2:00 a.m. after sampling every beverage at three different parties, your attitude control still has a lot more going for it than most small satellites'. For instance, you know where down is (the direction your house keys just went into the snow). For a satellite, down is a difficult concept. If a satellite were to drop its keys, where would they go? Granted, the topic is academic. We don't fly satellites with keys, but since when has that stopped anyone from studying an interesting problem? Satellites, having hypothetically dropped their keys, should have absolutely no problem. First, there is no snow in orbit to lose the damned things in. Second, the keys continue to move along with the satellite and, from the satellite's reference frame, go nowhere. But the satellite senses itself to be in a zero-g, no gravity, environment. So down is tough.

Can you walk a straight line at 2:00 a.m. New Year's morning? How about almost straight? Again, you are better off than satellites. Telling which direction you are going in a satellite is subtle. Very subtle. No wind. No sequential ads for chewing tobacco or Burma Shave, and no roadside to post them on. On the other hand, no Ladybird Johnson to get all upset about roadside signs. Lots of pluses in space. Lots. But forward is tough. From our satellite's point of view, it's not moving at all. Next time you fly on an airplane, walk into the restroom and close the door. How can you know how to point in the direction the airplane is going? Think of it as another step in the slow process of gaining empathy for satellite guidance and control.

For these and other reasons (like the fact that in 1957 getting into space, period, was more important than knowing where up, down, or frontwards was), the first satellites were unstabilized. Like Exhibit A here:

Un-stabilized

Sputnik— Khruschev's first attempt at Star Wars.

What Sputnik and our early satellites did about the attitude problem was—I can relate to this—they ignored it. They put a bunch of antennas in all directions, which made it easy to build neat models of satellites. Get a golf ball, some Elmer's® glue, and some colored golf tees. Glue the tees to the ball all over its surface, and, voilà, Sputnik. The satellite is clueless about where it is pointing and doesn't care. It worked fine then and it works fine now. The simplest solution is to ignore the problem. If you can do that, you can save money, time (which is money anyway, isn't it?), and payload weight for important things like ant farms and commemorative stamps.

Read dress-for-success books. They'll tell you pointing down is a very authoritative gesture. Looking down is handy for things like tying shoes and finding those damned keys. And pointing directional antennas down at the earth is handy for receiving weak signals from downed airplanes blown off-course en route to Red Square, or taking pictures of clouds over the earth to show on the six o'clock news. There is only one way to be unstabilized—like there's only one way to be dead. But there are lots of ways to be alive, for example, George Bush, George Harrison, Boy George, and there are lots of ways to point down.

Gravity Gradient (GG) uses the fact that the earth's gravity gets weaker as your distance from the earth increases. Thus, a satellite with a weight stuck out on the end of a long boom sort of hangs on the end of the boom, like little iron filings all lining up in a magnetic field. The satellite doesn't care if it is down and the weight is up, or if the weight is down and the satellite is up, so a method to flip it over is needed. But it's vertical. The variation in gravity over the length of a boom, 20 to 100 feet, is minute, and so is the gravity gradient torque. About the equivalent of the torque of a lightweight (flea weight) flea standing on the seat of a playground seesaw. Because no other stronger forces are present to upset the system, this minuscule torque orients the satellite vertically.

Gravity gradient satellites are popular because the stabilization is passive. No propulsion or gyroscopic systems are needed to maintain the vertical orientation. But the stabilization is weak and hence small oscillations about the vertical, up to +/-10 (degrees), are common. If anything starts the satellite rocking, like its initial separation from the launch vehicle, GG has no damping to attenuate the oscillatory seesaw motion. Thus dampers need to be added. We'll get to dampers later. Suffice it to say they add an incremental complexity to an otherwise simple stabilization solution.

Gravity Gradient

Hanging out on the long boom of the gravity gradient (GG).

I could go on. And on. But let's wrap up the first part of this primer with one more passive stabilization technique. Next time we'll take a look at systems that require the satellite to actively do things like spin itself or spin wheels or turn magnetic fields on and off to use the earth's magnetic field to create torques.

There isn't no wind in orbit (run that through your grammar checker). Just very little. Air density decreases exponentially with altitude. At space shuttle orbit altitude of 300 km, or 160 nautical miles, aerodynamic drag is about 100 times stronger than gravity gradient effects. By moving the aerodynamic center of pressure behind the mass center of gravity (in other words, by adding the equivalent of tail feathers to an arrow) the satellite points into the wind. Since the wind is created only by the motion of the satellite, like creating wind in a car by driving forward, pointing into the wind is the same as pointing in the direction you are going. In warm weather, dogs riding in cars often use the same technique to point their noses out the window in the direction they are going. The relationship between these applications of weather cocking is determined only by moving to the East (Tibet, not Baltimore) and thinking hard for 20 years. While you dial up your travel agent, the figure below shows an aerodynamic stabilized satellite.

Aerodynamic

Aerodynamically stabilized satellite.

The Chemical Release Satellites built for the Air Force are examples of this design approach. Stabilization pointing frontwards is needed to find the satellite from ground or airplane-based telescopes. The tail fin doubles as a radar reflector to increase the radar cross section of the satellite to make initial acquisition easier. An optical beacon (strobe) is carried that the stabilization system keeps pointing frontward. Don't you feel Domino's Pizza and Federal Express have cheapened this neat concept by using it for their delivery vehicles? Write your congressperson.

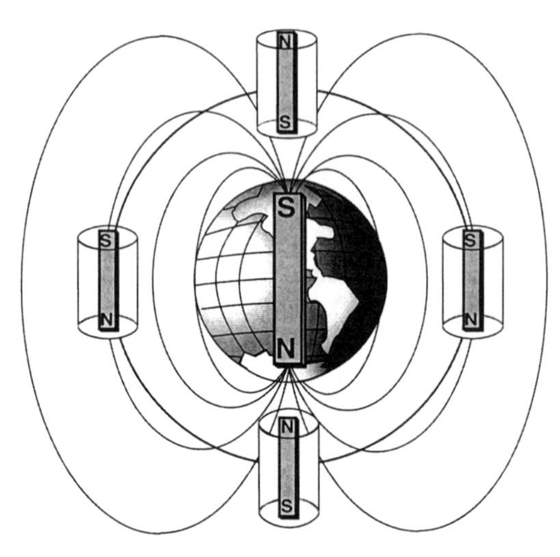

Magnetic stabilization.

The next part of this subject, spin stabilization, requires deep thought, possibly 30 years in Tibet. In the meantime, if you see Providence, tell her all is forgiven.

11.1 Active Control

Besides expressing tremendous empathy for the difficult conditions satellites live with every day of their lives. Chapter 10 reviewed many, though not all, of the popular stabilization systems that require no active control of the satellite. These included no stabilization (heavily favored by Existentialists), gravity gradient, and aerodynamic methods. In addition to these, a number of schemes using permanent magnets have been tested.

Using a permanent bar magnet fixed in the satellite, one satellite axis is forced to align with the earth's magnetic field lines. The sketches above and below illustrate this stabilization, which causes a twice-per-orbit flip in polar orbit.

Permanent magnet stabilized satellites have no stability around the field line and tend to roll about it, which precludes pointing at the earth. One exception has been exploited by Amsat. A dipole antenna can be aligned with the magnet axis and hence is aligned with the magnetic field lines. Except for near the magnetic poles, these lines are roughly parallel with the earth's surface so that the side lobes of the dipole are earth-oriented. This provides about 2 dB of antenna gain which, for the price of a bar magnet, could be a bargain.

Permanent Magnet

Permanent magnetic axis alignment of satellite.

I hesitate to leave passive stabilization. It works; it's inexpensive; but it tends to be overlooked and underrated because big satellites have largely stopped using it. Also, it helps rid your organization of pesky digital and automatic control engineers who are, at best, expensive, and at worst a negative influence on corporate decorum, what with their drinking, personizing (we used to say "womanizing"), and so on.

But I digress. If you find the beauty of active control beckoning you from beyond the ken (where it's always been for me), then read on.

The most complex active control configuration to implement is, paradoxically, the easiest to relate to. Shown below is a three-axis stabilized satellite.

The three-axis stabilized satellite, right side up at the North pole, would appear upside down (head on the ground, legs in the air) at the South pole.

If you look at it from, say, a chaise lounge fixed in space, preferably close to a blue-green ocean inlet dotted by wind surfers plying a light trade wind from the East, a three-axis stabilized satellite would maintain a fixed orientation relative to you. If your chaise lounge were near an X-ray pulsar (rotten luck, I'd say) or an interesting interstellar gas cloud, you could appreciate an important feature of three-axis stabilization: it can stare at a single point in space.

Note that, without some moving around, it isn't going to maintain a fixed orientation relative to the earth. Imagine the satellite right side up over the North Pole. Its legs, if it had any, would be pointing down toward the earth. It is fixed in space, so as its orbit brings it around the South Pole its legs are now pointing away from earth. From an earth observer's point of view it is now upside down.

On the other hand, if you are a satellite in orbit around the equator, your up-downness relative to earth is unchanging, which is the preferred way to stabilize the antennas of modern communications satellites. In continuous equatorial orbit, their three-axis stabilization allows them to point continuously at their service area.

Small disturbances, however, tend to rotate the satellite away from its desired attitude. This small motion is usually sensed by gyroscopes. The most common way to deal with the rotation is to design a three-axis stabilization system that uses reaction wheels. A small rotation forward is countered by spinning an internal wheel in

The three-axis stabilized satellite, right side up at the North pole, as in the previous drawing, would appear upside down at the South pole.

the same direction as the sensed upset rotation. The equal and opposite reaction of the rest of the body (conservation of momentum) causes the satellite to slow its forward rotation proportionately to the amount the wheel spin speed is increased.

By continuously sensing rotations on all three major axes and controlling the spin rates of three wheels, the body can be held inertially fixed. For a while.

Let's enjoy a brief intermezzo on bias. What does bias mean anyway? It means all things being equal, they're not. In an unbiased system, half your bike rides are in tail winds, half in head winds. Your shopping cart rolls straight without applying wrist-breaking torques on the handle. Your trousers, straight from the tailor, have legs of equal length. You think satellites have any better luck than you do? The things that push on them are biased, too. They go forward more than back. Or back more than forward. The sign (+ or -) doesn't matter. What matters is that the disturbance torques applied to satellites are biased. Why? Sometimes because satellites aren't symmetrical. Sometimes because life is like that.

Living with the hand Murphy dealt us, in this case, requires handling bias. If external forces keep pushing the satellite forward, we have to keep speeding up the forward spin on the reaction wheel to counter them, faster and faster. The wheel is spinning forward and if we slow it down, equal and opposite forces push the satellite forward. So we can't slow it down. Every time nature gives the satellite another push forward, we have to increase the forward spin on the wheel to counter it. But this can't go on forever, because the reaction wheel can spin only so fast. When it's spinning that fast, it is called saturated. Now what? The wheel is spinning forward as fast as it can and we get yet one more disturbance that pushes the satellite forward.

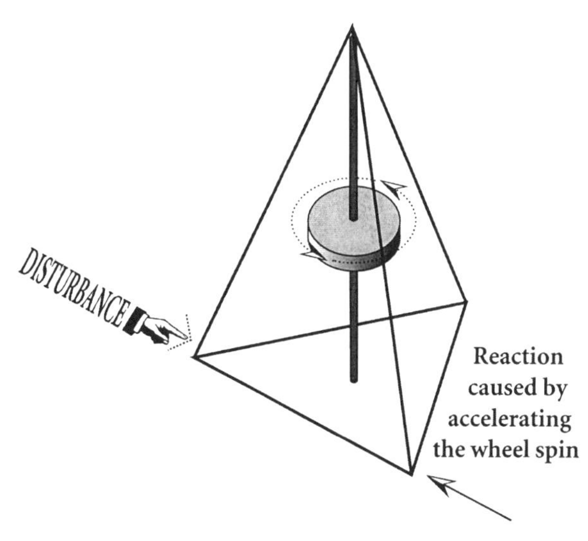

A stabilization system to counter small disturbances.

Kind of leads you right to the brink, doesn't it? Our satellite is caught between duty (pointing in one direction) and the immutable forces of nature, biased disturbance torques. You might wish to stop here so as not to ruin the ending when this is made into a 49-week docudrama on PBS and everyone in the office talks about it every Monday morning.

Or you might want to hear about propulsion. A lot of three-axis stabilized satellites have lots of very small rocket motors, around 1 pound force of thrust (for you metric sticklers, about 4.5 Newtons) distributed around them. Probably a hundred times you asked yourself why giant satellites have these tiny motors. Okay, maybe once or twice you've wondered. Even if you've never wondered, I'd say it's an interesting question and probably worth trying at cocktail parties and office lunches.

The boss decides the group needs a little more adhesion so all 37 of you pile into 19 Ford Escorts, minivans, motorcycles, and Porsches and head out to the local Taco Shop, which used to be a health club with a one-lane lap swimming pool. You are seated at a table three feet wide and 25 meters long and your group bonding experience amounts to facing one person you know only too well and have been trying to avoid for six months. The guy you've been wanting to BS with about that PBS docudrama is down where the diving board used to be. You're in the four feet. Well, here's the answer. Ask that true zero whom Fate has cast into the middle of your bonding lunch what all those little bitty rocket nozzles are doing on big, three-axis stabilized satellites. Here's the answer:

When that forward spinning momentum wheel saturates, or gets near to saturating, or just plain gets tired of spinning pretty fast day in and day out, one of these rocket motors is fired, which nudges the satellite the opposite direction from the biased disturbance forces. In our case, where natural forces are pushing the satellite forward, the motor is used to push the satellite backwards. Now the wheel has the opposite problem—the satellite is rotating backwards. The wheel slows its forward rotation to counter this backward motion. By applying small impulses with the motor and seeing the reaction wheel respond, the wheel can be returned to about zero spin rate. And that is one of the things small rocket motors on big satellites are for.

Another way to desaturate momentum wheels that doesn't require propulsion is to use magnets. A satellite in earth orbit is influenced by the earth's magnetic field and can generate torque using this magnetic field. An electromagnet coil is turned on when its orientation relative to the earth's field results in a torque in the right orientation to rotate the satellite in the desired direction to desaturate a reaction wheel.

The magnetic system is not always the solution to the desaturation for several reasons. In the very high orbits used by modern communications satellites, the magnetic field is weak and highly variable, making it very hard to use. Even in low-earth orbits where the magnetic field is stronger, the field lines are not always aligned well for producing the torques required to desaturate the reaction wheels. Despite the ability to use the magnetic field for the stabilization task, rocket motors could be necessary for trimming the satellite's orbit. Designers may opt to get double duty from the motors, using them for so-called station keeping propulsion and for stabilization. Both propulsion and magnetic torquing have been successfully employed with reaction wheels to achieve three-axis satellite stabilization.

11.2 Spin Stabilization and Conclusions

Now all can benefit from the highly enriched Pearls of Wisdom (POW) I gleaned during endless Popsicle® breaks taken at my previous place of toil with the sharpest engineer who's ever condescended to talk with me, Richard Warner:

POW #1: Thermal vacuum chambers are available for simulating space thermal environments. Anechoic chambers can show you how antennas are going to work on orbit. Shake tables and centrifuges verify structural strength. But there are no 0-gravity chambers to play with torque-free motion.

There it is—the wisdom of the ages right here. Who goes into a thermal vac test and changes nothing? Who does shake and antenna tests and doesn't see anything worth improving? Not me; probably no one. But we routinely design and build complex satellite control systems with various sensors, controllers and actuators without any testing besides computer simulation and analysis. We expect the resulting systems to stabilize, point, and track satellites because there is no room into which we can throw the satellite and have it rotate and translate freely without disturbances from, say, hitting the floor or the walls. Typically, we can simulate sensor environments and measure actuator response. But no one can watch a gravity gradient stabilized satellite approach equilibrium in a laboratory. This unfortunate fact is, by the way, quite often used as a lever by those scheming G&C engineers to extract ever more expensive computing equipment and salaries from their nearly bankrupt employers.

Which brings me to **Pearl of Wisdom (POW) #2:**

Every silver lining has a cloud.

Armed with our newly achieved enlightenment, spin stabilization is easier to treat. Torque-free motion is the way an object behaves without external disturbances like friction from touching surfaces and air resistance. In torque-free motion, a spinning body spins forever. Its spin axis—the direction the axle of a spinning wheel points— would never change direction. Of course, even if an object doesn't spin, in torque-free motion and without initial rotation rates, its axes never change their orientation in inertial space.

The difference between the torque-free motion of a spinning body (for example, a spin-stabilized satellite) and a non-spinning body (an object in orbit without any rotation) is one of those subtle physical things that many people would rather quit physics in high school than really get to know and love. Instead, they become bond traders and surgeons and CEOs of major corporations. But from their private jets and expensive homes overlooking Puget Sound, they know that something, some subtle physical insight, is missing from their lives. It is this void, Freud tells us, that motivates Donald Trump to put his name on skyscrapers, used 727s, and blimps. You can avoid all the time and trouble involved in amassing these uncountable millions, only to see a substantial fraction of them squandered on taxes and Jaguar XJ-Ss, on diamonds, and single-malt Scotch.

It is much more elegant simply to understand that when a torque impulse is applied to a non-spinning body, the body begins to rotate in whatever direction the torque has twisted it. Thus, a satellite initially pointing at, for example, the sun, soon loses this orientation after a torque is applied. It slowly rotates with no particular orientation, which, if you really want to look at the sun, is counterproductive at mini-

A non-spinning body, when subjected to a torque impulse, will begin to tumble and continue to do so.

A non-spinning satellite tumbling after torque is applied.

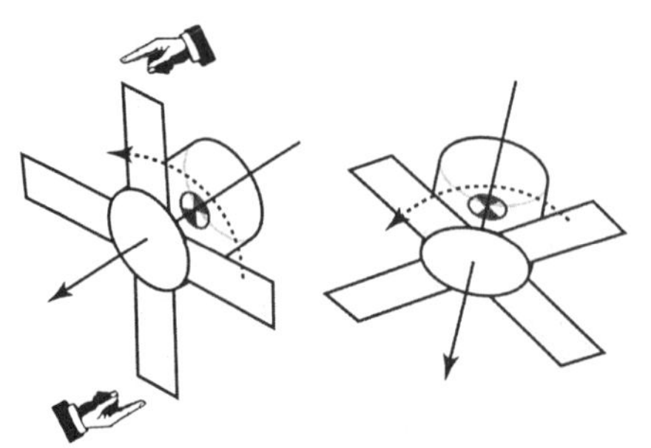

A spinning body, when subjected to a torque impulse, will precess its spin axis to a new orientation and maintain that orientation.

A spinning satellite with small deflection after torque impulse is applied.

mum. The situation is illustrated stunningly below.

It is even more elegant to realize that a spinning body, suffering a small torque impulse, responds only by having its spin axis, the direction its "axle" is pointing, change by a small, fixed angle. Thus, if a satellite is initially pointing at the sun and a disturbing torque impulse is applied, it still points in a fixed direction, deflected a small, unchanging amount away from the center of the sun. The more angular momentum (faster spin and higher spin axis inertia) possessed by the body, the smaller this disturbance angle is. The illustration shows the difference.

Now, isn't that easier than spending your time buying and selling major corporations? What good is a corporation once you buy it? What if you bought USX? Nobody really knows what USX makes anyway, and it really is only interesting cocktail party talk for a short while. Then you have to sell and buy again, maybe AMF this time. Then take up bowling. Spinning satellites, on the other hand, are neat and have real value.

This value stems from several attributes. Amsat's OSCAR IV and several following it spin so that their antennae, oriented on the spin axis, present gains to the earth when the satellites are near apogee. Thus they save power and reduce antenna requirements. Also, they all avoid a satellite's worst enemy—tan lines. Your know that if you fall asleep in the sun, your tan is uneven. Horrors! A true tanning buff (no pun there) wants to rotate slowly like a rotisserie so that he or she is evenly exposed to the sun. Satellites like this too. It keeps them from getting too hot on one side and to cold on the other. Thus, spinning reduces temperature gradients while helping hold a constant, inertial attitude.

ALEXIS, a spinner with a different attitude in many ways, always faces the sun. Its sun-facing surface is mostly solar panels that don't mind, in fact rather enjoy, continuous solar exposure. Its payload, on the other hand, hates to face the sun. A spinning satellite can provide this yin/yang experience because if a disturbance torque steers it a few tenths of a degree away from the sun it can be slowly dragged back by applying torques from its magnetic torque coils' interaction with the Earth's field. There is no hurry to make these corrections. After suffering a disturbance torque, a non-spinning ALEXIS would slowly, continuously rotate. Eventually the payload would face the sun and become unhappy, confused, and possibly damaged while the solar panels go into shadow and stop making electricity. A very bad state. But since it is spinning, the same disturbance torque now creates only a fixed, small pointing error, say 0.1 degree, and that's all. If the G&C system (like the engineer who built it) is out to lunch 22.3 hours a day, No Problem! When it awakens, it does not find a disoriented satellite, just a (ho hum) 0.1 degree pointing error. It corrects this error by applying an opposite torque with its magnetic coils.

One of the potential disadvantages of magnetic torquing is that the correct torque is not always available. Because of its stiffness, a spinner can afford to wait for correction opportunities even if they occur only a few times a day.

Spin stabilization is used in a range of applications. Spinning up rocket upper stages before firing makes the rocket point in a nearly constant inertial direction despite inevitable disturbance torques produced by small thrust asymmetries. Geosynchronous communications satellites, particularly from Hughes, have a spinning body that provides a stiff platform for orienting highly directional antennas. The antennas themselves cannot spin because they have to stare at a region on earth, so Hughes uses a large bearing and the antennas sit on a de-spun portion of the satellite.

Another variation on the spin-stabilized satellite is to move the spinning portion into a small box somewhere on the satellite and leave the rest of the body unspun. This configuration, shown below, has the same response to a disturbance as the all-spinning satellite.

The advantage is that only a momentum wheel inside a housing, which looks temptingly like a computer hard disk, spins while the rest of the satellite does not. Configuration power needs to be supplied to get the wheel going. This power varies roughly with the amount of stiffness you need, which in turn depends on the satellite mass, how tightly the attitude must be controlled, and the size of anticipated disturbance torques. To be reasonably stiff, a 250 lb. satellite at 400 mile orbit requires only a few watts to spin a wheel weighing a few pounds.

Stabilizing effect of a momentum wheel.

11.3 Conclusions

In this chapter I have irreverently but accurately glossed over every major type of space vehicle stabilization system used today. All without more than a few smirks and furtive grins from the real G&C types I have to deal with regularly. These systems have included:

- Unstabilized
- Passive Stabilized: gravity gradient, aerodynamic, some spinners, magnetic
- Active Stabilized: spun/despun, some other spinner, three-axis

It should also be intuitively obvious, as they say, to the most casual observer that these themes have lots of variations. On a welcome note, in this book you won't see any more attempts to address these myriad concepts. Having lived through this voyage into the virtually unknowable, the only cloud enclosed in your silver lining is the gap that finishing this opus might leave in your reading. Have you tried the yellow pages?

Chapter 12

Memory Systems of Spacecraft —or— Memory — What Is It Good For?

—with Richard Warner

According to Seinfeld, the original title of *War and Peace* was "War—what is it good for?" Talk about rhetorical questions—no wonder the publisher dumped it. The song (by the groupWar) of the same name notwithstanding, War is good for killing your enemies, plundering, raping and absorbing wealth, maximizing your popularity among your legions, and raw ego satisfaction. It's a great technology driver, and paves the way for cultural imperialism and opening up new markets. If War didn't exist, surely Coca Cola or IBM would have to invent it (hark! a war cry of the '60s when we thought they really did.).

Satellite memory systems are good for something too—but somehow they lack the death/sex/violence cachet of War or X Files. While the infoworld has focused on computing power—8086/80286/80386/80486 and Pentium all have become household words—the manipulative power of a fast microprocessor is worthless without data rapidly available for manipulation. In 10 years, the amount of memory we think we need in our computers has leaped from 32k or 64k (k being roughly 1000 bytes) to today's 6M to 64M (M being roughly 1 million bytes). A factor 1000 increase. The case can be made (though not in this book) that TV ratings need a good war here and there, and that solid state memory like that found in PCs created the small satellite revolution.

To be small and simple, ergo, cheap to build, and to be cheap to launch, microsatellites tend to be in Low Earth Orbits (LEO) from where the earth appears rather hugely close by (an object 12,000 miles in diameter only 300 miles away looks something like a beachball an inch from your eye).

At that range, you can't see a whole lot of the beachball or, in the case of the LEO satellite, the earth, and from the microsatellite's perspective, it can't communicate with what it can't see. Flipping over all the cards, most small satellites only see the ground station that links them to the people that paid for them to frolic carelessly in the heavens about 25 minutes per day, coincidentally about the same amount of time most dogs are active.

So what do they do the rest of the time? Of course with dogs, the options are few—eat, lounge, sleep, rip up furniture. With satellites, the amount they can do during the 23 hours and 35 minutes of solitude afforded them depends expressly on how much memory is on board. With no memory, they are like dogs except they're easier on the furniture. They can soak up solar energy and charge batteries, and they can sleep. Now even a cheap satellite in orbit with a ground station will run you a few million dollars, and having that investment sitting idle 98.3% of the time could be equated to spending a couple of $100M on something that would work 100% of the time—like a large geosynchronous satellite in orbit over the equator at your longitude. To be cost effective, small satellites need to be awake much more than dogs, and employing a large on-board memory, they can do that.

Maybe you want to map the whole earth, looking for spectral signatures of oil near the surface or undersea. Ideally, the satellite would remain awake all the time and take hundreds of digital images of the earth's surface in a few key wavelength bands. What happens to all those images until the next pass when they can be sent down to the ground? They get stored in the spacecraft digital memory.

Maybe you want to collect status messages from sensors all over the world. You might be monitoring ocean temperature, or just keeping track of the positions of containers on ocean going vessels and trains, or the status of pumping stations and electrical distribution centers. The satellite collects these data to send down on the next pass over the ground station—and meanwhile stores them in the spacecraft digital memory.

Maybe the satellite images astronomical objects. Many of these are not bright, requiring integration over very long times. Their images are slowly built up through repetitive addition of photon counts into the spacecraft digital memory.

None of these applications are possible without large, fast, low power and mechanically simple devices for mass storage of digital data. The most capable small satellites can downlink about 128 kBytes (about one million bits) of data per second. With an average of 1500 seconds of contact time per day, 2 Gbits (2 billion bits) of data is typically targeted as a data storage capacity, since that's all the data which could be downlinked in a day's worth of contact with the satellite.

12.1 History

Classically, satellites either had no storage or they used electromechanical devices like tape recorders. There are still many satellites in orbit, quite costly sophisticated ones, with less memory than an average home computer. Because satellites evolved before large memories were available, their missions have been built around real time control and data relay. They require no more memory than does a telephone or a television set since they do not have the job of storing data, but rather relaying it as quickly as it is received. They do not execute stored commands, but rather respond to real time commanding from the ground.

Small satellites, because of their tight volume, mass, and cost constraints, and because they have found wide application in providing digital data store and forward (mailbox and e-mail) service, have really driven the development of the newer generation of solid state memories. Working at AeroAstro, I (RW) have been involved in the development of digital memory systems for several small satellites.

12.2 What's Available in Satellite Memory Devices?

The classic data storage device for spacecraft for decades has been the flight tape recorder. These were reel-to-reel devices like you would see in a mainframe computer installation, complete with motors, capstan drives, springs, and guide wheels.

Its main disadvantages include large mass (ten kgs), high cost (close to $1M), reliability problems due to the moving parts, and lack of flexibility in terms of access to the data—random access to specific portions of the recorded data is virtually impossible. Anyone who has tried to find their favorite track in the middle of a cassette tape knows this problem! Other moving media devices have been considered: computer hard disk drives, magneto-optical disk drives, and VCR-type cassette drives. These devices offer large storage capacity, but suffer from reliability concerns similar to those of conventional tape drives, and they tend to access data serially. You can't go in and read any file you'd like, play back from front to back, or back to front.

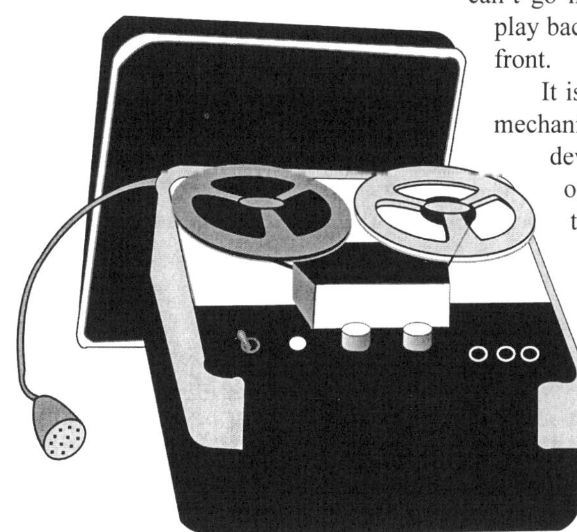

A reel-to-reel storage device from the Jurassic Era.

It is also worth noting that moving mechanical parts inherent in these devices can generate small but occasionally significant vibrational disturbances to a small satellite platform. Spinning devices like the tape spools or the metal platters on which modern hard drives store data add angular momentum, which can make attitude maneuvers more difficult. These forces are insignificantly small in your lap top, but the satellite lives in a much quieter environment. The stabilization force on some small satellites can be swamped by the force of the weight of a housefly.

Also, though some of these terrestrial computer components have been toughened for portable use, i.e., with laptop computers, they are still quite sensitive to vibration. Most difficult to handle is the harmonic, periodic loads of the launch. Just because a hard disk can withstand a drop from the tabletop to the floor does not mean it can stand the strong, prolonged, highly resonant vibrations of the launch vehicle and the tests designed to ensure the satellite will survive the launch.

Bubble memories were quite vogue for a few years in the `80s. Their main feature is that once data are written, they are not erased even if power is lost, unless they are written over intentionally. Bubble acts like random access memory (RAM) in a computer, but holds data even without power. But bubble memories have proven large, costly, heavy, and quite slow. Lacking acceptance in the terrestrial computer industry, they have not kept pace with other solid state devices. These technologies, particularly RAM, designed around semiconductor memories like those found in computers, offer many advantages. In fact, they are the only memory type that have gained acceptance in the small satellite world. While other technologies have been looked at, particularly use of hard disks and optical disks, solid state memories have become so compact, reliable, and power conserving that the memory does not drive the spacecraft design, and the motivation to find an even better solution, if any even exists, is weak. So for now and the foreseeable future, RAM will be the choice of most small satellites and small rockets. They are the topic of the rest of this chapter.

12.3 Anatomy of a Solid State Memory System

The diagram below shows the three major components of a solid state memory system. The memory array, which provides the actual storage, is almost always expandable and configurable for particular applications. The interface section provides the path into the memory system; data are written and read from the memory through this path. The controller manages the memory array, keeping track of where the data are, what portions are used, and what portions are available for more data.

12.4 Semiconductor Memories

Three different types of semiconductor memories to consider for a spacecraft memory system are: dynamic random access memory (DRAM); flash electrically programmable read only memory (flash EPROM); and static random access memory (SRAM).

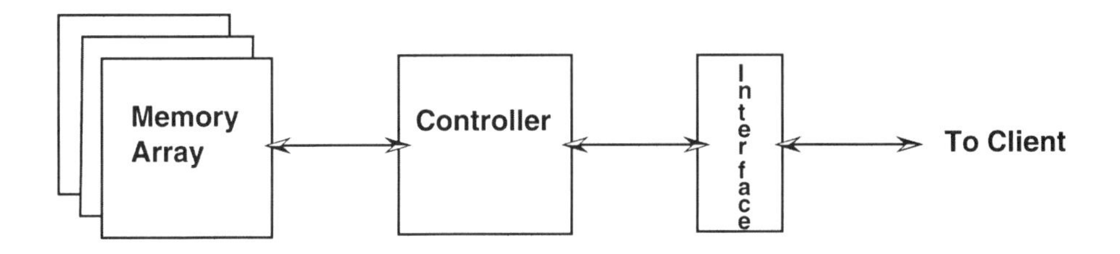

Solid State Memory System.

12.5 DRAM Storage

DRAMs are the semiconductor storage devices most commonly used in all computer systems for terrestrial applications like your Mac or PC. They have the advantage of the best density and lowest cost per bit. In the past, they were not selected much for space applications because they require frequent refreshing, which requires power. They have also had a low radiation tolerance. But power requirements have dropped a lot as manufacturers of laptops and other portable devices like cellular phones have innovated to combine low power with lots of fast, cheap memory. There has also been a serendipity in the radiation area. While this is not significant for terrestrial applications, radiation tolerance is vital for most LEO satellites. It has turned out that the processes necessary to get good manufacturing yields of very dense DRAM happens to yield quite radiation tolerant parts! Also, faster processors mean the satellite computer can check for and cleanse the memory of errors quite frequently, so that they can be caught and corrected. The more frequently you scrub, the higher the error generation rate that can be tolerated. Thus, DRAMs have become the standard technology for spacecraft memory.

12.6 EPROM Storage

Flash EPROM is used in some palmtop computers and many electronic gadgets like digital musical instruments. Storage is permanent even if the device is unplugged from the circuit, yet it can be electrically bulk-erased and reprogrammed for a limited number of times, typically 10,000 to 100,000 write cycles. The main advantage of Flash EPROM is that once you write it, the data are permanently stored without requiring electrical power until you purposely erase them. Flash EPROM is dense, though not as dense as DRAM, shows promise of good radiation tolerance, and can be completely shut down indefinitely for the minimum possible power requirement. It is also expensive—ten to one hundred times more expensive than DRAM, though component cost is rarely a driver for a spacecraft (2 Gbit of storage might cost $5000) and the savings in power may easily justify that cost. For some applications, these devices promise to be an excellent match. Though they have not yet seen wide application in spacecraft as bulk memory, they are used to store critical data like the operating system and spacecraft state vectors.

12.7 SRAM Technology

SRAM or Static RAM, was the overwhelming technology of choice just a few years ago. Static RAMs offer low power consumption, high data transfer rates and reasonable density, but now have a limited commercial market base, which is causing their cost and density to lose out in competition with DRAM.

The current technology for DRAM is 64Mbit parts. Each part holds over 64 million bits of information. Historical trends were a quadrupling of capacity for a given part type every two or three years, though that pace may be slowing. Still, given that most microsatellites are aiming for about 2 Gbit, only 32 parts will do the job. They could be packed into a 4x4 cm circuit board using surface mount. But there are repackaging companies that stack the silicon wafers into "sugar cubes". The entire 2 Gbit made up of 32 wafers can be housed into not much more than 1 cubic centimeter. Thus the actual memory is rapidly becoming an insignificant consumer of mass and volume.

12.8 Radiation Effects

One of the more significant differences in operating a memory system in a space environment compared to a terrestrial environment is the effect of radiation. Radiation has several effects with different consequences. Long-term exposure to radiation results in a gradual degradation of electronic components. This effect is referred to as total dose. Bias currents increase and thresholds change. At some point, the device stops working. This problem is solved with a two-part strategy: first, by selecting components with intrinsic hardness; and second, by shielding, which means packaging the components in a metallic enclosure to reduce their exposure.

Energetic ions are atoms and molecules missing a few electrons and hence carrying a net positive charge. These ions, present in the LEO environment, can embed themselves into circuit element substrates and cause some types of components to latch up. Essentially, the component suffers an internal short circuit. The strategy for dealing with latch-up is again two-part: careful selection backed up with "circuit breaker" protection circuits. Often, if the latched-up part has its power cut and restored, the part can be restored to normal. While latchup can be destructive to the part, it is rare and there are materials that are virtually immune to it. If power is cut quickly enough, requiring special circuitry that monitors for sudden increases in current flow, latchup damage can be prevented.

Much more often an ion does not cause latch-up but instead causes a single event upset (SEU). SEUs occur when a circuit changes state, so that a stored 1 becomes a 0 or vice versa. If SEUs occur in the actual memory components, the data are corrupted; if they occur in the control circuitry, the memory management is disrupted.

12.9 Error Detection and Correction

SEUs in the memory array are handled by storing more bits of information than strictly necessary—essentially by storing redundant data. These allow reconstruction of the correct data even after some corruption.

The simplest example of this is parity. For every byte of data in memory, the system allocates and maintains an additional bit of parity information. If the total numbers of 1s in the data is an even number, the parity bit is set to 1. If the total number of 1s in the data is odd, the parity bit is set to 0. When the memory system is given a byte of new data to store, it calculates the current parity value for that byte. When the byte is read out of the system, it can be verified that the parity bit still agrees. This way, theoretically, if you suffer an SEU and a bit "flips," you'll know something's awry.

This scheme has numerous shortcomings. Parity is not a way to correct corruption. It only tells you if the data are good or bad, not how to fix them. Likewise, a byte with multiple bad bits is likely to be identified as good.

More sophisticated EDAC (error detection and correction) schemes store a number of parity bits for each data word, which allows correction of single bit errors and detection of many multiple bit errors. For example, a common scheme uses an algorithm known as 32-bit Hamming encoding. For every 32 bits of data, 6 additional bits of parity, or check, are generated and maintained. Imagine a situation where a memory array begins with correct, uncorrupted data. Over time, SEUs invalidate bits in the array. The memory now contains bad data. When the first few bits flip, the Hamming code can used to detect and correct the error, thus returning the memory to perfect fidelity. However, if nobody "looks" at the data and EDAC bits, these random bit flips start to add up. After a while, all of the data in the array may be corrupted, and therefore useless. No EDAC scheme can help you.

The solution is to periodically "scrub" or "traverse" the memory, going through each location, verifying the data, and correcting bad locations. How often scrubbing is required to avoid accumulating too many errors to affect a full correction is determined from a probability analysis of the frequency of bit errors. If you have an idea of the upset rate (which is a hard number to quantify) for the orbit of your mission and the RAM technology you are using, you can calculate the total error rate (the rate of accumulating errors which can't be scrubbed using the error correction schemes in place). This calculation is based upon the frequency of traversal, the number of memory bits, and the capabilities of the specific error detection and correction (EDAC) codes used.

For small memories, scrubbing is relatively easy. The microprocessor controlling the memory reads each block of data and uses the EDAC bits to fix occasional errors. However, as the memory grows in size, the task gets harder. Of course, twice the memory requires twice the speed in the processor to scrub the whole memory. Most larger memory systems have dedicated hardware specifically designed to speed the EDAC process.

Another important error correction function of the memory is to "map out" bad memory parts. Mapping out can eliminate potential errors due to hardware problems. If a memory chip, for example, should fail, the processor can eliminate that part of the memory from its address space. It just stops using the bad parts. Software sensitive to repetitive errors in specific addresses can be included for doing mapping out tasks.

12.10 Transfer Rate

The transfer rate tells how quickly data can be moved into and out of a memory system. A typical application might require the input rate to be fairly low, on the order of 10 Kbit (10,000 bits) per second, and the output rate to be higher, say 1 Mbit (1,000,000 bits) per second. This is typical because while data may trickle in slowly as the spacecraft orbits the earth, it must be downloaded rapidly during the brief period of time when it can communicate with the ground station.

Because DRAM is very fast, the transfer rate is not limited by the memory technology, but by the operating speed of the interface.

12.11 Interfaces

A wide variety of power and data interfaces to memory systems are available.

For the power interface, most systems can be configured to accept either unconditioned 28V (or any other voltage) or conditioned 5V supplies. From a systems point of view, the first option is easier; however, the second option can be slightly more efficient because the spacecraft may have a lower number of larger DC/DC converters, rather than converting locally at each device.

Data interfaces can be either serial or parallel. Serial interfaces are better from a cabling and configuration point of view; parallel interfaces allow faster data transfer rates and generally simpler support software.

Typical serial interfaces include simple asynchronous dedicated links, usually based upon the RS-232 standard, and multidrop serial links, for example, MODBUS, based upon the RS-485 standard and avionics standards like MIL-1553.

Parallel interfaces are most often specifically designed for a particular application.

12.12 Power Consumption

Static RAM uses very little power when quiescent; thus, the size of the memory array has little impact on the power consumption. You simply leave most of the memory in its quiescent state, and only activate the portions you need to read or write. What does drive power consumption is the data transfer rate and activity and the scrubbing rate for error correction. Dynamic RAM does consume power during every refresh cycle, however with current technologies, even two Gbyte carefully designed may only consume a watt or two.

12.13 Testing, Quality Assurance, and Reliability

One nice aspect of memory systems, compared with altitude control systems or deployables, just to pick two examples, is that it is quite possible to do a thorough job of testing on the ground. Minimum weight deployables require the absence of gravity to work exactly as they will in space. Attitude control systems would ideally like to see the surroundings exactly as they look from space, and exist without disturbance torques from gravity and drag that are a part of being on earth. But memory systems are really only sensitive to launch loads and vibrations, radiation, and thermal conditions. A well-devised "shake and bake" test plan can provide high confidence in system integrity.

Radiation testing is more problematic, but it is a fairly mature, well understood art. Radiation testing an entire memory system would prove quite difficult; instead, parts are tested at the component level. Once the parts are characterized, we can simulate the bit flips expected in space and ensure the EDAC actually works.

Electronic components are available in a wide range of screening levels. The basic parts are the same, but the higher screening levels are subjected to additional inspection and testing. Some of the levels, in ascending order of supposed quality and definitely cost, include commercial, Mil-883C Class B, and Class S. You typically find one order of magnitude increase in price for each step up the screening hierarchy. Note that you are NOT getting an order of magnitude increase in part quality for your money. You are getting more traceability and documentation and some increase in the probability that the parts will not fail prematurely. Usually.

Typically, small satellite programs are cost-constrained. In many cases, small satellites are built with lower screening grades than would be used for a large conventional satellite. The thinking here is that testing and part replacement is so easy in a small satellite that system level screening will weed out bad parts. Also, small satellites simply have less parts to fail, so they can accept lower reliability specs per part.

Memory systems by their very nature offer a high degree of graceful degradation. Generally, they do not fail at once, but rather individual SRAM chips fail. Note that this is one disadvantage of denser RAM. More eggs in fewer baskets. It can be very worthwhile to consider a lower cost part and accept the potentially higher failure rate, with the knowledge that this system can easily map out failed parts and continue to operate with only slightly reduced capacity.

12.14 Additional Processing Tasks

It is also worthwhile to consider that the memory controller in many systems is resident in the general purpose microprocessor that might have considerable extra computational power besides what's needed for the memory maintenance function. In some applications, this extra capacity can be used to advantage. Data could be prioritized so that important information is downlinked before mundane stuff, or data compression algorithms could be executed in background mode. Check with your memory vendor about support for these kinds of enhancements.

12.15 Vendors

Seakr Engineering builds a number of different memory systems, some with Shuttle flight heritage. Texas Instruments has also provided both complete systems and innovative components for space memories. AeroAstro has built memory systems for it's own satellites and others. But as density of memory, processor power and integration levels all increase, many spacecraft developers, and processor developers, are building their memories into their products, rather than purchasing them as separate components.

Shown in the photo on the following page is a 1990 era memory unit built by AeroAstro for the Spectrum X-Gamma mission. While today the components and packaging are much smaller, the basic elements are unchanged. Most of the cards house banks of memory chips. One or two additional boards are microprocessors responsible for addressing, reading and writing to the memories, and executing the EDAC algorithms. They also maintain EDAC statistics to test our models of error rates with actual flight data. The memories use an RS-422 serial interface and draw about 3 watts for over a gigabyte of storage.

12.16 The Future?

SRAM, though intrinsically rad hard and low power, has continued to follow the evolutionary path of bubble memory. It is not cost competitive with DRAM, and its power advantage has been eroded by advances in DRAM technology. DRAMs with 64 Mbits on a single chip are now available. This means that the entire spacecraft memory, all the data that could be downlinked in a day's worth of contact to a single ground station at 1 Mbit/s downlink rate, will fit in a single "sugar cube" package of

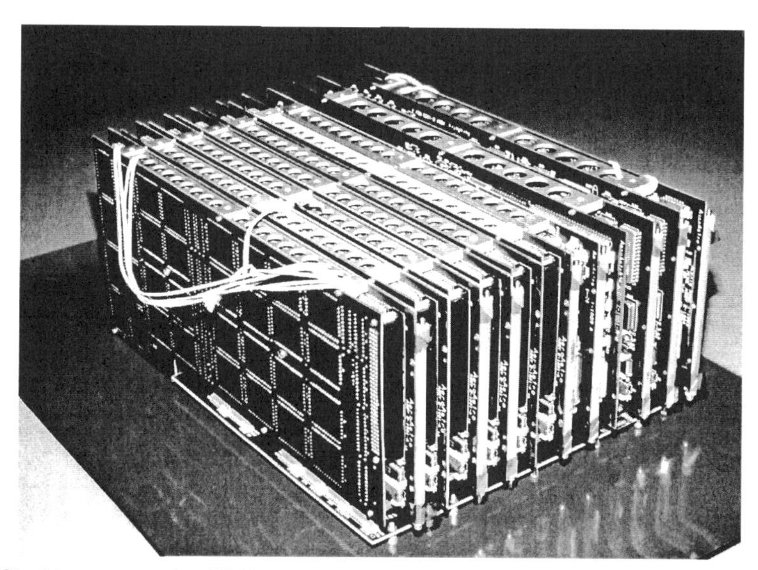

A 1 Gigabit memory using SRAM technology designed by AeroAstro in 1989 (cover removed).

much less than 1 cubic inch (16 cc). Power consumption, not counting the spacecraft processor, which is necessary whether the memory is there or not, can be below 1 watt. Thus until there are significant advances in the downlinking of data, and excepting some special case missions where many days or weeks might go by without a downlink opportunity (interplanetary missions could be one such case), memory is not a major size, mass or power driver in microsatellite design.

Chapter 13

Mechanisms: The Nuts and Bolts of Small Satellites

13.1 Moving Parts

Just one typical integrated circuit, one of those tens or hundreds of little plastic bug-like creatures planted on their 32 steel legs on a typical circuit board inside your satellite or your home computer, has typically 500,000 or even 1,000,000 active elements: simple switches (transistors); capacitors; resistors; logic gates; memory cells; each one capable of switching state millions of times a second. It is remarkably unremarkable that these innocuous, seemingly static things change state, in other words, do things, easily 1 billion times per second. A malfunction of a single gate, with a dimension of a few millionths of a meter, can be the end of a multi-million dollar mission. You say you built in redundancy? Congratulations. Instead of sweating the failure of one in ten million elements, your reliable, redundant satellite is now vulnerable only if two failures occur out of twenty million active devices.

As far as I can tell, the steely-eyed macho satellite engineer has no problem accepting this risk. He'll stake his career on 10 million devices, each one of characteristic dimension 3 micron, all working fine at a state change rate of a billion per second, despite the reliability this implies for each one of those 99¢ DIPs (Dual In-line Pin) devices. But try telling Mr. or Ms. Cool that you want to deploy a boom with an antenna or magnetometer away from the satellite 5 or 10 feet. Or maybe you want to deploy the secondary mirror of a Casegranian telescope, or a solar collector or solar panel. Two moving parts, maybe three. You'll get two reactions: naked fear coupled with naked greed, for only by spending prodigiously can you assuage the fear. Something as "complex" as that could take millions of dollars to design, build, and test, and it will be the least reliable part of the whole space mission. That's what program managers will tell you.

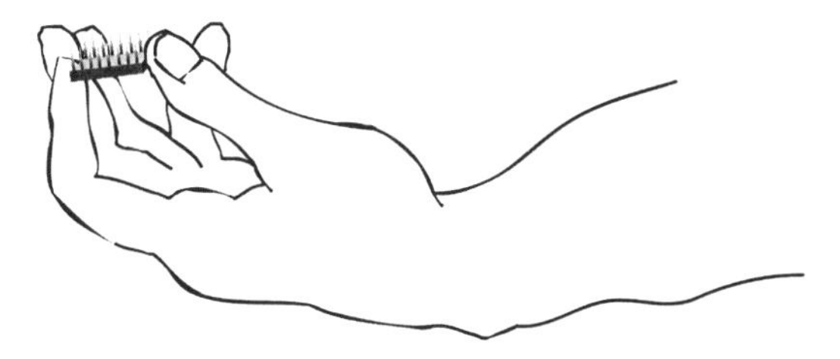

An IC chip with up to a million active elements—but no moving parts.

These moving parts are examples of deployables, which are things that are mechanically unfurled, moved out, or unfolded after the satellite gets on orbit. They are guaranteed to unnerve even the coolest designer. First comes denial.

"You really don't want to do that," the systems engineer explains, trying to remain rational.

Then anger.

"I thought you said you wanted a reliable, cheap satellite..." The designer imagines a black smudge on an otherwise perfect flight success record.

Then comes recrimination.

"I should have stuck with a simpler career, brain surgery or President of the United States."

Finally, acceptance.

"This is going to cost you big bucks."

Oh, the mighty dollar. Our ultimate weapon.

Hey, what's all the fuss? You've got a billion transistors on board, and all I'm asking for is one simple hinge. The difference is that none of those transistors physically moves. They are conductors, semiconductors, and insulators in and around which electrons flow. A deployable, like an antenna or sensor on a boom that extends on orbit, means physical motion and all that goes with it. Nasties like flexibility and poor alignment, leading to increased friction, jamming, and galling. It means force and restraint, momentum and potential deformation. Thermal inhomogeneity leading to dimensional instability. Vacuum welding and the problems associated with non-outgassing lubrication. Fine structures that can resonate and break. These are the lions and tigers and bears of spacecraft mechanisms. They have conspired with Murphy over the years to destroy billions of dollars in space missions.

13.2 Shall We Confront Our Fears?

On earth, we use deployables every day without fear or spending millions of dollars. Every morning I used to hop in my Buick, when I had a Buick, and switch on the radio. A whip antenna automatically, electrically deployed about 3 feet out of a 4" cylinder in the trunk, powered by a tiny electric motor. Tens of millions of cars have these, and they work great. If you break one off somehow, you are talking $179 to replace it. More than a McSalad, but not a big factor in a $20M spacecraft mission.

So why aren't we flying these marvels of reliability? This is where we space guys get former Senator Proxmire's Golden Hammer Award. But not 'cause we want it. Let's follow the life history of that simple antenna deployment mechanism as if it were to be applied to spacecraft. For one thing, antenna pieces are made out of plastic that outgasses, meaning that in space they get rough and stiff, and the evaporated constituent materials could recondense elsewhere on the satellite, like on the surface of

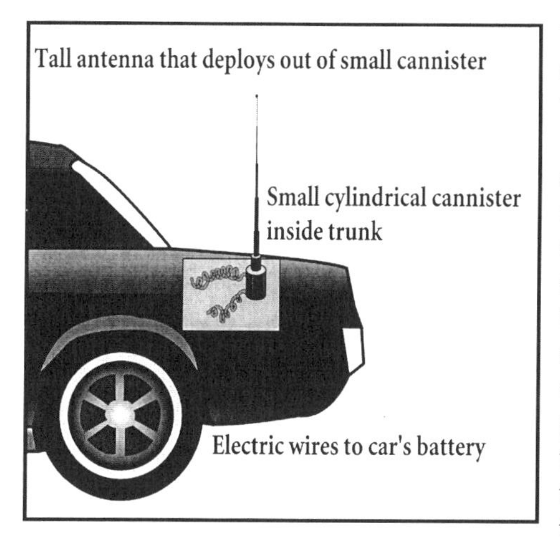

An earthbound deployable.

the optics. OK, so we replace that stuff, assuming there's a replacement with suitable properties.

Next problem is that the Buick antenna has lots of metal-to-metal contact—surfaces that have to slide over one another. On the ground these are lubricated with oil or grease, but these materials can't withstand the space vacuum. They just evaporate away. Two metals touching each other generally do nothing on earth, but in the space vacuum, two metals pressed together can weld themselves together, their crystalline structures intermingling to form a bond. This technique was for a while even used by Peugeot to make cold welded aluminum bicycle frames, using a terrestrial vacuum chamber to join the various structural members, but you don't want your deployable antenna to try to weld itself together in orbit. So lubrication is a big issue.

The Buick antenna's motor depends on air circulation to cool it while it works, a feature not to be expected on a satellite, and it too is lubricated with space no-nos.

Space is a benign force environment, but getting there isn't. That Buick antenna is pretty tough, but can it survive 50 degrees below zero Fahrenheit (-40˚C), and +175˚F (+80˚C)? Those are extremes that the mechanism might see on orbit, or on the way there. Not necessarily in the rocket ride, but in planning a mission, we have to plan on contingencies - what ifs. And a what if could be like this: What if the spacecraft stabilizes itself in some weird attitude and your mechanism is left in the shade, radiating its heat away to the black vacuum of space and it cools to -40C. Will it survive? Will it deploy at that temperature? If not, and if that mechanism is needed to work in order to get power or communications going, that could be a mission killer. G-loads as high as 21 times the force of earth's gravity (21g) are encountered in some launches, and qualification tests might go to as many as 30g in vibration.

By the time you make all the design changes to use a motor that won't overheat and will operate without conventional lubricants; find metals that won't weld themselves together in vacuum and plastics that don't evaporate and that stay subtle and flexible over the full range of temperature; then beef it up so that everything can withstand the thermal and mechanical environments, you discover you can't actually build what the Buick engineers built.

As soon as you modify even innocuous small things, which the motor, lubricant, and basic structural material are not, you have a new, untested product. Which leads us to Dilemma #2—that when you head off to build your own mousetrap, you will discover that the reliable and cheap Buick antenna actually took many years and many millions of dollars to get right. GM went through hundreds of prototypes and tested each of them for thousands of hours before fitting one on your car. Probably in the first year on the market, GM installed one million antennas, and got a million vehicle years of test data upon which they based a few additional design tweaks, which resulted in the nice reliable unit I got on my Buick. The reason it only costs $179 is that the development cost is spread over millions of cars every year for many years. If $50M were spent to build that deployable antenna and get it right, and it's used on 10 million cars a year for 5 years, that's $1 per car in R&D cost. You, on the other hand, want one deployable, and presumably you don't want to spend $50M to get it, nor do you have years to wait for it. There are companies that sell these spacecraft deployment devices, partly to spread the cost over many satellites. But they sell one or two or maybe five a year, so they still typically cost millions of dollars.

In the real world, building a self-deploying antenna, or a solar panel on a hinge that simply folds out, or a rocket payload fairing that splits in two and disappears (in other words, leaves the vehicle) without leaving any pieces behind and without hitting the rocket or its payload, all turn out to be hard problems. We live in a society bathed in clever consumer devices, things like 6 GByte hard drives with discs the size of a silver dollar that can withstand the laptop computer they live in being dropped four feet onto the floor. We think these ought to cost $300, and they do because millions of them are made every year and their R&D costs are well distributed. And they are the nth generation of decades of continuous product investment and improvement.

Like the days between autumn and the end of the year, your choices, when you need to deploy something aboard a small, inexpensive spacecraft, come down to a precious few:

1) Find a mechanism that has already flown and worked and try to finagle a good price for it;

2) find a terrestrial device that can work in space;

3) build something from scratch and try to beat the odds.

Small satellite developers play all three of these options. The weakest is the use of a previously qualified device. If it was built for a major mission, it is out of your price class. A few meters (10 feet) of deployable boom fully qualified for a major program can cost $2M. Or even $10M, which is more than the total budget of most small satellite development programs. Still, many of us have deployed things, and it's not a bad idea to call around and see what's on the proverbial shelf.

Terrestrial devices are tempting indeed. They work, and you don't pay any development cost for them, so they're cheap. But before you try to fly one, go through the following list of questions as a minimum:

- Can the terrestrial part withstand the temperature, vibration (including not just a shock, but harmonic loads), and static load environment of my mission?
- Does it contain magnetic materials that can interfere with spacecraft instruments or create disturbance torques on orbit in the presence of earth's magnetic field?
- Will the materials it is made of outgas? Will they survive in space?
- Does it require lubrication? Can I find substitute lubricants and if so, will they work?
- Is vacuum welding possible?
- Does it require air to cool it? Will it sustain large temperature gradients without immersion in air that could cause warpage or failure?
- Does it contain any microelectronic parts that might not be radiation hard and that will therefore fry in the increased radiation environment of space?
- How reliable is it really? In space nobody is around to jiggle the knob or fuss with a balky connector. It works or it won't.
- What are its momentum characteristics? For instance, does it have rapidly spinning components that can cause the spacecraft to tumble when they spin up? Does it have angular momentum that can interfere with spacecraft stability?
- Flight safety, particularly if you launch on the Shuttle: Does it have any glass or other materials that can fracture and leave small pieces? What type of insulation is on the electric wires? Will it evaporate in vacuum and cause a short and potentially damage other components or maybe even drain the spacecraft battery through the short?
- How much power does it need? To a Buick, 250 watts is no real big deal—just look at how many 15 amp fuses are in its little fuse box. But that can be ten times the total power budget of a small satellite and even for a transient load, could drive the size of the power distribution system. Since most deployables only work for a few seconds, your major concern might only be instantaneous current capability of the electrical system, but you have to think about how to

switch it, and without large transients that can harm other equipment. Also, make absolutely sure it goes off and stays off!

- Can it be tested? If you have a device that can be adequately tested and demonstrated on the ground, that's a big plus. Beware of components too flimsy to work in the earth's 1g environment and at 1 atmosphere of pressure, since they are hard to prove workable at all.
- Finally, if you have to modify it to make it meet the space criteria, or to perform somehow better in space (for example, more angular precision in the way it's pointed once deployed), you should assume you are starting basically from scratch, since it will not work exactly like, nor will it have the reliability implicit in, the original device. It was the replacement of one simple wire that caused the recall of 50,000 Saturn automobiles. Remember, therefore, to budget for a significant test and qualification program.

In the words of a wise old technician I apprenticed under at a German wind tunnel: "Alles nicht so Einfach." Nothing is as simple [as we first think].

Finally, there's from-scratch development. All of the above disparaging comments notwithstanding, do-it-yourself can work. It has worked in the past, otherwise we wouldn't have the flight heritage designs we all try so hard to keep using today. Some pointers:

- Look at all the criteria listed above for using devices originally designed for terrestrial application.
- Make sure you test it over as wide a range of conditions as possible.
- Get a few uninvolved reviewers to look at your design, development, and test plans.
- Make it stronger, harder, bigger, and in general more overdesigned than you think you need to. You can always refine it in the next generation.
- Allow plenty of time (6 to 12 months) after the part is "finished" to make improvements and corrections.
- Design the spacecraft systems so that the failure of the home-brew part does not terminate the mission.

13.3 What's out there?

Following are a few deployment aids you might consider:

Pyrotechnic bolts. These favored classics have been around for many years. Space deployables have to withstand large launch loads, and then suddenly be free to move. One way to accomplish this feat is to use nice strong bolts fitted with explosive charges, often moving bolt cutters, to break them. They are reliable, but not particularly testable, because once tested, they can't be reused. You don't want to be in the same room when you initiate them. Pyrotechnic devices often impart high shock loads. They also require special circuitry to ensure that they absolutely will not fire prematurely. In other words, they are normally shorted to ensure that current doesn't flow through the initiator. Finally they aren't particularly cheap. A single bolt is $600 to $2500. The fact that they are probably the most commonly used actuator, despite all their disadvantages, speaks well for their advantages, which are reliability and a very solid, instantly removable structure.

Hot Wax Actuators. A relatively recent, though now flight-proven device, electrically heating and thus vaporizing wax captive within a small, stainless steel cylinder forces the motion of a piston that can be used to open a latch, release a coverplate or door, or push a moving mechanism on its way. Hot wax actuators are more expensive per unit than pyrotechnics, but they can be used and tested repeatedly. They have no special safety hazard, and they produce no shock loads. On the minus side, it can take 30 to 60 seconds to achieve a deployment, and the concomitant electrical energy requirement is high. Also unlike pyrotechnics, the hot wax actuator continues to draw current even after deployment is successfully achieved, so the spacecraft must include logic to terminate actuation. A final disadvantage: if anything goes wrong, Space News will probably report that the failure could have been anticipated because the deployable was held on only with wax! Hey, this actually happened to me. We live in a media ocean, and you've gotta think about swimming.

Melting wires, or I should say "vaporizing"? You can constrain a solar panel or antenna with a simple wire, part of which is heated electrically and vaporized, thus releasing the wire constraint. These are very inexpensive, safe, and shock-free. Some people worry about the vaporized metal and where on the spacecraft it might redeposit.

Sublimation. Hold two parts together with a plastic molded part made out of a material that outgasses 100%, meaning that it evaporates completely in space vacuum. When that happens, the two parts separate. This technique has been used to achieve very good thermal isolation by removing the mechanical supports of the isolated vessel. Again, redeposition of the sublimated material onto sensitive surfaces is a potential issue.

The deployment mechanisms themselves include:

Hinges. Hey, they're simple; they usually work.

Carpenter tape. An old AMSAT technique, antennas are made of metal carpenter tape rolled up. When released, the tape straightens out and voilà! An antenna. Most recently seen on Orbcomm, the technique has been used by amateurs for over 20 years.

Stacer. This word refers to a ribbon of metal rolled into a truncated cone. The metal ribbon wants to unwind into a long rod but is constrained by a pyrotechnic bolt. Stacers deploy with plenty of gusto. In fact, they can puncture arctic ice! Cheap but not hugely reliable, stacers rotate as they extend, making them less than the first choice if you need to tightly specify the deployed object's attitude. Stacers have been used to create low cost gravity gradient booms.

AstroMast and alternatives. AstroMast is one example of a whole world of deployable structures. It is a small trusswork collapsed into a box. On command, the trusswork assembles itself and a boom is extended. The process is reversible, which the Stacer is not, which can be handy.

All kinds of shapes have been deployed: beams, flower petals, complex radiator geometry, multiply folded solar panels. Some of what has been produced is truly amazing, but that does NOT mean you want to try them. Most require years of development time and tens of millions of dollars to design, build, and qualify.

13.4 Testing

Space engineers appear obsessed with testing. Vacuum chambers, shake tables, anechoic chambers, radio ranges, all the high tech paraphernalia of preparation for life at 7 km/s, high above the atmosphere. Many parts of satellites are really quite easy to test. Radios could care less if they

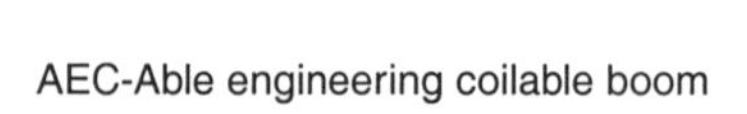

AEC-Able engineering coilable boom

are in zero gravity or not, ditto (remember before searing right wing talk radio when normal people used to say ditto and there were machines that were named accordingly?) computers and batteries. Yeah, they need to be tested, particularly to ensure they can withstand the thermal and radiation environment, but the absence of gravity, hard to simulate on the ground anyway, is not a major issue for electronics.

But it sure is for mechanisms. Deploying a 10 kg (22 pound) mass at the end of a flimsy 10 meter (33 foot) boom might not even be possible on the ground. Many space booms can't support the mass they support in space in any orientation on earth. They can't push it upward, because their motors and springs are too weak. They can't push it downward because the momentum of the deployed mass causes the structure to fail, and they can't do it horizontally because the side loads are too great and will bend the boom. Big programs try to get around these problems in clever, albeit expensive, ways. For instance, they create very flat, 15 meter (50 foot) long stainless steel tables with thousands of tiny holes machined in them through which compressed air is pumped. These massive air hockey stadia are virtually frictionless horizontal surfaces and stuff gets deployed on them. But even they are not perfect. For instance, in deploying a boom horizontally on such a table, the boom is supported on the side, preventing distortions that could occur in space where there is no table working with gravity to keep the whole thing nice and flat.

This simulation stuff is all pretty interesting and creates lots of challenging jobs, but for low cost space work, it's useless. You can't afford one of these tables; you can't even afford to use one for a month, but it's interesting to know that if you could, you would. And you'd still have questions. Just in case you think everybody in this business is hidebound and limited by their own imaginations, I know of one 0-g mechanism that was tested horizontally with helium balloons used to offset earth's gravity. Problem there is helium balloons have a lot of aerodynamic drag relative to their lifting force, hence they have trouble moving as fast as the mechanism and hence they produce a retarding force. In the end it worked, but not without significant fiddling around and pledges that we would never try that again!

The other side of the testing coin is that we really don't test things at all. Not compared with the commercial world. How many testing hours does that Buick antenna accumulate in its first year on the road? Say there's a million of them sold on all the GM cars, and each radio goes on and off twice a day, so that's two deployments, 2,000,000 a day, 365 days, for 730,000,000 deployments in the first year. You can build some failure statistics on that. How about testing a new drug? Take 1000 people and test it, or 10,000 people. That's a hugely expensive test program. But it's nothing compared with the number of Tylenol doses sold in a year, which is billions worldwide.

For most products, the field test is a final step in the testing process. Final testing might result in the occasional recall, but that is hard to avoid, given that real-world testing is millions of times more exacting, or at least more exhaustive, than the best lab test. Aerospace mechanisms don't get the opportunity to be tested in application. They have to work on their first time out.

13.5 A World of Mechanisms

Not all mechanisms are deployables. A very common mechanism is opening access doors. Sensitive optics or materials need to be kept in vacuum or otherwise isolated from the integration and launch environments. There are many, many sad stories of doors that refused to open on orbit, even after significant test programs on the ground. Some causes of failure have been launch vibration load, temperature extremes, and vacuum welding problems.

Other mechanisms include camera shutters and rotating carousels. Gyros and momentum wheels are also mechanisms, though they are often purchased as sealed units. The spacecraft designer doesn't need to be too concerned with their operation other than to be careful handling them. The bearings, lubricants, and assembly techniques developed over many years to make these devices reliable shouldn't be taken too much for granted. A momentum wheel seems simple enough until you actually try to build one that will survive the launch and operate for many years on orbit.

The same can be said for computer hard disk drives. Problems flying these in space include:

- They make angular momentum that can interfere with the satellite stabilization.
- Their lifetime is relatively short compared with many space missions.
- They require an air environment to float the heads and for cooling, so they need to be in an airtight enclosure sealed for the duration of the mission.
- Though they can absorb shock, the are not made to withstand the extended, possibly harmonic vibration of launch.
- Their electronics are not radiation hard and hence they will not work in most orbits, even Low Earth Orbit (LEO), without major design.

A side comment on these devices is that modern satellites tend to be downlink, not memory, limited. For example, even with a 1 Mbit per second ground station, the typical 20 or so minutes of daily contact time to a LEO satellite from a single earth station amounts to only about 1 Gbit (10^9 bits) of downlink data a day, spread over several contacts (see Chapter 12 on memory for more on this). Thus, for many missions 1 Gbit of memory is plenty. This much memory, and more, can be built into less space using RAM than using a hard disk, avoiding a complex mechanism and a potential failure point.

13.6 You Turn Me On: Rock & Roll and Explosive Bolts

A certain coincidence has apparently gone unnoticed in the popular satellite press until now. (If any satellite press is popular; I mostly read it because it's my job.) Satellites (the man-made kind) and rock and roll got started at the same time. As products of the '50s, they share certain commonalities, not the least of which is that people remain fascinated with them today for largely unfathomable reasons. Just what was so great about hula hoops and waitresses on roller skates at drive-ins? But in historical perspective, this coincidence explains why people who build satellites are obsessed with turning them on, a more modern, '60s kind of concept.

Think about the turning on part. A little satellite could weigh 50 kg (110 lb). Maybe 10 of those kg (22 lbs) are batteries. Sometimes it takes a month to get a satellite launched. And sometimes, as in the case of the Shuttle, you install the satellite six months before it gets into space where its solar panels can receive sunlight and recharge the batteries. Ever leave batteries in your favorite flashlight for six months and then turn it on? Zilch, right? Add the fact that NASA likes to launch satellites with their batteries discharged to make sure they don't do anything unexpected, and you begin to understand the problem.

If you haven't built satellites, you might not believe that hitting the ON switch occupies the time of many MIT Ph.D.s, but it does. Fundamentally, more than just battery discharge is involved. For safety reasons, satellites are usually launched while turned off, so their radio transmitters don't go on and interfere with the launch vehicle guidance system, for instance. Never thought of that one, did you?

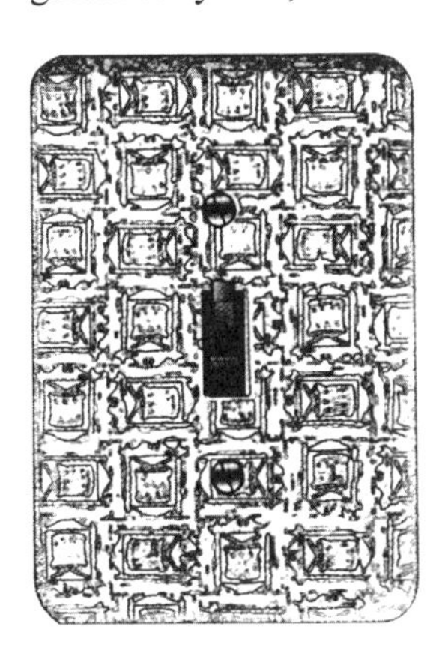

You know how you're not supposed to use the cellular phone when going through a road construction area where they're using explosives? Well, guess what a solid propellant rocket looks like? About 25,000 kg (55,000 lbs.) of blasting cap. If your downlink radio happens to resonate into that mass of plastic explosive, the launch crew is definitely NOT going to have a nice day.

So we're talking serious OFF here. Back to rock and roll. Besides OFF, the satellite is seriously ATTACHED to the launch vehicle. A ride into orbit on a typical launch vehicle has a lot in common with a real blood and guts rock and roll concert—serious noise, vibration, and physical abuse. The ride quality is 10 times worse than a roller coaster Bart Simpson wouldn't touch. Test loads are often 20 times the force of gravity, so your 50 kg satellite has to be bolted down as if it weighed a ton. Then, after a few minutes similar to being bolted to one of KISS' big speakers on stage at LA's Fabulous Forum, your satellite is supposed to spring off lightly and amble quite civilly into its orbit.

One solution to these problems would be to build launch vehicles that aren't so sensitive to the occasional, highly unlikely, radio getting turned on and that don't shake their payloads within an inch of their lives. Until that becomes a reality, engineers spend a lot of their time Turning On, and to quote a favorite rock and roll cliche, Getting Off.

The de facto standard Getting Off technology is the Marman Ring, which is named after a real person who was not, contrary to often-seen spelling, a Mormon. He invented the method sketched below for making a really good mechanical joint, and then severing it simply, quickly, and symmetrically. By symmetrically I mean so that it doesn't come off tumbling (a tough problem for the satellite attitude control system), but straight. Axially. Probably a more generic term for this kind of separation design is "manacle ring."

The Marman, or manacle, ring separation system is actually composed of several parts. The rocket and the satellite use tapered rings, actually truncated cone shapes with a very flat, mated surface. The two mated surfaces are surrounded by the third ring, called the Marman band or manacle clamp band. This very springy metal belt is bolted tightly around the joined satellite and launch vehicle rings. When the belt is correctly tightened, the two ring faces are held tightly together. If the belt is severed, the two surfaces are not attached at all. In practice, the belt itself is not severed, but rather the bolts that tightened it around the two rings. In fact, the band is often made of two halves bolted on both sides where the halves are joined.

Manacle ring separations mechanisms are used both for separating satellites and other payloads from vehicles, and for separating the stages and other parts of launch vehicles.

The bolts that hold the halves together are severed by pyrotechnic (explosive) bolt cutters. If either bolt is cut, the band, which is very springy and wants to straighten out flies off, flies off and the satellite is separated. Usually the bolts are fitted with two cutters each, so if any of the four cutters works, the satellite is separated. Many rings are fitted with

Marman or manacle ring.

springs that push the satellite away. Other times the launch vehicle backs out of the way. Some satellite carry propulsion systems that fire after the Marman Ring releases, and the satellite accelerates away.

Marman or manacle rings are to satellite separation as Mick Jagger is to rock and roll—sort of instant credibility. Nobody doubts Mick is the real thing. Tell any tough reviewer that you're using a manacle ring separation mechanism, and they're happy. Unlike Mick, who really is the real thing, the Marman Ring isn't constant as the Northern star, which Joni Mitchell once pointed out. Like her, over the years it's been reinterpreted many times, and the real physics behind its operation are somewhat subtle and beyond the current scope.

Make sure the Marman Ring you base your satellite's life on really is a Marman Ring and not just a circular something that oughta work. Potential problems include deformation of the rings under the stress of being tightened, which tends to held them together without the attachment belt; vacuum welding (remember that grease is good for that 50s haircut, but really messy in space where it redeposits all over your clean satellite); and injuring the satellite as the band goes flying off when it gets cut. Another problem is space junk—holding onto the pieces of the Marman band and the bolts so that they all stay with the rocket.

Bristol Aerospace of Winnipeg, Canada, is the leading supplier of manacle ring separations mechanisms (the real thing). Bristol has flown over 1,1000 manacle ring mechanisms without a single flight failure.

There are other ways, of course. People have used the bolt cutting trick alone, for instance. Bolt the satellite to the launch vehicle and when you want to get off, cut the bolts. Trouble is, all the bolts have to cut, and all at the same time. But this does work. Slip rings cut by a circular charge, and lots of other ideas, are used in special circumstances. The Marman Ring isn't perfect; it's just well understood. It takes height, and that band is really moving when you cut it. It also requires explosive cutters, which means lots of safety qualifications. Still, that's where the technology is.

So you got off, but so far you didn't turn on, which is where aerospace engineering really goes paranoid. You really DON'T want it to turn on prematurely, but then it HAS to turn on. So half your brain is busy making sure it won't turn on, and the other half is making sure it will. This is like Spy vs. Spy, layer after layer, first an "inhibit" to keep it all off, no matter what, then another actuator to make sure it goes on, no matter what.

Typically, the action starts with a switch. Sometimes the switch is actuated when the manacle ring disappears. But that is sometimes too fast. If, for instance, the satellite wants to deploy solar panels when it gets turned on, you want to make sure that the rocket and the manacle ring are out of the way first, and that means letting a little time go by first. So OK, a switch and a timer. No big deal, right? But what if the timer doesn't go on? What about a second timer—maybe set the second one for a long time, like two weeks, and start it when you kiss the satellite goodbye on the launch pad. Why two weeks? What if it starts raining, they don't launch and they don't let you back at it? If it rains for five days and your timer is set for three, maybe the solar panels deploy inside the rocket. Fun.

Now about that switch. You're going to trust your baby to one little microswitch? Of course not. Put four in there, all in parallel. But then the safety people, who don't care if your satellite fails just so it doesn't hurt anything else when it does, say, "Gee, if any of those switches shorts prematurely, does the satellite go on and blow up the rocket?" So each switch has to go in series with another switch with a DIFFERENT actuation mechanism. Maybe it looks for sunlight or feels for the launch vehicle instead of for the Marman Ring.

Well, all of these arrangements could, and do, get tedious. You get the picture. With as many solutions as there are satellites and launch vehicles, you can bet that every satellite designer has agonized over this simple little conundrum. But this is good news for you. Don't reinvent the wheel. Go talk to your neighborhood satellite jock who is building her or his wheel, and paint it your own color. One final note on getting off: Deployment mechanisms kill more satellites than rocket failures or failures on board satellites. So don't shy away from putting real thought and lots of testing into this critical part of the mission.

The world is full of mechanisms: landing gear on aircraft, elevators and escalators, the animated characters at Disneyland. None of them have to survive a difficult environment like space, and all of them are the product of more testing than you can do on your pet mechanism in a lifetime. All of them failed at some time in their history. But, through a long process of refinement, the ones on airplanes and in amusement parks get fixed and improved and developed into the highly reliable devices we have today. Without that luxury, using existing, proven devices, overbuilding everything and making sure you can test at terrestrial conditions, along with following all of the guidelines for making mechanical structures for space is your best option. None of which is to say it can't be done.

We have deployed enormous structures in space, and even bigger ones are planned. Many of them have never failed to work perfectly, though did you ever wonder about that Apollo picture of the US flag planted on the Moon's surface, apparently waving in the breeze? Some breeze!—a deployment mechanism that didn't quite work as planned. Right in front of our eyes in one of the most famous of all space pictures, possibly one of the most famous images in the history of humanity, glares out at us a little lesson on deployable mechanisms. They sort of mostly work.

Chapter 14

Batteries Not Included

What is included: *useful information on electrical storage for small satellites sprinkled liberally upon a canvas of occasionally entertaining comments at best generally peripheral to the subject at hand...*

Think about this simple phrase: Batteries Not Included. It's printed on the side of almost every package on sale at Toys-R-Us. Of course it means there are no batteries inside the item. It also means that you are going to need batteries before you use your toy. Since not you but that little 8-year-old monster of yours is going to unwrap the item and life immediately becomes nothing short of unbearable until the batteries somehow do mate up with the toy, you need to buy batteries. Not just any batteries—the right size and number. D? C? A? AA? AAA? 9V? Alkaline? NiCad? Transistor? Gold Top? How many? One, two, three, and some gadgets require 8 or 12! Used to be you could at least hypothetically live in a world without all these choices by moving to the Soviet Union, where the upper bound on battery types and sizes would closely approximate 1. Well, we all know what happened to that utopia... Society without zillions of battery models is just not viable. That's the clear lesson of the cold war.

That option no longer available, you find yourself without recourse, standing in aisle 1b at Toys-R-Us, donning those reading glasses that make you look and feel like your mother-in-law, and scanning the fine print: Four D-size batteries required. You still don't know what kind of D-size, so you pick the cheapest, no, the second cheapest, lest the kid label him or herself Child of Misers, and off you go.

When you get home, your loving spouse tells you that you should have gotten NiCads (Nickel Cadmium batteries, aka rechargeable) because they're ecologically more desirable, or at least you should have snagged the disposable kind with reduced mercury. You slink off to bed, feeling a bit beaten by the whole experience, wherein you pick up your *Newsweek* only to find that Cadmium has replaced polyunsaturated fat as the bad guy of the year.

All that, dear reader, to run a little red fire engine fitted with a light and a siren whose sound, impinging on your eardrums already weakened from years of blasting Metallica on your (battery-powered) Walkman and ghetto blaster (is English a great language or what!), is shorted directly to that little spot right behind your eyeballs that no aspirin tablet has ever been able to reach.

14.1 When You're Away

We know why Walkmen, ghetto blasters, little red fire engines, and Nintendo games need batteries. They need electricity and normally you don't plug them into the wall. Satellites, whose batteries, believe it or not, are the subject of this chapter, have a taste for electricity too. As a homemaker/owner you probably think you own an infinity of extension cords (then why, you might ask yourself, can you never find one when you need one?). But you don't have enough extension cords to plug into your orbiting satellite! Even if you do, they'll get very messy as they wrap around the earth once every orbit.

14.2 Let the Sun Shine

Most satellites use photovoltaic (solar) cells arranged into neat panels to convert sunlight to electricity. This setup is nice when the sun is shining upon your panels, but rather inconvenient when the earth ambles between you and old Sol. Most small satellites end up in low earth orbit (LEO). Most LEO orbits are eclipsed from the sun by the earth for about 40 minutes out of each 100 minute orbital period. You need batteries if your panels are momentarily not pointing at the sun while your satellite does some attitude maneuvering or if you want to keep the satellite on during launch while it's enclosed in the rocket fairing, or after insertion in orbit before the solar panels are deployed. Some satellites have very high peak power loads. You turn on a big radio transmitter only when passing over your ground station in Lima, Ohio, which happens 20 minutes a day. Batteries allow you to store the trickle of power from your solar cells all day long, and use that electric power in a big blast when you need it.

14.3 Have You Ever Had to Make Up Your Mind?

You will be glad to know that you probably face fewer choices in battery shopping for your satellite than for your kid's next weapon of mass discomfort. Spacecraft batteries don't sell nearly as well as consumer products. Research has proven that the annual cell output of the US' leading space battery manufacturer is less than the number of cells purchased in one busy hour at any single Toys-R-Us. Also, space application imposes lots of constraints on the battery design.

Ignore the few satellites that have been orbited without solar panels. These operate for only a few hours before the batteries expire or do not have any electrical systems. An example was the Echo passive reflector satellites flown in the '60s. And the moon. Satellite batteries need to be rechargeable. A very few make electricity from nuclear power, an option that would cost millions, except that it requires sign-off right on up to the President of the United States, which is safely relegated to the "not likely" file.

The only other thing in consumer products like NiCad (Nickel Cadmium, two fairly common metals that make up the electrodes of the battery), for some applications, for many years has been lead acid. Car batteries, which use a liquid, or aqueous, acid and lead as the electrodes, are of course rechargeable. They are the most commonly used lead acid batteries.

Besides NiCad and lead acid another rechargeable is available for satellite builders: Nickel Hydrogen (NiH2). Battery manufacturers only recently discovered the reason NiH2 has not been popular in small satellites. It's because there is no neat acronym like NiCad. Seriously, you think MacDonnell Douglas sold many planes before they went to MacDac? No way, José. They actually tried NiHy (for the battery, not the airplane company), but they were sued by a company that makes chocolate soda pop. All of this motivated development of a new battery called Nickel Metal Hydride, which has a nice acronym, NiMH. I always thought that stood for the National Institute of Mental Health. Wrong, but at least I remembered the acronym. Lithium batteries have been available to consumers for a long time, particularly for long life applications like watch batteries. Many laptop computers and handheld video cameras use lithium ion rechargeable batteries. This technology is slowly transitioning to satellites, where it will provide four to ten times the storage capacity of NiCad.

14.4 How Do They Stack Up?

Satellite batteries operate in a demanding environment. Our typical LEO satellite goes into the earth's penumbra (shade) 14 or 15 times a day over 5000 times a year, a huge amount of charging and discharging. To get good life for the satellite, we want them to withstand maybe 20,000 cycles. If you could get 20,000 cycle NiCads for the kid's toys, and recharged them every single day, your 8-year-old would be ready for retirement before the batteries would be!

Launch costs range from $1000 per kg to $100,000 per kg, so we also want the batteries to be light, small, and not to leak. You probably haven't seen a battery leak since you stopped buying those cheap paper ones that wrecked the seven-transistor radio you paid $6.95 (1963 dollars) for back at Lafayette Electronics. Note: That's roughly $55 today. Those little guys weren't such a bargain, were they! But because the batteries could be exposed to the vacuum environment, leakage is a problem in space.

Under certain overcharge conditions some batteries might need to vent gas, typically hydrogen, which could be hazardous. Under thoroughly abusive conditions, lead acid car batteries can vent gas, a preferable alternative to exploding.

NiH2 batteries are the lightest option now in wide space application, storing about 50 watt-hours of electricity per kg, compared to 35 WH per kg for NiCad and for lead acid. Besides its acronym, NiMH's attractiveness stems from its light weight, about 10% lighter than Nickel-Hydrogen.

This number is particularly important because, to maximize lifetime, satellite designers allow their batteries to routinely use only about 15% of their capacity.

For example, a pretty capable small satellite might have a 60 W continuous power requirement. Going into penumbra for 40 minutes results in 60 W x 40 minutes = 2400 W-minutes or 40 W-Hours of battery drain. That should be just over 1 kg of NiCad, and less than a kg of NiH2, right? Wrong! Multiply those mass numbers by about 10 if you want lifetimes measured in months or years and not days or weeks. It is not unusual to require 15 kg of NiCad batteries on a spacecraft bus weighing less than 100 kg. That can make batteries the single largest consumer of your precious mass budget.

14.5 You Say Tomato, and I Say Tomahto...

In social circles hep to battery nomenclature, it is important to differentiate between a "cell" and a "battery." When you buy a 9-volt battery (9V), it is really a collection of 6 tiny cells, each with 1.5V. A cell is a single cathode and anode, and typical single cell voltages are 1.2V for the rechargeables and 1.5V for disposables. Up to 3.8V for Li-Ions.

It is not really a good idea to plan to use these things at 1.2V, i.e., as single cells. To produce a peak power of a modest 100 W, you'd need 83 amperes, which would mean using cables about twice as thick as your automobile jumper cables.

Commercially available collection of six tiny cells.

Anyway, because most electronics need higher voltages, like 6 to 28 V, people tend to prefer higher voltages. To produce these voltages, we combine cells with smaller voltages into a group to create a battery with a higher voltage. When you shove 8 D cells into the old ghetto blaster, you are creating a 12 V battery out of 8 cells, each of which provides 1.5 V.

Now you know why they call them D-cells and C-cells that each put out just 1.5V, but 9-V batteries. And now you can make cocktail conversation about batteries even at highly chic aerospace gatherings without fear of embarrassment. In other chapters we addressed the motor vs. engine, perigee vs. perihelion, and satellite vs. spacecraft, but for now stick to batteries and stay away from

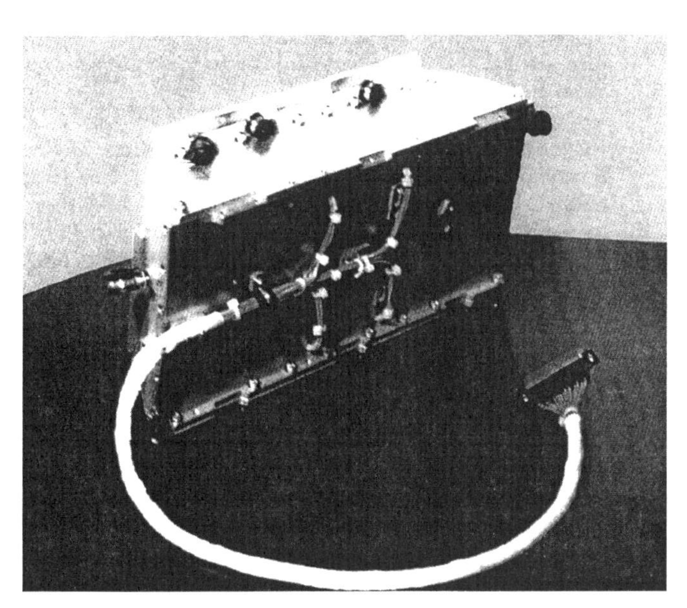

NiCad battery for the HETE satellite, containing 24 NiCad cells.

orbits and engines. If asked about non-battery subjects, just excuse yourself and say you've got to run out to the car and use the cellular to check in with the babysitter.

William Safire, you might have noticed, I ain't. The point of the Cells vs. Batteries discourse is that to keep current (amperes) low, you need several cells in series. Most spacecraft use 28 V. Some use lower. But just as an example, a 28 V electrical bus requires about 24 cells, figuring that as they discharge they get down below 1.15 V per cell. The smallest NiH2 cells are 30 Amp-Hour (AH). Configured into a 24 cell string, this is 720 Watt-Hours (WH).

Even assuming a depth of discharge of only 15%, that is 108 WH, appropriate for a spacecraft requiring about 165 W continuously. Most satellites carry at least two batteries consisting of two sets of 24 cells, thus providing enough for a 330 W continuous drain, which is about 10 times more than the most power-hungry small satellites. Indeed, two 24-cell strings of NiH2 batteries would have a mass of 45 kg, possibly as much as the entire satellite. In practice, most small satellites depend on 4 AH or 6 AH cells, and 30 AH NiH2 cells are just way too large for them.

NiH2 batteries have one additional drawback for small satellites. The cells are round because the internal hydrogen gas pressure is high, typically about 65 atmospheres (950 psi). Rectangular prismatic NiCads pack much more densely. But to round out the picture, NiH2 batteries also have excellent lifetime records and probably are capable of more charge and discharge cycles than any other space battery.

What about lead acid? It was bypassed for a long time for space application for several reasons. Clearly, flying a liquid acid like a car battery, including ports for gas venting overboard, looked like a bad idea for a satellite that needed to be tested in all attitudes. The acid would pour out! But lead acid batteries with a gelled acid do exist, like the gel formulation used in those leak-proof paper batteries that killed your seven-transistor radio back in '63. From that experience, you know lead acid can still leak. Most of today's pricier consumer disposable batteries are in metal cases to prevent leakage, as are the lead acids used for satellites. These so-called sealed lead acids have been used in a variety of field applications, including small satellites.

Lead acid cells are slightly heavier than NiCad compared to AH, but their voltage is slightly higher too, so in terms of power storage they are comparable. If you shop consumer batteries, you'll notice that NiCads don't last as long in your walkman as disposables, but they are also much lighter. Comparing a disposable C-cell with a rechargeable, for each cell the disposable or lead acid lasts longer. For each kg, they are comparable.

14.6 Qualify, Qualify, Qualify

There are no space-qualified lead acid batteries. You have to buy commercial units, test them, and take your chances. Lead acid batteries don't last as long as space-qualified NiCads, but they cost about $6 a cell vs. $2000 a cell for space-qualified NiCads. They come in cylindrical packages to provide a pressure sealed vessel to contain a small amount of generated gas, so nominally they do not vent.

Another low cost option is commercial-grade NiCads. If it worries you to use the same cells in your satellite as you do in your cordless screwdriver, then this approach is not for you, and neither is any lead acid variety, for that matter.

Commercial NiCads, like lead acids, run about $6 a cell, have power densities comparable to lead acid, and also come in cylindrical packages that look like the D-cells at Toys-R-Us.

14.7 This Year's Model

Last year's ecologist was pushing NiCad batteries for home use. Why throw out all that metal when you can buy a charger and recharge the same batteries hundreds of times? We can charge and discharge thousands of times in space applications where the depth of discharge is more limited.

This year Cadmium being introduced into the environment causes a lot of concern, not just from disposal issues, but because of the waste product of their manufacture. Our industry historically has believed that, as such a small part of the pollution picture, our designs shouldn't be torqued around by pollution effects only important on a much larger scale.

Nonetheless, every industry is now required to be ecologically responsible, and ultimately the future of NiCad technology could come into question. Lead acid, NiH2, NiMH and Li-Ion are our available alternatives. Lithium-ion is coming. In the meantime, in consumer application it is not clear whether the mercury-free dispos-

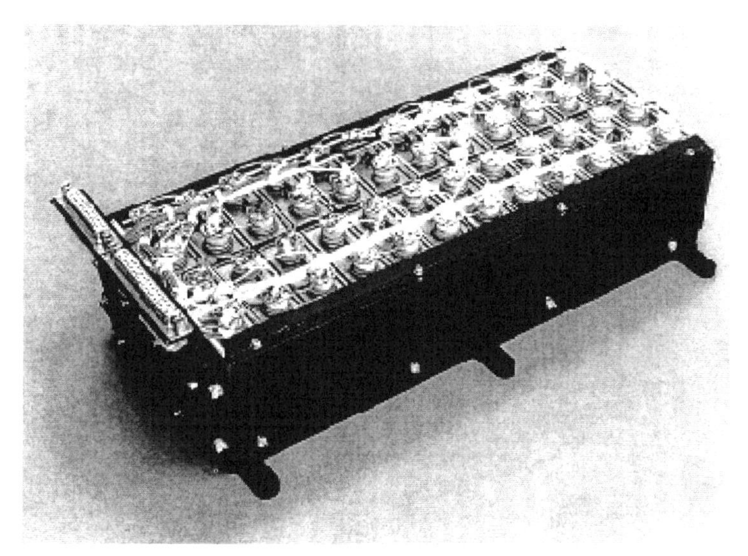

The Swedish Freja satellite used this space-qualified NiCad cell stack.

ables are ecologically more or less sound than NiCads. For right now, a rational buying choice probably should depend more on economy than ecology.

14.8 Qualification II: the Sequel

Commercial cells have been used in space application for decades, but, like recycling, it has only recently become fashionable. Also like recycling, there are plenty of ways to do this incorrectly. Batteries must be carefully screened and matched. No matter how carefully you prequalify batteries, commercial units are more failure-prone and also do not survive as many charge and discharge cycles.

In addition, commercial units have a shorter shelf life because of diffusion of materials within the battery. This difficulty might be slowed by cold storage, but once any battery is placed in service, even if you don't charge and discharge it, it is aging. The biological clock ticks faster in commercial NiCads than in space-qualified units. Even if your mission lasts only a year, if the satellite sits in testing and in a warehouse, er, cleanroom, waiting for integration and launch, all that time counts against battery lifetime.

A space NiCad is a different animal than a consumer NiCad. The materials used in battery manufacture, and the manufacturing processes themselves, differ between commercial and space units, a contrast with many electronic components like ICs or discretes that might differ only in screening level. For this reason, application of commercial NiCads, while promising to save potentially hundreds of thousands of dollars, must be accompanied by prudent design and testing. Usually a larger number of commercial cells are used in order to ensure redundancy against a cell failure.

14.9 Solar Arrays for Small Spacecraft

The primary source of power for small LEO to geosynchronous spacecraft missions lasting from a few days to several years is solar arrays. They are an ideal source of power for spacecraft with long life and very high reliability. Several issues are key in determining the design of a solar array, including spacecraft configuration, required peak and average power levels, operating temperatures, shadowing, radiation environment, illumination or orientation, mission life, mass and area, cost, and risk.

14.10 Required Power Level and Mission Lifetime

The three key design considerations in sizing a solar array are mission lifetime; average power requirement, and orientation relative to the sun. Because of the degradation of solar cell performance over time due to radiation (refer to Radiation Degradation below), the solar array must be sized to meet the power requirements of the spacecraft at its End of Life (EOL), which, however, causes an oversupply of power to the spacecraft at the Beginning of Life (BOL). For missions with a lifetime of more than ten years, sources of power other than photovoltaics should be considered because of the degradation of the solar cells.

14.11 Operating Temperatures

A solar cell's performance and efficiency are affected greatly by temperature. The power characteristics of a cell are usually determined or quoted at 25 to 28 C. Each type of solar cell has a different temperature coefficient (% decline in efficiency / increase of 1 C). Silicon cells typically have a temperature coefficient of -0.46%/°C, while gallium-arsenide cells run at about -0.22%/°C.

14.12 Spacecraft Configuration

Because a single cell's power output is very low, cells in an array are usually arranged in strings. Cells are placed in series to reach the required voltage and in parallel to reach the required current. When a cell is not illuminated, it acts as an open circuit; therefore, the shadowing of a single cell in a string causes the loss of the entire string. This problem can be reduced by several methods: actively pointing and tracking solar arrays, using diodes, and arranging cells in a "ladder" network. On two- or three-axis stabilized spacecraft, tracking and pointing the solar array can minimize the effects of components, antennae or structure that might shadow the cells. Diodes may be used to bypass groups of solar cells in a string to help prevent damage to cells that are not illuminated. A ladder places several parallel strings in a series parallel network. In the case where a cell within a string is not illuminated, this ladder network offers the remaining cells an output current path, therefore reducing the degradation of the output.

14.13 Array Configuration

There are two types of solar array configurations: planar and concentrator. Either configuration can be body or panel (deployable) mounted. Planar arrays are by far the most often used arrangement for a solar array. The use of panel arrays is applicable to three- and sometimes two-axis stabilized spacecraft. Panel-mounted arrays may or may not be pointable to increase the power output of the array.

Body-mounted planar arrays, which are used on spinning or tumbling spacecraft, reduce the need for and expense of tracking panels. Using body-mounted panels produces a less efficient sun incidence angle that in turn reduces the array's efficiency. Body-mounted panels also operate at a higher temperature because they cannot radiate excess heat from the spacecraft into space, which also reduces the efficiency of the arrays.

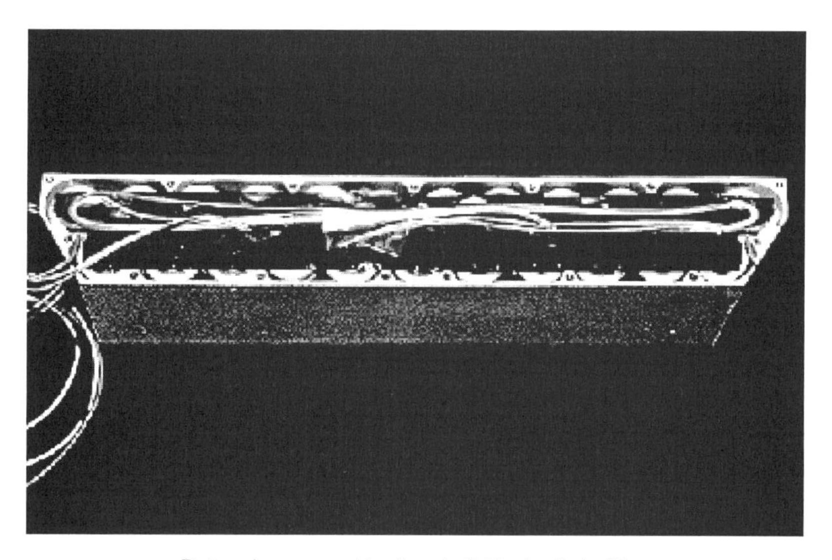

Battery box opened to show individual cells inside.

14.14 Energy Conversion Efficiency

A cell's energy conversion efficiency is defined as the power output divided by the power input. The solar illumination intensity (1358 W/m2) is the input value for the power input to the planar array. For example, an array with an efficiency of 16% produces 217 W/m2.

14.15 Radiation Degradation

An inherent weakness of solar cells is their loss of effectiveness due to radiation in space. The degradation of energy conversion efficiency over the life of a solar array can be caused by several other factors such as thermal cycling, damage from particle impact, and material outgassing, to name a few. By far, however, radiation is the largest factor in the degradation of solar array performance. For a silicon array, the degradation in performance might be as much as 3.75% a year with 2.5% of that caused by radiation. For gallium-arsenide cells, the degradation is reduced to 2.75% a year, with 1.5% caused by radiation.

14.16 Illumination and Orientation

Orbital parameters such as the sun incident angles, eclipse periods, solar distance, and the concentration of solar energy influence a solar array's illumination and intensity. The sun incident angle is the angle between the normal angle to the solar array's surface and the sun vector. The output power of the solar array varies with the cosine of the sun incident angle, and is a very important consideration in designing the solar array to provide adequate power.

Chapter 15

Bring 'Em Up Clean

Planning to be the proud mom or dad of a budding little satellite? Dr. Sparck answers some new parents' most frequently asked questions on how to get Junior off to the healthy start she or he needs to ensure a long and productive life on orbit.

***Q:** Dr., I've heard that developing young satellites need to be kept in a clean room 'till launch. What exactly is a clean room?*

Dr. Sparck: There is no single definition of a clean room. What some parents consider clean is, to others, unacceptably dirty. Generally, a clean room has the components shown in the sketch below. All clean rooms have controlled access, including a staging area for donning special clean clothing. The idea is to minimize introduction of contaminants such as particles of hair, skin, lint, and dust into the clean environment. All tools, test equipment, and materials brought into the clean room must be thoroughly cleaned first. Special ceiling, wall, and floor coverings that do not generate dust also reduce the generation of contaminants, particularly drywall and paint dust.

An air circulating system introduces filtered air into the clean room and keeps the ambient pressure above the outside air pressure, preventing dirtier air from entering through doors and crevices. The air is moved rapidly in the clean room in a laminar stream from the ceiling toward the floor, so that newly generated particles are rapidly carried away. Clean rooms are classified in terms of the number of particles per cubic FOOT of air. The most common level is class 10,000, meaning that there are on average 10,000 particles present in each cubic FOOT (about 1/30 cubic meter) of air. This level is about 100 times cleaner than a typical office environment. Some clean rooms used in aerospace fabrication achieve levels below class 1000. For some assembly operations, even cleaner environments are sometimes needed. Generally a clean chamber (glove box) or hooded tabletop is used, since achieving levels of class 10 to 100 is almost impossible if a human operator is present.

***Q:** My home computer, Walkman, and television all operate in a normal environment, and our house is no clean room! What's special about satellites that requires such particular care?*

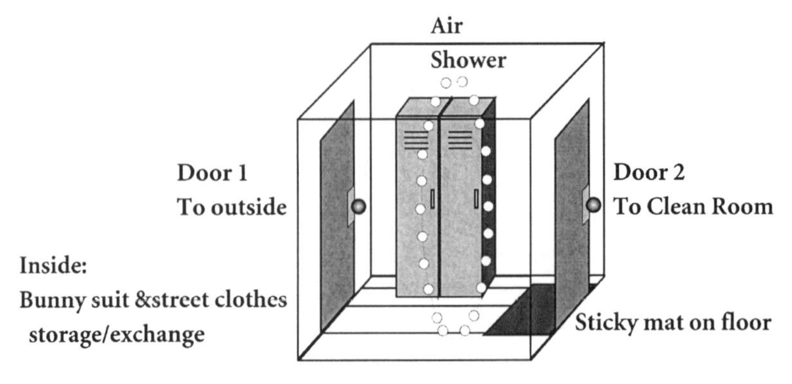

Entrance to clean room.

Dr. Sparck: Young satellites most sensitive to dust are those with optics. Sensitive optical surfaces including special filters could be damaged by normal cleaning and hence must be covered or stored in a very clean environment. Van Der Waals and electrostatic attractive forces result in a gradual buildup of foreign material on lenses and coatings, but this buildup happens much more slowly in a clean environment. A special concern with UV (ultraviolet) optics is contamination by "organics" - organic molecules. Common hydrocarbons like waxes, lubricants, and cleaning solutions might be present in the air either as tiny droplets (aerosols) or as vapor that can condense on optical surfaces. The high UV absorbtivity of many hydrocarbons has created a huge market, such as skin cosmetic products that prevent sunburn. But if your spacecraft is sensitive to organics, special care needs to be taken in selecting paints, coatings, sealants, cleaning solutions, filter materials, clean room clothing and floor, ceiling and wall coverings that do not outgas UV-absorbing vapors.

To reduce weight, some spacecraft use extremely thin photovoltaic (solar cell) cover glass, too thin to withstand normal cleaning procedures. This glass must therefore be kept in a clean environment at all times. Historically, clean room environments have also been maintained to prevent small particles from potentially short circuiting fine electronic wiring.

Q: *But Doctor, some satellites are built in people's basements, garages, and living rooms. How do I know what's right for my family?*

Dr. Sparck: We all want what's best for our children, and it's true that dust, dirt, and debris can cause failures and performance degradation. But keeping a child in a completely clean environment isn't the answer for every parent. Clean room upbringing is expensive. Besides the cost of building and maintaining the clean space, satel-

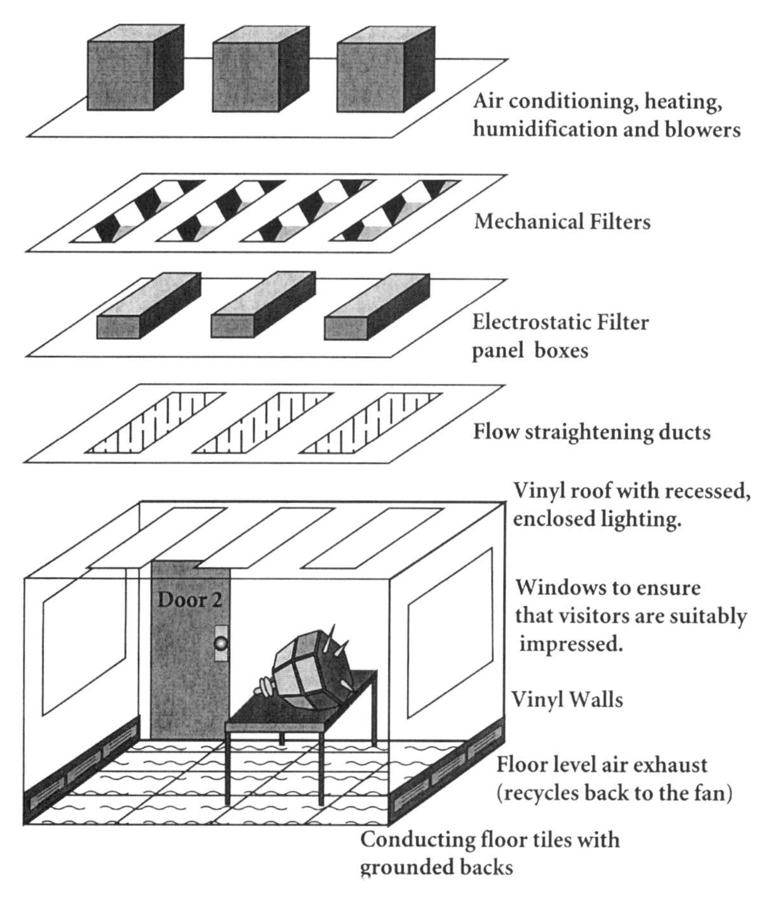

A diagram of the Complete Clean Room.

lite care is much less efficient. Every tool that goes into the clean room must be swabbed, usually with alcohol using lint-free cloth.

The few minutes required to suit up each time you enter the clean room add up. Often workers find themselves changing into and out of clean room hats, masks, gloves, overalls, and boots five or ten times a day. This ritual is tiring and time consuming. There is a temptation to make do with inadequate or inappropriate tools rather than go through the changing process for the 11th time during a long shift. But many tools simply can't be used inside the clean room, including most heater and refrigeration units, which require purchase of special gear.

No little satellite lives its whole life in one room. Eventually it has to go to vibration and thermal vacuum testing, range tests for the radios and antennas, and the launch site. If you bring up the child to thrive only in a clean room environment, all of these operations are slower and more expensive.

On the plus side, the clean room is a barrier to more than just dust. When your cousin, Butterfingers Bobby (known as Bull at the family reunions for the time he accidentally kicked over both flaming barbecues, demonstrating once AGAIN that elegant place kick of his that won the Big Game down at State U 15 or was it 17 years ago?), shows up to bounce the new baby on his knee, you'll be glad to have the air-lock between him and your pride and joy.

Clean rooms are the Prams of satellite rearing. Politicians, reporters, financiers, and traditionally minded engineers believe a gleaming clean room to be the sine qua non of professional satellite development. When they see it, they'll believe you've really arrived!

Many small satellites have no optics at all, or simply use covers to protect the optics. No small satellites have used photovoltaics too fragile for cleaning with a lint-free cloth and alcohol. The hazard of dust contamination of electronics is remote, and can be controlled in many other ways. Cost-conscious parents can raise a fine satellite without the benefit of a clean room.

Q: What are these other steps we can take to keep the satellite clean outside the clean room environment?

Dr. Sparck: The most important step is care. Serious contamination occurs both in and out of clean rooms through human error and facilities problems like leaking ceilings, insect invasions, spilling liquids, cutting wires and letting the shreds fall onto the satellite. The cleanest clean room isn't going to help you if you shear off a bolt head and it and its shreds fall into the electronics box!

Most electronics can be cleaned. Periodic swabbing with approved cleaners, for example, alcohol using lint-free cloth, keeps dust and debris down. The same is true of the spacecraft structure. Keep it clean. Use only cleaning, lubricating, and coating materials known to be approved for spacecraft, which usually means they have no organic contaminants, don't produce dust, and don't outgas. I know the can says it's made for rockets, but save your WD-40 for squeaky hinges on the Dodge, not on your mini-Delta.

Conformal coating is an option. It's a thick, plastic-y material that you paint onto circuit boards, but make sure you clean them first. Special cleaners are available that will ensure the boards are clean enough to guarantee adhesion of the coating. Conformal coating provides an impervious coating and protects your components. Most conformal coats are so tough, they provide some mechanical support to components that might be in danger of being overly stressed in vibration testing and launch. But don't count on it.

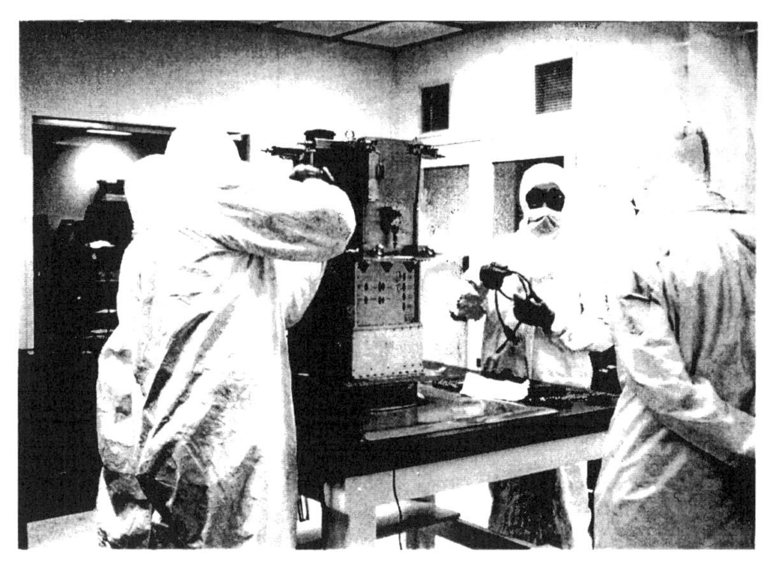

A clean room that Mother would approve.

Something else you can count on, too, is that conformal coats are so tough that you can't get them off when you need to. If you fry a component on your board, no guarantee says you'll be able to cut through the coating and replace it. If you attempt to slice the coating off, you could damage other components and circuit board traces in the process. Despite this irksome drawback, conformal coating is used in a large fraction of satellite programs. One last caveat, however. The conformal coating only protects the surfaces you coat, and that can never be every surface. It does not eliminate the requirement for care in handling and working around satellites.

The oldest and most reliable means of ensuring cleanliness, particularly in sensitive optics, is covers. Covers are more effective than clean rooms because there is no air circulation to introduce new dust particles onto surfaces. Also, they can protect against the occasional dropped screwdriver, sneeze, and flying wire shaving. But covers have their drawbacks too. First, you have to remove them before flight. Murphy has reared his ugly head several times on this basic precept so I will not, for the sake of the memory of the dead and the careers of the living, divulge any specifics. Covers built to remove themselves in flight using mechanical actuators can fail. At any rate, the cover needs to come off, either for a test or in orbit. If the rest of the satellite resembles Pigpen, the optics do too, thanks to outgassing and diffusive motion.

The combination of careful handling, cleanliness, and using covers is the most common cleanliness system in small satellite practice today.

Q: Besides dust and dirt, what else do I have to be concerned about?

Dr. Sparck: The Number One hazard to the developing young satellite is not dust and dirt. The glamor of the clean room environment with the gleaming satellite surrounded by people in white clean room garb has brought cleanliness into the public awareness. The real killer is static electricity or, more exactly, electrostatic discharge (ESD). Particularly in dry environments such as heated indoor spaces, charge potentials of many tens of thousands of volts can build up between objects. Discharge of this voltage can destroy sensitive electronics. Maybe worse, it can cause latent damage unapparent for many days or even months of operation; then, without any other abuse, the component mysteriously fails, possibly during on-orbit operation where the mission might be compromised or lost.

Q: But what can I do about static—ESD?

Dr. Sparck: A lot! We know a lot about ESD, and it can be completely prevented. The first line of defense is to properly condition the air in areas where the spacecraft and its electronic components are handled, assembled, tested, and stored. Relative humidity should be controlled to between 45% and 65%. While higher humidity inhibits ESD, it can also cause condensation and possibly corrosion. But humidity control is just a start. ESD-sensitive components must be stored in conductive plastic (pink poly is one common name, but the heavy gray-toned plastic ICs come wrapped

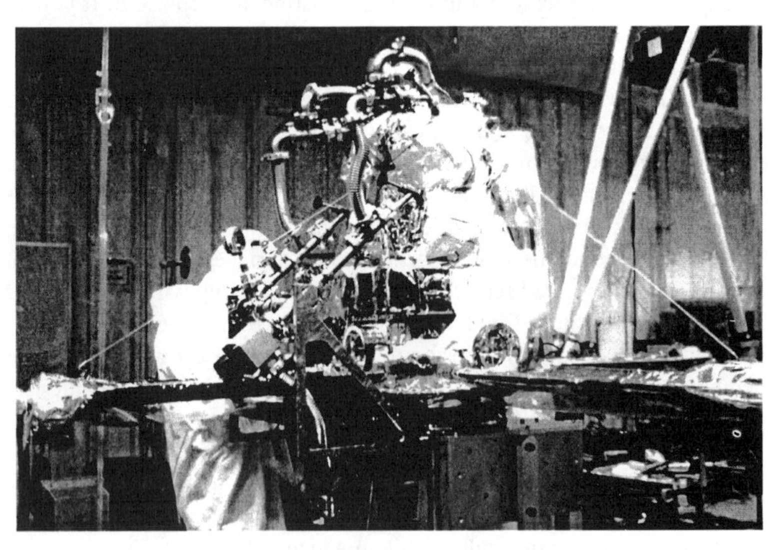

ALEXES satellite being mated to Pegasus rocket is protected by a portable clean room created with clear plastic curtains.

in is also conductive). Just because plastic is pink or gray does NOT mean it is conductive. Make sure your plastic is conductive, and also that it does not outgas or cause organic contamination. Many packing foam materials are available that are antistatic and non-outgassing. Use ONLY these.

Absolutely critical is human handling of components, boards, and systems, for example, electronics chassis. You must be grounded at all times when handling electronics. Grounding is usually accomplished by wearing a flexible metal bracelet with a wire attached to a grounding post. Commercial products are available that show a green light when proper ground path is provided, red when the strap is not grounded. Rooms where electronics are handled should be static-protected. This means the floor should be electrically conductive AND GROUNDED. Both vinyl and carpet floor coverings are available in conductive materials. They should be laid down over a conductive substrate, typically a copper sheet that itself is grounded. ESD furniture should be used. These are chairs and tables built of conductive material that are grounded to the floor at all times. They are surfaced with materials that do not tend to create charge when you or your clothes brush against them.

Beware of all cleaning procedures. They usually involve a process of rubbing a gas (like an air jet) or a cloth against the material to be cleaned. Like rubbing a party balloon on your hair to build up charge to make it stick to the wall or ceiling electrostatically, rubbing electronics with a cloth or a stream of air can build up tens of kilovolts of potential during cleaning. Air supplies and their canisters and nozzles must be grounded and only conductive cloths should be used for cleaning.

With these simple steps—humidity control, use of appropriate packaging materials, ground straps, conductive floor covering, care in cleaning and use of ESD protected furniture—you can protect your electronics from ESD. Remember, ESD damage is insidious. It can weaken a part without causing it to fail immediately. And, ESD-induced degradation cannot be detected.

Q: *OK, we know to be careful about cleanliness and about ESD. Anything else?*

Dr. Sparck: Only a few things, but they're important. Don't forget all this discipline when your satellite goes out to play. Wrap it in clean, non-outgassing anti-static materials. Make sure the satellite is grounded to shipping containers, and that they are vented only through adequate filters. Young satellites are prone to damage through mishandling in transportation. Shipping containers should include shock mounting and so-called shockwatch indicators. These record whether the maximum g-loading (acceleration) applied to the shipment exceeds a preset value. If the shockwatch indicator is tripped when you receive the package, the satellite should be thoroughly tested for possible damage. If it is not tripped, things are probably OK. The worst environment for a satellite is the salty air typical of many of the world's launch sites, including Cape Canaveral (Florida), Kourou (French Guyana), Barking Sands (Hawaii), Wallops Island (Virginia), Kagoshima (Japan), and Andøya (Norway). A responsible parent reviews the accommodations the satellite will have at integration and at launch to make sure exposure to the corrosive salt environment is minimized.

Personally, I believe most satellite damage isn't the result of any of these sophisticated hazards. How many of the cameras, watches, radios, and tape players you've owned have broken from vibration, static discharge, corrosion, what we would think of as intrinsic failures? Compared these with extrinsic failures caused by, say, dropping them overboard from your two-person sailboat, wearing them out in the rain, running them over with your car while pulling into the garage, or putting the batteries in backwards. These devices seem more reliable than we are!

When handling the satellite, use caution, allow a minimum number of people present at one time (because they distract each other), follow procedures, and NEVER rush or work to exhaustion.

The satellite is no more reliable than you are in handling it. If you don't make a big mistake, if you don't drop it, connect the plus lead to the minus pole and vice versa, drop things into it, spill things on it, or just bump into it (ever see a solar panel cover glass crack itself?), you've eliminated the major causes of failure. Use of protective covers is, last time I checked, not against the law either.

If any of these things do happen, and at least one of them will (probably many more than one in the course of development of a small satellite), your best defense is honesty. Report the event to the project team, analyze what damage might have occurred, and test to see if in fact there has been damage. An extra disassembly and cleaning is a preferable alternative to flying a satellite with a nut and washer loose inside it somewhere, waiting to lodge in a mechanism or short an electrical lead.

Chapter 16

Choosing A Launch Site

Launch Sites are real estate, and like all real estate, location is the critical ingredient. A very wise professor of mine once proved the existence of God (when I said he was very wise I wasn't kidding). He explained that if you look at the energy required to get into earth orbit, a function of the earth's mass and the gravitational constant, and compare it with the energy of chemical bonds which we break and remake to create rocket propulsion, it turns out to be just barely possible to get into orbit. Slightly weaker chemical bonds or a deeper gravity well and we'd be locked down upon earth's surface. Slightly weaker gravity or tighter molecules and going into orbit could be as easy as "launching" a Cessna 150. Only a God interested in challenging us could have engineered the improbable circumstance of barely-possible space travel.

If there is a God, religions tell us he or she is Omniscient. This fact is borne out as well by orbital mechanics and geography. You can launch from anyplace. But, some anyplaces are a bit more equal than others. The most popular single orbit is geosynchronous—the orbit with a period identical to that of earth's rotation and with a plane including earth's equator. Satellites in this orbit appear "stationary" in one spot over the equator. Billions of dollars are spent every year to put satellites in GEO.

16.1 Cleveland? The Equator? Florida?

Orbital mechanics dictate that all orbital trajectories lie in a plane which includes the earth's center. Thus if you launch from, say, Cleveland, your basic orbit will be in a plane which includes Cleveland and the earth's center. This is not an equatorial orbit, and your satellite will never appear to be stationary over one spot. In order to get to GEO, you wait 'till the orbit crosses the equator, then burn lots more rocket propellant to turn your satellite parallel to the equator. If you start from Cleveland, you may need as much as 50% more propellant energy to make that turn.

But if the site you pick happens to be itself on the equator, you will launch into a plane that includes the equator and the earth's center, and voilà, you're on your way to GEO. Plus, if you spin a basketball on your fingertip, you'll notice that the equator moves really fast, while the poles don't have much velocity at all. A rocket launched due East from the equator gets a 5% free boost from the earth's rotational velocity compared with a launch from the pole. 5% isn't much, but remember that getting to

Possible launch sites that are not Cleveland.

orbit is a marginal proposition anyway, and that 5% can translate into 25% more payload. (The Eastward velocity of the Earth has to be canceled to reach polar orbits, giving far northern and southern launch sites the slight advantage for them.)

Not to oversell the equator and create a land rush, devastating the earth's tropical rain forests and emptying Europe and North America of their populations, but equatorial launch sites have a third advantage. If you launch away from due East, say slightly left of North, or Northeast or Southeast, you can achieve any orbit inclination you want, from 0 degrees (equatorial) through and beyond 90 degrees (polar and sun synchronous) without energy-wasting turns. Launch from Cleveland and your orbit will always be inclined somewhere between the latitude of Cleveland (41 degrees, 28 minutes, in case you had to know) and polar. Unless you pay that big energy hit to turn while traversing the lower latitudes.

The Space Shuttle launches out of Florida at about 28 degrees. That's why it's orbits are always 28 degrees inclined from the equator, or higher. The Shuttle can't afford the energy hit to turn Eastward as it crosses the equator and still carry a meaningful payload.

It turns out there are a few other key requirements in picking your launch site. Historically, rockets occasionally go to some location in between the pad and orbit. These we refer to as failed launches, and we try to site launch facilities so that if a failure occurs the probability of hitting a person with a rocket fragment is less than one in a million. Thus most launches are from deserts and coastlines.

16.2 Rockets Are Big

They're heavy things that have to be trucked, trained or, transport-aircrafted to the launch site. The engineers and technicians who build them, the managers and money people the engineers and techs work for, and the political dignitaries who allocate the money and tend to own the real estate the launch site occupies, all face the fundamental dilemma of rocketry. Lemma #1 states that nobody wants a launch site in his or her own back yard. Lemma #2 is that, as United Airlines has only recently admitted, incessant travel is no way to live. People just don't want to spend 11 days on travel to get to the launch site, particularly if that travel is through category IV virus infested rainforest via riverboat and Range Rover.

Admitting that we will have to travel to someplace deserted, human ingenuity, ready for any challenge, invented a euphemism. Infrastructure. While often (mistakenly) interpreted as the existence of railroad lines, telecommunications terminals, and radar for things like bringing in rockets, communicating to ground stations, and trajectory tracking, infrastructure actually means a jet airport with a decent Red Carpet Club, a safe city with hotels that have decent air conditioning and a reasonable bar, reasonably good food, and cable television.

The list goes on. Rockets are sort of dangerous besides the threat of a failed one achieving a trajectory that intersects your present location. For instance, they can be used to launch warheads, and any orbital rocket has the power and guidance precision to launch a warhead to anyplace on earth. Politics dictates that launch sites are preferred on territories known for their lack of proclivity for launching warheads, at least at us (and us is defined by whomever is paying for the launch site). The probability of a revolutionary change in government has also to be factored somewhere into the political calculus.

16.3 Weather, Ecology, and Another Proof of the Existence of God

Weather is also a factor. As a kid interested in sun, sand, and eventually bikinis, I used to think Florida had great weather. 'Till I started hearing about all the Shuttle launches delayed for weather. Thunderstorms and tons of rocket propellant are not a good mix. Neither are high winds buffeting a huge airframe shaved of all extra mass to enable 1% or less of itself to finally get to orbit. Dust and sand are not friendly to precision equipment aboard rockets or their payloads, and cloudy, rainy skies inhibit visual and radar monitoring of the climb-out which is necessary for range safety.

Finally, we have the ecology to think about. Even though in the grand scheme of things rockets amount to a subtrivial amount of scrap metal being dropped on the ocean and combustion products injected into the atmosphere, they have a huge PR impact. Nobody really notices when 1,000,000 different Americans light off their 2 cycle lawn mowers, snowmobiles, and skidoos, but set off one rocket, and you're on TV worldwide. And ecology is more than whether a titanium, aluminum or carbon composite cylinder or three lands in the Coral Sea and Great Barrier Reef. Whose sacred burial grounds might you be overflying? What rich fishing ground might be the livelihood of fishermen claiming that loud rockets scare away that livelihood?

Getting back to the nagging question of the existence of God, pull out your trusty globe and find all the land masses possessing the following qualities:

- on or near the equator
- water or fully uninhabited desert all around in an arc from West of due North all the way clockwise to West of due South
- no large population center within 50 miles (sometimes rockets blow up on the pad or turn in the wrong direction)
- politically stable and friendly (to your side, whichever side you're on)—fitted out with infrastructure including daily jet service
- good weather most of the time
- no major ecological, religious, commercial or territorial conflicts at the site or for hundreds of miles around the above mentioned arc.

Just as there are only a few molecular bonds which—if we're careful and go to obtuse extremes like throwing away 90% of our rocket to help the last 10% put about 1% of the total mass in orbit—there are just one or two locations that meet most of these criteria. In the whole earth, just one or two? Why not a whole bunch of choice locations? Why not none? QED, as physicists say.

16.4 Some Good Non-Equatorial Launch Sites

The fact is, good equatorial sites are rare. And all of them are far from the governments which ultimately invest in all the costs of equipment, land improvements, buildings, transportation and personnel necessary for launching rockets. Hence a lot of the world's sites are far from the equator at places like Vandenberg, Kodiak Island, Wallops Island, Tanegashima, Plasetsk, and Andøya. These places cope either by specializing in non-equatorial launches, which are now becoming more popular since the rise in popularity of LEO communications constellations which are not geostationary and in fact want high inclination coverage to serve the entire world, or they accept the loss of payload intrinsic in the turn maneuver to head East upon crossing the equator.

Every launch site is a compromise, trading off issues of rocket efficiency, accessible orbits, convenience, safety, reliability, political risk and gain (including national pride, favor exchanging, and other irrationalities), ecological and cultural invasion, and of course cost. The complex of pluses and minuses, plus the ever changing ways in which we use space, and hence the relative desirability of different orbits for science, communications, defense and remote sensing, all combine to explain the apparently almost random distribution of launch sites around the world.

Chapter 17

Satellite Constellations

It has only been a few years since aerospace engineering was in a great slump widely predicted to be a permanent steady decline. Geosynchronous comsats were widely believed to be losing the competitive battle with fiber optic cables, the end of the cold war meant a decrease in military missions, and the public was becoming sufficiently glazed over about space that billion dollar space science missions were going to be history. Darkness was falling over the land. Last aerospace slump, at least we had an energy crisis that needed to be solved, or a war. Here we were, last decade of the millennium, going out of the space business.

All three of those ominous factors turned out to be true. But we're building lots more satellites now than ever before in human history. Small satellites, and particularly constellations of multiple identical satellites, are the main reason. Our society's increasing mobility coupled with the ever increasing cost of installing new fiber and other terrestrial networks, conspire to foil the fiber optics, and takeover from geosynchs is also no longer even predicted. And while giant interplanetary science missions are, at least at this writing, out of vogue, they have been replaced by a larger number of missions flown by ever shrinking satellites.

Satellite constellations were, until Motorola's announcement of Iridium, widely believed to be science fiction. How would anyone launch so many satellites—and how could they be controlled and coordinated? Motorola moved the topic out of the space groupie press and into the *Wall Street Journal* with its announcement of a plan to place 77 small satellites in low earth orbit to provide global cellular telephone coverage. Several other firms then proposed constellations of small satellites to provide new and/or improved communications services.

Why a constellation, or a duck, for that matter? Way back before the big Bon Jovi disappointment, man envisioned two ways to communicate with satellites. One was to put individual, large, complex satellites into very high, 24,000-mile "Clarke" orbits that appear stationary in the sky, also known as geosynchronous or GEO orbit. The other was to put many smaller satellites in low earth orbit. Because satellites in low orbit circle the globe about every 100 minutes, they move in and out of sight quickly. (Refer to the articles on Orbit Mechanics for a weird but largely accurate treatment of satellite orbits.) Those who remember Telstar know the problem. It worked great for about five minutes.

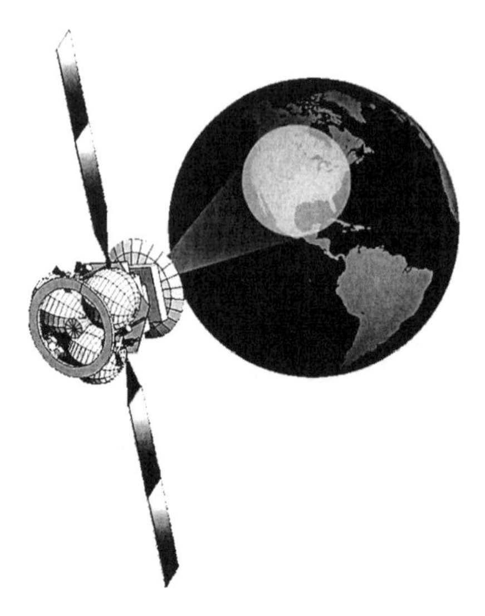

Large communications satellite at GEO.

The way to get over the disappearing satellite problem is to fly a large number of small satellites. Each one provides a few minutes of service to any one particular user. Then its role is taken over by another cluster member. An earthbound analogy is the cellular phone. As you drive from Costa Mesa to Mesa Blanca or from Cerritos to Cerillos (or, for that matter, Pawtucket to Pawkatuck), you move from cell to cell. You never realize that your phone is switching its attention from one cell to another. A cluster of satellites is like a cellular system except that you, the user, are relatively stationary, and the cells are circling the globe at four miles a second.

"So what?" you say. Hey, I'm getting to that. Large satellites in GEO have a few disadvantages. Take, for instance, the point of view of the poor soul trying to finance one of these collosaltrons. You go to the bank and tell them you need, say, $200M in small bills, all of which may, if some well-meaning technician has a bad day in the clean room and leaves a rag in a fuel line, be essentially destroyed and their remains sunk to the bottom of some nameless and deep ocean trench. (This happened.) The thought of such risk irks those buttoned-down bankers, a notably nervous lot, of late, who then send our friend the satellite entrepreneur to the insurance companies. They don't like betting all their marbles on one launch either, and life gets expensive, varying to impossible.

The small satellites, destined to become part of a large constellation of their brethren, are put in orbit one or a few at a time, thus requiring several launches and an explanation of why launch companies think clusters are The Way to Go. A launch accident or a failed satellite is a 5% perturbation on the program funding and hence no big deal.

Another advantage often touted is that low earth orbiting (LEO) satellites are, well, low down (don't really need an ivy education to get that far, eh?). Hence, you don't need that 8' diameter status symbol parked in the back yard to tell your neighbors you've got a satellite receiver. You can reach the satellite with an antenna about as complex as the one on your FM transistor radio (like that nostalgic imagery?). To be fair to Mr. and Ms. Big Satellite, this assertion is not rigorously correct. Big satellites have big antennas, and they can use spot beams to give very good service to small terrestrial antennas.

Ultimately, the little guys in the white hats win because of another subtle point. Being 20 or 30 or 77 in number, they can be close to everyone on earth simultaneously. That would be more than just a lot of spot beams on a big satellite. It would be impossible. A single geosynchronous satellite doesn't see the whole earth. You'd need about three to do it, assuming you could put a lot of agile antennas to work on each one, not an inexpensive proposition. The teeming multitudes at the North and South Poles (Mrs. Claus, the elves, and their nuclear submarines parked under the ice) would still be out of range unless they use those big dishes that don't transport well on sleighs (tangle in the bells) or on submarines (attract sharks and rust a lot).

When you're alone in the car with them and the radio is blaring to avoid eavesdropping, the old-timers (ten years older than I am) will tell you a story about the engineers at Bell Labs who pioneered low earth orbit communications links in the 60s. They carried the day technically, but politics prevailed. The federal government threw its weight behind large geosynchronous systems as a way of justifying the large rockets needed so that our guys with the white scarves could beat the Russian guys with the red scarves to the moon. Of course that lunar base has been a longstanding cornerstone of our nation's defense, so nowadays it's considered well worth opposing compelling reasons to use small, low earth orbiting satellite clusters. Heresy, of course. Our government would never meddle in commerce and technology, but I wanted to represent the views held by a few thousand older satellite folks.

Well, one way or another, big geosynchronous satellites became the mainstream and weze guyz became a technical curiosity. Counting up the clusters announced, the numbers seemed incredible when they reached 175 satellites totalling over $3B in project funding. (And they said small satellites were just a hobby!). Then along came Teledesic, backed feebly by Bill Gates and Craig McCaw and the total started to approach 1000 LEO satellites. Arthur C. Clarke predicted the application of geosynch, and he also predicts that LEO will be become so crowded that it won't be usable—everybody flying into each other in a massive orbital traffic jam. Could he be right again? The numbers make it look that way, and industry practice and legislation

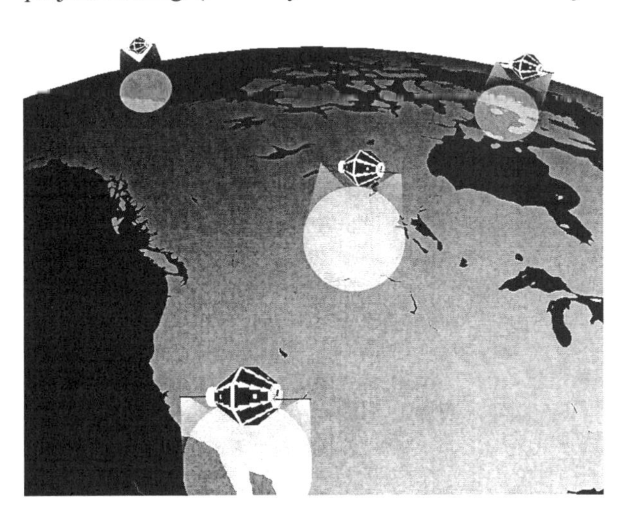

LEO Satellites

are at work to ensure those satellites leave LEO and get out of the way as they reach the end of their life, as GEO satellites do already.

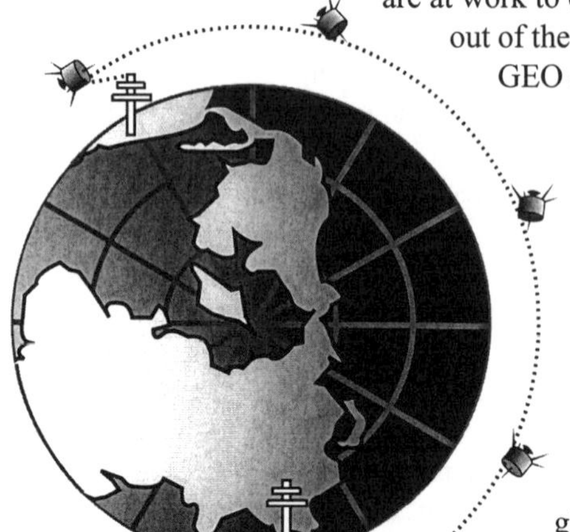

The simplest cluster configurations: to carry a message from a remote location, the nearest satellite of a cluster receives your message and carries it immediately for downlink to the addressee when it passes overhead.

17.1 What Are These Satellites Doing To Keep Busy Up There?

Orbcomm, LEO-1, and E-Sat are a few of the companies aiming for the low cost end of the market with message forwarding services. They would like to see a small box in every car's glove box with a connection to the radio antenna that every car already has. Break down by the side of the road? Of course not! Cars don't break down by the side of the road any more than people die in cemeteries. They break down in the middle of the road and then, except during snowstorms in Washington, DC, move to the side of the road so as not to get hit or just to stay out of the way. No matter. The so-called Little LEOs will allow me to set the button to message #11: "Broke Down By Side of Road," push Enter, and my message pops up at some AAA or police headquarters along with the latitude and longitude of my rusted hulk. (The car, that is.)

Or I'm out hiking and break an ankle wrassling with a bear on some rocks. No problem! Set the little box in my back pack to message #83: "Broken Ankle, Road Rash, and Bear Claws," and the park service gets the message along with my position and I wake up in a helicopter with a paramedic casually nursing my wounds and furiously attempting to verify my VISA credit card limit. With ground terminals selling in the range of $1000, these systems intend to service millions of users with a basic, reliable, immediate means to send coded messages, possibly even a short e-mail, fax or pager message, worldwide.

Well, almost worldwide. Technology and politics clash again. The satellites are glad to provide worldwide service. Governments aren't. World radio regulations, devised for geosynchronous satellites with their narrow spot beams and unchanging coverage, are not ready for a satellite that covers the whole earth. It came up for the first time at WARC (World Radio Allocation Conference) in 1992. So-called mobile satellite (I have never understood this moniker. Are their immobile satellites?) service will probably keep lawyers and politicians busy for decades, while the world's satellite builders and markets wait for an elusive resolution.

Once again I digress. Next up the food chain from the low cost message and position relay satellites is Ellipso (aka Ellipsat), so named because the satellites are in a slightly elliptical orbit to maximize their time over North America. This is a concession to WARC since the service was only to be approved, at least initially, in the U.S. Though the satellites will cover the earth, they'll only operate over the U.S., maybe Canada. Kind of a waste, but they can spend their minutes over New Zealand recharging batteries. With plenty of time for R&R, the satellites can be built lighter and cheaper so it's not a total waste. A big part of OrbComm's pitch is focused on how the system is configured so that it is impossible for the satellites to carry messages across national boundaries. A complex system of earth station "gateways" and on-board software is designed to emasculate the power of their global cluster, a concession to the inability of politicians to catch up with technology.

Ellipso and Loral's Globalstar concepts are shown conceptually below. They allow you to pick up your cellular phone anywhere. Maybe you're not in range of a cell, for example, you're rowing your boat from Block Island to Narraganset dragging most of your mast and sail in the water behind you. Switch your phone to overdrive (satellite) mode to reach one of the 24 satellites as it passes overhead. It relays your call through the Cranston East cell just as if you were in Cranston (you can always dream). The system augments the existing cellular network by getting you into it even when you're out of its normal coverage range. Since it is only a link into the existing phone network, it is relatively inexpensive and simple—if you can call a billion dollars and 24 satellites inexpensive and simple.

To put inexpensive and simple in proper perspective, there's Motorola's $3B Iridium. Shown on the next page, it is visually similar to the Orbcomm / KITComm / LEO-1 architecture shown here.

But! More than just an enhancement to cellular service, Iridium provides its own long distance transmission network. Note that the signal from your car, truck, or bicycle telephone doesn't go through an existing

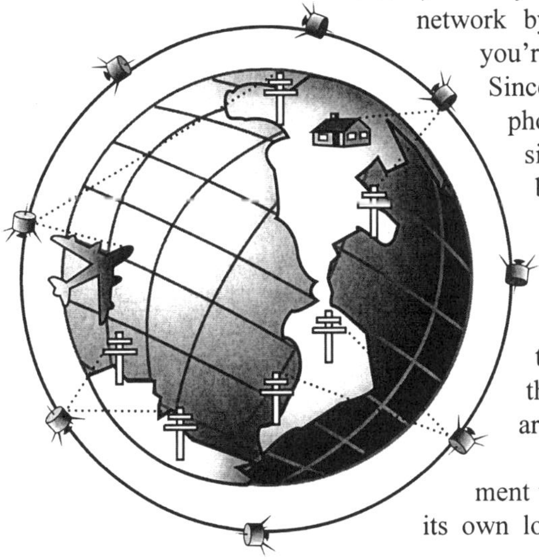

More capable clusters have enough satellites to give continuous service linking users to the land-based telephone system.

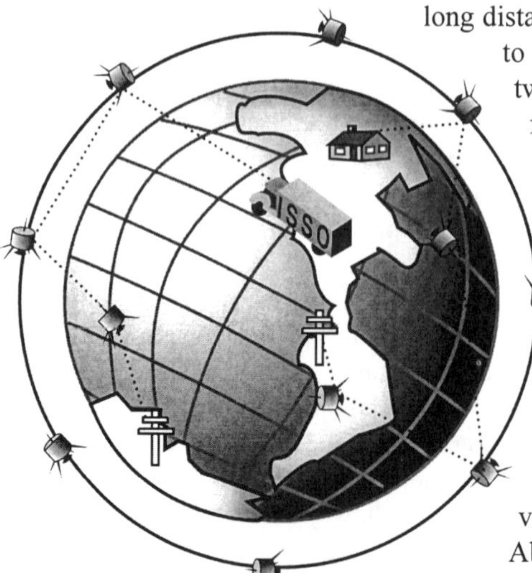

More complex clusters like Iridium relay communications allow cluster members to link two users anywhere on earth without terrestrial links.

long distance phone network. If, while bicycling to work at the shimmering AeroAstro twin towers with the commanding view from the Blue Ridge to the Potomac, you decide to phone the AeroAstro overland express plying the back roads of Piedmont, your call is routed up to the nearest cluster member, then relayed (cross-linked) from satellite to satellite until the call reaches the cluster member passing over Piedmont. Then it goes down either directly to the truck, or to the local phone service.

About a week after announcing the proposed Iridium system, The Wall Street Journal did a special interest story on national telecommunications systems, noting them as important export products for many smaller countries. Well, Iridium and all the LEOs, Big and Little, are stirring up that captive market. When you threaten steady revenue streams, you're likely to get noticed.

The last chapter of "Saga of the Satellite constellations" has definitely not been written. Science missions are now looking at constellations to do simultaneous measurements mapping time various fields—like earth's magnetosphere. Or simultaneous imaging of the entire surface of the sun—or the earth. Nobody knows which systems will succeed or how the world will regulate, tax, and utilize them. But we're about to find out. And the prospect of punching a button on your wristwatch, which is spending the afternoon on the beach with you in Shirahama, Japan, and dialing up your boss, who is, as it turns out, rowing his sloop into East Greenwich Harbor, Rhode Island, for repairs, became a reality in the 1990s—the decade that started with the end of the space era—thanks to constellations of low earth orbiting small satellites.

Part II

Missions and Management Reliability

Section 1
Missions

Chapter 1

The Smallest Show on Earth or, Tom Thumb in the Big Top

Circuses, you may know, are no longer politically correct. After all, those poor lions and tigers and bears were intended by God to roam free in order to hunt and kill each other without interference from men and women who build cages and snap whips and shout commands at them.

My college roommate considered that all synthetic things were in fact natural. He reasoned that, because people and their brains are natural, they are in some sense just a conduit for building molecules, structures, and devices that would have taken too long to happen haphazardly without us to act as God's vehicle. (Whips and cages might be included in that set of molecules, structures, and devices.) In other words, we carbon-based thinking machines merely speed up what would have happened anyway during a long random walk. We are, in fact, living proof that God is just as impatient as we are—which makes sense since we are made in God's image, paranoia and attention deficit disorders presumably included.

1.1 Comparing Aerospace to a Circus

It's too bad really, the decline of the circus in America, because aerospace itself is a circus. (And without real circuses around, everybody will miss the metaphor.) If you didn't go to a circus when you were young—but you undoubtedly did because they weren't politically incorrect back then, and parents who send kids to college, like yours, tend to also take them to circuses—you might fail to see that important fact. Then you'll be liable to think that what we aerospace engineers do is a serious profession pursued by people who studied hard in college and now ponder daily the deep mysteries of life and the universe. As the Wizard of Oz said, "Pay no attention to that man (aka Geek) behind the curtain!" This phrase now appears in the *American Dictionary of Slang* as an acronym credited to the software industry, Panambic. But let's all step up now, and get on with the show!

And now, La-dies and Gent-le-men, to the Three-Ring Circus!

Grab some popcorn, cotton candy, and overpriced souvenirs and head for your seat under the Big Top. It's time to feast your wondering eyes on the Greatest Show on Earth. Hubba, hubba, hubba, in Ring #1, you've got what still passes for Man In

Space. (The PC Police still haven't been able to launch Crew in Space or Man and/or Woman In Space into the mainstream, or even into the American Dictionary of Slang.) The act in this Ring is tough to mount cheaply, mainly due to the fact that if one of those humans in space is killed, the President of the U.S. will most likely fire the head of NASA. That's a supreme risk no bureaucrat will undertake, though there are plenty of astronauts who would.

Nope, Ring #1 is the Ring of Gold. No wonder it takes a million signatures to sign off on a shuttle launch. The huge facilities necessary to train astronauts on a small fraction of the software and wiring already built into the shuttle's navigation and control systems, the sex appeal that forces divisions of the circus into all fifty states—and now almost as many countries—so we can all lay claim to contact with space-faring people—all of these have to be factored into this high-wire act.

And now, may I have your attention please! In Ring #2! Pre-ZEN-ting the Colossaltron in Space. (And a tip of the hat to Robert Close for the term.) Colossaltrons can't be cheap—otherwise they wouldn't be colossaltrons. When NASA asks for ideas for a $1 billion or $2 billion mission, you propose a $2 billion mission. If somebody came up with a $2 billion mission that only cost, say, $10, NASA would have a $1,999,999,990 surplus, which Congress would gladly reclaim. Which is just no way to run a government agency.

Don't blame NASA for this seemingly fallacious argument. Economists will tell you that if we came up with a way to avoid spending hundreds of billions of dollars every year on cars, and fuel for cars, and parking for cars, and guardrails, and so on, and if we could transport ourselves by clicking the heels of a pair of ruby slippers purchased at K-Mart for $8.95, we wouldn't save hundreds of billions of dollars at all. Instead, we'd plunge our economy into a Depression that would make the Ice Age seem like a mild winter in Boca Raton.

There are places to do $10 missions at NASA—in the Kennedy Space Center's coffee shop, for instance. But we have to face the facts. Colassaltrons are very popular—almost as popular as People In Space, but for different reasons. Colossaltrons employ colossal numbers of people, and put colossal amounts of money into Congressional districts for constituents to spend, for instance. And, though a bit less efficient than discovering an American who can go out and win the Olympic Pentathalon, colossal space missions contribute to national pride and to America's standing in the world—always a tenuous thing, even in the best of times.

Ring #2 is the Ring of Silver, because it's the actual hard currency being spent that's the real point of the Collossaltron. Nobody on the ground understands why peering hundreds of millions of light years into space matters, or even why that's the equivalent of looking back that far in time. It's sort of an article of faith. But the jobs, the orders, the ringing exclamations of Congressional excitement that peal forth when the most complex Collossaltron ever built actually works, is what it's all about.

And now, La-dies and Gen-tle-men, for Ring #3, our ring—the Ring of Cardboard! Building, even launching, highly-reliable low-cost satellites is amazingly easy. There's no sense in reading books about it (this one excepted.) My sister, though as bright as they come, has never picked up either a thermodynamics text or a freshman physics book in her life. But she could launch a low-cost satellite. Bottom line: You don't need a ring of Gold or Silver. But to mount your Cardboard act under the Big Top, you do need one item that's very hard to find: a mission. Just what do you do with your El Cheapo satellite once you launch it?

Let's review. The fact that people are in space routinely is an end in itself. We're told it's part of pursuing our destiny to explore the planets and stars, the natural migration of our restless people off of this planet and onto others.

Similarly, nobody understands a Collosaltron, nor can they appreciate what a million light years or a billion dollars really means in terms of their drive to Starbucks or the cost of a decaf latté once they've gotten there. But they do understand that if it's big, it must be impressive. For this reason, no politician is going to back forking over, say, a mere $3 million just because it's a lot less than $3 billion. $3 million is a lot of money, sure, but for sound bites on the Evening News, it doesn't have the sex appeal of $3 BILLION. Saving money is simply not as newsworthy as plowing through huge amounts of it. And $3 million certainly doesn't have enough pork in it to pick up more votes in the next election.

Let's face it. What is a circus without the bright lights, the midway, the wild animals, and the gaudy, glittering costumes? Well, it's just another cheap carny show, a dime a dozen. Who would notice?

Riding the wild Collossaltron.

1.2 The Need to Find Uses for Satellites

For decades, computers had a similar problem. If what was called a computer was a room-filling Gargantua (aka Collosaltron) petted and fussed at by very intense looking men attired in pocket protectors and narrow ties, well, it was impressive and worth developing for its own sake. But once a computer got small enough to put on a tabletop, looking something like a breadbox with pimples, we suddenly had to have a use for it. And even if one of these newfangled personal computers cost just one part per million of an Eniac, few individuals were going to spend money on an Altair box just to play tic-tac-toe and print out their names thousands of times until, finally tiring of the sport, they hit a button and stopped execution.

Luckily for the San Francisco Bay Area, in previous times renowned for virtually nothing except Alcatraz, somebody out there invented word processing and spreadsheets: the very first "killer apps" for PCs. All of a sudden, people needed these otherwise ignoble boxes, festooned with unwieldy connectors and running hopelessly mysterious and unreliable operating systems. Now, thanks to Internet web pages featuring Dilbert and real time weather radar maps, we don't just need personal computers. We need computers with color monitors, fast processors and modems, tons of random access memory, and lots of hard disc space to store the latest Web browser upgrades.

If we could come up with a killer app for a pebble in Low Earth Orbit (LEO), I could build you a satellite for 10¢, though I'd try to convince you to pay at least $100,000 for it after space qualification. So the real issue isn't "can we build cheap satellites?" It's "can we find uses for them?"

Lest you think this is just a rambling philosophical discourse, remember that there is plenty of money out there to build satellites. The only question is, can you give an investor a huge return on invested capital without having to make a major technical breakthrough that's the equivalent of, say, sustained fusion? Or can you provide a country with a crucial service—security, crop improvement, education—with a satellite whose cost is within that country's means?

1.3 Cheap Satellite Ideas

As you can see, these books of mine are not strictly about "how to build cheap satellites." They're about missions that can be done and done well in Ring #3: the cardboard realm. Luckily for my standard of living, a few such missions do exist, and their number and variety actually seems to be on the increase. So when somebody comes to you looking for a cheap satellite, your answer is easy: "Yes, I can do it." But the hard part of your job is to find a mission you can easily fulfill. Here are a few suggestions:

"And for his next death-defying feat, he will actually put a passive re-flec-tor in SPACE!"

Fly a mirror, or some other target, in space. Yes folks, step right up. For less than the productive output of an average American worker over his or her lifetime, you too, can put a passive reflector in space. But it's not as trivial as you think. Your space mirror may lack radios, a guidance system or thermal controls, but the payload still needs to stay shiny in the harsh environment of space, and somehow withstand first being launched and then being ejected from the launch vehicle. And somebody may ask you to make it fairly lightweight while you're at it, so it doesn't take too much money to launch. If they want to get fancy, they may ask you to prove it won't spin itself up (say because of solar pressure or thermal disequilibrium), or they might want deployables to make it bigger.

Another mission that's typically cheap is messaging. Ham radio people have been doing store-and-forward of short messages for 20 years. The trick was for entrepreneurs to realize that there are commercial niches for sending non-real-time messages, such as monitoring the height of water behind a dam. Nobody really wants or needs an hourly phone call relaying the current depth of water in a lake.

Most people don't realize that the science world is full of instruments that don't need very fine pointing, consume only a few watts of power, and don't generate gigabits of data per microsecond. Latch onto them and extol their scientific virtues to your potential customers. Similarly, there is no added complexity, from the satellite engineering point of view, in replicating something ten or a hundred or even a thousand times. But for some instruments, having a thousand of them in different orbits could prove really interesting. Such instruments could be deployed for instantaneously mapping the three-dimensional magnetic field over a volume extending from 200 miles to 200,000 miles from earth.

How about technology demonstrations? You don't need Collosaltrons or astronauts to prove that some electric propulsion device functions reliably enough to trust on a major space mission. The first solar sail will be rigged to a cheap little satellite.

So now you know the trick of small, low-cost space. This isn't rocket science. Well, even if it is, it certainly isn't brain surgery, or nuclear physics, or talk-show-hosting, or whatever the hardest job in the world is thought to be this week. What happens is that, human nature being what it is, we get impatient and decide to try to sneak the absolute most complicated mission we can into the smallest, cheapest box possible. Little kids do this and we call it "not knowing when to quit," or, more urgently, "enough is enough already!" Take my advice—avoid exotic challenges at any cost—things will work better.

If you want to be a hero, build a cheap satellite, make it work, and still have time for bungee jumping off the Big Top. Space people have a tremendous propensity for finding new ideas for missions, but we've been a failure at creating profit-making enterprises out of any but a very few of them. Just find a very simple mission to do. The small, low-cost satellite part will happen naturally.

Let's look at where new ideas in some other fields come from, and see what we could do better...

Chapter 2

Telepresence: Paul Bunyan Takes a Hike

Did you know that the National Institutes of Health has now channeled about half its budget away from mundane pursuits like finding a drug to prevent the spread of AIDS or cures for cancer or multiple sclerosis? After all, we've already cured tuberculosis and polio. And cardiac bypass surgery is pretty well worked out. Hey, these NIH people are sharp and they need new challenges to stay at the leading edge. So fully half their budget is now devoted to inventing a handheld electronic device about the size of a Walkman. We only know four defining facts about this thing: its outward appearance includes a circular patch resembling a small speaker; it makes sort of an eerie humming sound; when a trained physician passes it over the prone body of a sick or gravely injured patient the person is instantly and completely cured without drugs or surgery; and most importantly we know such a device is possible because the brilliant minds that brought the world Star Trek have revealed it to us. NIH admits it's unfortunate that, during the decades that will be required to build this fantastic gadget, millions of Americans may die of diseases we could have found cures for in the meantime using mundane techniques like genetic engineering and analytical drug design. But once we realize Gene Roddenberry's vision for the future of medicine, the sacrifice of maybe 10 million lives and $100 billion will seem almost insignificant.

What's really interesting to me is how significant trends in one branch of science quickly migrate to other realms. For example, the FBI has been buying up all the remaining DeLorean sports cars. Rather than expending taxpayer dollars wearing out shoe leather collecting evidence and rounding up suspects, they too have learned a high tech trick from genius visionaries in Hollywood. How much easier it will be to convict a thief, murderer or drug runner by carrying a video recorder back in time and space to the scene and moment of the crime and just documenting the whole thing as it happens. No more criminals getting off free, no more DNA matching and petty squabbling over evidence. And no more year-long multimillion dollar OJ-style trials clogging the courts. Everything the FBI needs to get started on this fantastic revolution in crime busting was revealed in the movie "Back to the Future." Naturally, access to the substantial resources of the US government will obviate the need for kludges like using the clock tower in town square as a lightning rod to generate the 1.21 gigawatts necessary to trigger the Flux Capacitor. But the basic principles are

Catching bad guys the old-fashioned way.

clear from the movie. All that is left for our technocrats and phenomenologists (Washingtonese for engineers and scientists) is to build from fiction's blueprint.

Does it sound crazy to you that the FBI would curtail conventional, proven crime fighting activities to try to convert an out-of-production sports car into a lightning powered time travel machine based on a sci-fi movie? Or that we would give up searching for cures for cancer, AIDS, and heart disease to invest billions and billions of dollars in an electronic gadget used to cure fictional TV series space travelers of the future? Well, the good news is America's top health researchers and crime fighters have better things to do with their scarce resources than to be led by the nose by creative but technically ignorant writers. The bad news is the same can't be said for the US aerospace community.

For instance, other than a legacy from ancient cartoons—published before most of us were born or before a rocket had actually put an object into orbit—why would anyone bother to build a launch vehicle that looks like a souped-up jet airplane? Right down to the stubby swept delta wing, heavy retractable landing gear, crew compartment outfitted with windshield and joy stick, this paragon of aerospace pop culture is the US Space Shuttle. NASA is proud to serve neither science nor a questionable human destiny in space but rather to kowtow to popular fantasies as ridiculous as "Back to the Future's" car or "Star Trek's" handheld universal healer. Are we to believe the Shuttle needs a captain with a steady hand on the stick? Hundreds of millions of dollars were spent on that joy stick and the control systems and sensors that service it. But it is impossible to hand-fly the Shuttle or any other rocket into orbit. The stick, vestigial from craft that plow through earth's dense atmosphere, makes as much sense on a rocket as a propeller on a beanie.

NASA sidetracked a tiny bit more of your money to make a really neat poster showing the fiery launch of the Shuttle with the catchy caption "Going to work in space." The analogy is clear. Going to work means stepping into a carriage and burning fuel. A car, for instance. Or a macho FWD truck exempt from federal fuel efficiency standards, so you can light a bigger fire in its belly. Up the scale another notch,

maybe you fire up some high-bypass turbojets and fly or ride a fighter or transport airplane to work. The highest rung on America's commuter ladder is our astronaut core, firing up millions of pounds of rocket propellant to get seven or eight people to their offices in low-earth orbit. The message? Space is just like air which is just like land, and our future is just like our present, except that our buggy whips are made of Kevlar instead of leather and our suits are Nomex instead of cotton-polyester blend.

But there's an alternative. Let technologists guide technology. And let fiction writers create fiction. Like 25 million other Americans, I often go to work by clearing yesterday's debris off the dining room table, booting the computer and logging in to my company's e-mail network. I project my body across town, across the country or across oceans only because this uncomfortable, time and resource consuming, environmentally undesirable alternative is a sort of a low-tech patch for flaws in our crude telepresence infrastructure. Given the huge cost of placing and sustaining even one family, two millionths of a percent of America's population, in low earth orbit for about 0.05% the span of a normal working career, (estimates range from $500M to $2B for that family vacation in space) robots and telepresence are a good way to go to work in space.

But we treat uncrewed missions as stepchildren. We are almost ashamed we couldn't finagle the funds to send people to Mars and Saturn, like all the sci-fi writers tell us we ought to be doing. Meanwhile, back on our planet, AT&T Sprint, Apple, and Kinko's Copies are building large, new, profit-making enterprises based on people's appetite for staying home and letting bits—photons and electrons—do their traveling.

I know, that's not how they do it on TV. On the screen people jump into sexy humming rockets and dash off to the odd galaxy. Yes, but on the screen Mr. Ed talks back and Dick Van Dyke jumps into a chalk sidewalk sketch and enters a cartoon kingdom. Compelling as we find these characters, few research dollars are spent on talking horses or expeditions to sub-sidewalk worlds.

Let's leave our Buck Rogers-era baggage, handed to us not by aerospace scientists but mainly by cartoonists and novelists without technical training, on the sidewalk for a minute and plunge into a fantasy all our own. No rules—except physics. Freed from the burden of lugging our bodies with us, we need no food, no air, no pressure suits. No medicine, no radiation shielding, no huge swimming pools for training in weightlessness. Anyone can be an astronaut—kids, people with serious physical disabilities, grownups with jobs who are unable to pay the hundreds of thousands to millions of dollars necessary to put themselves into LEO for even a few seconds or a day.

In the past 50 years human biology has not evolved at all while electronic circuitry has shrunken by eight orders of magnitude—a factor of 100,000,000—in mass, volume and power required. In 1960 a human being was the lightest, cheapest flight computer you could buy. But today, people, myself included, build miniature spacecraft for less than the cost of one astronaut's space suit, spacecraft that can perform missions humans can only dream of and that foreseeable human spaceflight budgets make highly improbable. These space robots can sit in LEO and observe earth and its oceans and atmospheres, and they can robotically perform manufacturing tasks in space in an environment a million times cleaner than the Shuttle's or the Space Station's. They can orbit the moon or crash into the moon, if you'd like. Soft land on the moon and rove around looking for ice. Fly to and onto asteroids. Hang at the first Lagrange point a million kilometers sunward from earth and study the sun and the earth from an ideal perspective. Fly to the other side of the sun from earth and view the part of the sun we can't see. Or even leave on a multidecade trip to another star.

Yes, they can't zoom around at Warp 6, but what's the rush? When I do go to an office, I usually go by bicycle—about 1/2 the speed of a car. But time is relative. Time in a car is often boring and sometimes frustrating. Time on a bicycle is recreational and fun, and I'm glad the ride isn't over twice as fast. Letting our robots travel for us, we can go about our lives here on earth while they sit in a sort of suspended animation awaiting their encounters. Solar astronomers don't sit on their hands, isolated from friends, family and comforts of home, waiting for a solar eclipse. They live their lives, study things other than eclipses, and when the eclipse happens, they are there. Similarly with our robotic explorers, when they get there, we're there with them. Until then, we go about our business in safety and comfort.

This is disturbing news—that robots are better space explorers, better space colonists, better space researchers, than people. Welcome to reality. The fine art of cutting down trees is now left to machines, Paul Bunyan notwithstanding. Trains go faster than the fastest human, not to mention 747s. Computers are a lot better at routing telephone calls than banks of operators with phone jacks and wires. I'm sorry, but this is our job, us techno-geeks, to do things better, cheaper, and less expensively using our machines, our tools. Why should people in space be somehow uniquely qualified, whereas human bank tellers can be replaced by ATMs and internet banking?

A big draw for me into the world of miniature satellites was the personal involvement. Space is hard to be involved with. Something like one in every 50,000,000 people alive today has been to space. Working on large, conventional satellites, my role was a minute part of the whole. And I had absolutely no role in deciding where the spacecraft would go, what sensors it would carry, or what it would do when it got there. In the microspacecraft world, a few people get together and build exactly the robot they want, they find the launch, and they control it in orbit. You don't need a Ph.D. in aerospace engineering to build a 2-pound, 10-pound or 50-pound spacecraft. You need a collection of people among whom there is a vision, a desire, and a mixture of basic engineering and organizational skills, especially the skill to make do with what you can scrounge or afford. High school students have a role in one of the satellites we've made. Many college undergraduates have helped conceive satellite missions and then built the actual craft.

If you think telepresence just won't satiate your need to put your own flesh and blood in space, despite the discomfort, inconvenience, expense, and hugely poor odds, then just exploit microsatellites as scouts. Very few people, just two, in fact, got to be Lewis and Clark. But millions followed them into a new frontier. Astronauts are poor scouts. They are so laden with their own costly life-support systems, we spend all our time trying to find money just to send people to LEO, at best a first rung on the ladder to somewhere we actually want to live outside of earth. A microsatellite can go places and in the process uncover reasons for people to go there. Our society can afford to colonize the moon with the technology we have now. Or even Mars. But we can't justify the cost without the motivation. Queen Isabella and King Ferdinand did not sponsor Christopher Columbus out of a sense of duty to humanity's destiny, but rather to scoop the market on spice importation, which translates into wealth and its attendants, power and ego reward. The price tag to roam around the moon with people is just too high—as proven by our 20+ year lapse in revisiting the lunar surface. But we can easily afford the robots to do it—an amateur group with enough motivation could get there. And if the discoveries our silicon-based life forms make—water, oxygen-containing ores, exciting vistas and landscapes, are attractive enough, governments or possibly private enterprises will find the capital to exploit the new possibilities.

Let's not forget that plenty of scouts went to places people never really chose to inhabit. The north and south poles, for instance. Tierra del Fuego. The floor of the Pacific Ocean. Reality is, if the reward of being there isn't worth the cost, all the lobbying and political activism on earth won't change the course of our destiny—people won't want to go there. That we wasted only a few dollars, instead of a significant fraction of the world's GNP, to get a negative result on some destination at least allows us to continue trying—sending our robots to extra-lunar places—Mars, moons of the planets, asteroids, and candidate locations outside the solar system.

Microsatellites. Robots. Silicon-based life. Whatever you want to call it—it's a way we—meaning everyone who's interested—can be in space right now, doing the things we want to do there, at a price tag the average person/country can afford. And we can go in such large numbers, to such a wide diversity of destinations, with such a vast assortment of instruments and telepresence gadgetry, that the probability of waking up society at large to space's attractiveness, and to finding places worth visitation by people and ultimately to colonization, is the best it can be. If you're serious about moving into space, in Nike's words, Just Do It. Your silicon surrogate is ready when you are. And while you're at it, why don't we take a look at these ever-shrinking silicon surrogates?

Chapter 3

Being Disruptive
—or—
Lessons from the Ever-Expanding Backpack

A canvas backpack is my constant traveling companion. I affect one not only to maintain my carefully cultivated casual image. I'm also an innocent victim of technological progress. Ten years ago the only kind of luggable electronic device was a calculator—250 grams (8 oz.) max. My payload was mainly papers, pencils, and maybe a book—a tasteful leather portfolio sufficed. Today, I lug a 2.5 kg laptop, 1 kg palmtop, 500 g cell phone, a couple of kg worth of spare batteries, connecting cables and Zip disks, plus another kg or two of AC and cigarette lighter power adapters and power cables. To haul that 7 kg (15+ lb) load of electronics, which, by the way, still has not replaced any of the compliment of papers, pencils, wallet, credit cards, and other physical objects required of the modern road warrior, I need a good backpack.

This is not just a fluke irony of technology—that the shrinking of telephones and computers has loaded us with ever more junk to drag through airline terminals and into hotel rooms. It's also not an accident that as our machines work faster and harder for us, we work more hours and our free time is more frequently interrupted by work. It's easy to look back with nostalgia to before FedEx and fax machines, when "it's in the mail" could hold off a deadline for days, even a week, or at least long enough to finish dinner and put the kids to bed.

It is in fact the nature of new technologies that they are often handicapped in almost every respect compared with those they displace—except for one or two key features. A book is in many respects preferable to a text file resident on a computer. A book is portable, requires no batteries or charger, is easily and cheaply replicated, has a shelf life of thousands of years, and requires no special hardware to view. Notes can be permanently affixed with a simple stylus device (pen) and even a cheap $1 book supports high-res graphics and colors. A book can safely and without violation of FAA regulation be read on an aircraft even while landing and taking off. And, except if you dump it in a lake, a book is fairly indestructible. But a book can't be easily modified or electronically transmitted—text files on computers can.

Early aircraft had virtually no advantages compared with a passenger or freight train. Aircraft still are far less comfortable than trains; they were at first much less safe, and their schedules were unreliable, being subject to weather and an unreliable lightweight propulsion system. Air travel remains more costly, and at its introduction was often not much faster than train travel. Aircraft payloads even today are minute compared with trains' payloads. But what aircraft did provide was an alternative means of transportation that happened to suit new niches (e.g., flying mail across the Andes) not served by trains, and gradually, over 75 years or so, some (but not all) of their negatives were addressed.

Disruptive technologies, those that establish new paradigms in stable technologies and markets, follow a development path that is widely misunderstood. We think newer technologies are better across the board than the old ones—but that's usually false. Personal computers at their introduction disrupted the technology of main and mini-main frames. Despite that, they were significantly handicapped compared with their predecessors in almost every respect—speed, memory, I/O capacity, program size, programming language, and even cost measured in dollars per FLOP (floating point operation per second). This led the then President of Digital Equipment Corporation (now part of Compaq) to state there was no reason anyone would want to own such a [handicapped] machine for the home. The niches PCs would fill—including telecommuting, web-surfing, game-playing, e-mailing and music production, to name a few, were all yet to be conceived at the birth of the Altair and the Apple II.

Inside your glitzy new laptop are other disruptive technologies that are quite inferior to their predecessors. The tiny disc drive whirring inside it, with platters about an inch in diameter, is expensive, power hungry (per byte stored), noisy, and slow, its storage capacity limited and space-inefficient when compared with the big Terabyte (1000 Gbyte) drives on facility computers, or even the medium size drive on a desktop computer. But it fits in a laptop—that is its one and only advantage.

One more example. Is a Sony Walkman anywhere near as good as a stereo system? No Walkman offers CD, tape, and FM Stereo. It can only be heard by one person at a time, its fidelity is significantly handicapped by the small, lightweight headset and its space, budget, and power limited electronics, and its lifetime is usually short. But, you can't carry a pair of Infinity speakers, a Yamaha amplifier plus turntable, FM tuner, CD player, and tape drive on a bike ride or onto an airline seat.

So it is with tiny spacecraft. They produce much less electric power, downlink much less data, cost more per kg, point less accurately and in general do nothing as well as larger spacecraft. Their popularity owes to two factors:

1) They are Cheap. 2) Their development time is fast (they can be developed and launched in a year or two rather than the decade or two needed for large spacecraft).

In some ways, though, the Better / Faster / Cheaper paradigm is wrong when referring to microspacecraft. They may be faster, but per watt or per bit they are not at all cheaper (though per satellite they are), and better they definitely aren't—in fact they are worse, just as a walkman is worse than a big sound system and a cell phone is worse than one connected by wires (except that it isn't).

While economics have forced an industry-wide downsizing of conventional spacecraft, there is still a tendency to compare cost per bit or kg, or "sophistication" of missions between large and small spacecraft, and a need to somehow postulate that a $1M or $10M spacecraft can approach the capability of much larger ones. The truth is, it can't. No more than an Apple II could compete with an IBM 370B or a Pentium II Dell with the latest industrial supercomputer—except in price. But even there, in dollars-per-flop, the mainframe is the hands-down winner.

Rather than try to build the next Hubble stuffed into a Walkman chassis on a Cub Scout Troop fundraiser budget, AeroAstro and I have focused on the one key feature that has driven the acceptance of both the PC and the microsatellite—low cost. Not comparable capability at low cost, just low cost. In spacecraft, low cost must include low mass, because launch cost is only cheap if spacecraft mass is small. Spacecraft are generally much more useful after insertion into space than they are idling on the ground, hence their name.

Let's pretend we're Einstein or Schrodinger and engage in a Gedanken (thought) experiment—a quaint pastime no longer popular as it is subject to fatal degradation via nagging interruptions by ringing cell phones. Imagine that we have a 1 kg spacecraft producing just two watts of electric power, supporting only a megabit or two of memory and sporting a modest 9600 baud modem. (Younger readers will have to also imagine and accept as an article of faith that such a crude technology as a 9600 baud modem did at one time exist. People paid big money for it. They were introduced at $1000 when $1000 was enough money to buy a real racing bicycle.) The hypothetical microspacecraft does nothing as well as even a conventional small satellite. Its lifetime might be less than a year, its pointing accuracy is no better than a degree or two, and even a small payload could only be turned on during 10 or 20 percent of an orbit in order to live within its power budget.

But this waif of a spacecraft is better in one respect—it's cheap. Cheap to build and cheap to fly. Cheap to build because its very modest radio, based on existing L-band digital cell phones, its battery, adapted from a camcorder, and low bandwidth interfaces are all available at low cost and can fit together on one single circuit board—eliminating most of the components, wire harnesses, cables, structure, chassis, clips, tiedowns, and connectors that make even small spacecraft expensive to build and time consuming to troubleshoot. The 1 kg box would be quite cheap to fly as it comprises 0.01% of an Ariane IV payload in both volume and mass.

Immersed in our Gedanken experiment, we ask: "What would happen to this spacecraft in today's market?" Answer: "Nothing." It can do exactly none of the missions we currently ask our satellites to do—too little power, too little data rate and so

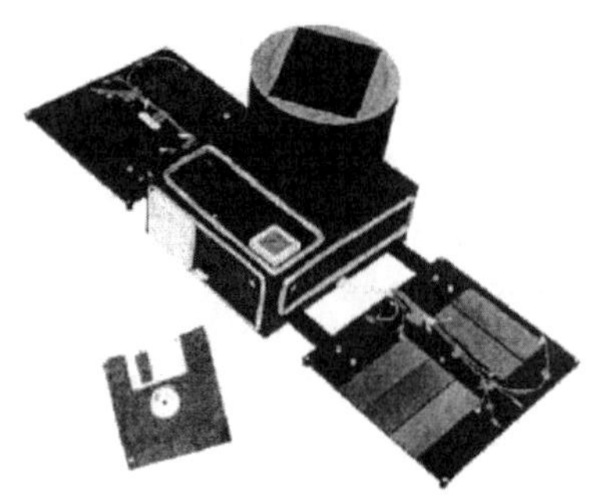

AeroAstro's Bitsy: The first McPizza in space?

on. Just as the first aircraft could do none of the transportation missions handled 100 years ago by trains and ships and the first PCs couldn't do what we did with those IBM-370Bs back in 1971. The aircraft carried too little payload and they were too cold, noisy, rough, and dangerous for commercial passengers. The early PC couldn't numerically solve the Navier-Stokes equations or search for 70+ digit prime numbers, which is what we thought computers were for.

Before we return to our normal nonthinking state, one more Gedanken experiment. Imagine the marketing group that proposed that a contingency of humanity migrate itself from the warm and fertile plains of northern Africa to the cold of Europe. There's a concept: substitute warm sunshine, a soothing zephyr and abundant food for a tightly wrapped animal skin, fire, and a club to kill a rather unappetizing critter as a substitute for sweet, colorful, and attractive fruits, vegetables, berries, and grains. Cold and lousy weather is worse. Not having abundant food is worse. But ultimately the migration happened—because clothing and fire sport one key feature—they enable humanity to populate the 99% of land mass that isn't warm and filled with edible plants. Of course, until the first attempts at those early migrations, there was no mission to the north, and no use of clothing to keep warm.

Truth is, the 1 kg spacecraft Gedanken experiment is a hoax. AeroAstro is already trying it out in hardware, as are several other government and commercial satellite building organizations. Our 1 kg, single board, zero wiring harness Bitsy spacecraft does not handle a single one of humankind's contemporary space missions—it meets no known need in space. Are Bitsy and its cousins at other organizations therefore a marketing and sales concept somewhat akin to concrete canoes, McPizza and Earth Shoes? About half of AeroAstro's own engineers have answered yes, but my answer is—Bitsy's success depends on finding a mission it can do better than any alternative. And as it turns out, there are several of these.

Bitsy's low cost allows flying hundreds of spacecraft in a quasi-randomly distributed constellation, ideal for three-dimensional mapping of tensors like the earth's magnetosphere and geotail. It is light enough to create a piggybackable module that can rendezvous from a virtually random orbit insertion with a malfunctioning spacecraft, taking pictures of it and "listening" for the electromagnetic hum of its processors and other on-board systems. It can be tossed out of an airlock to inspect damage outside a space station, reducing the need for and risk associated with crew EVAs, none of which are "real" missions today. But they might be what satellites are mainly used for tomorrow—just as web browsing and e-mail, what most of us do with PCs today, don't require the horsepower needed to map the flowfield around the Space Shuttle during reentry, and the number of people using massive computers to look for huge prime numbers is today insignificant compared with the number of high school term papers prepared on Pentiums.

And as technology continues to evolve, like the PC, aircraft, and radiotelephone (cell phone), Bitsy-like spacecraft will gradually become more capable. Deployable antennas and solar arrays will increase their on-board power and communications capacity, laser and GPS technology will allow Bitsy-class spacecraft to coherently detect radio and even light across baselines of thousands of km; and exploiting LEO comms constellations will bring the satellites within reach of your computer and modem at any time and location (yours and the spacecraft's) without need for a ground station. Big spacecraft will still find a few applications that only they can do, but for most people and most people's choice of missions, a microspacecraft will be the way we work in space. Smaller size = smaller cost, a logical concept worth discussing in some depth.

Chapter 4

More of Less is More: The Logic of Microspace

If you've surfaced much over the past twenty years, you may have noticed that everything is bigger. Today's cars, for instance, are huge. My whole car doesn't come up to the top of the tires of my neighbor's new SUV (Sport Utility Vehicle) behemoth. Our government tells us people are getting about 1% heavier each year, on average. A genuine New York H&H bagel rivals the bulk of a half-loaf of bread. A four lane freeway is sort of a quaint remembrance of simpler times, found only out in upstate New York and central Utah. My wife's closet is about the size of the studio we first lived together in. It would be indelicate to comment on her shoe size, but mine is well into the two digits.

For a guy fixated on making satellites smaller, these apparent signs of our growing affluence loom ominously. Stock traders and football coaches like to talk about "the fundamentals." Are today's fundamentals Big Dog XXL T-shirts and megamalls? Yes, but the fundamentals that I see are why I haven't switched careers into bigger satellites, 56" HDTV screens or subzero refrigerators which, in simpler times, could have housed a walk-in closet.

One simple fundamental is that space use, travel, and exploration are too expensive. Not too expensive for the people who already have the money to build and use satellites—people like the US Government, the phone company, and CBS. But too expensive for hundreds of colleges and thousands of college students. Too expensive for tax-paying people to ride into space like Alan Shepherd and John Glenn did on our nickel. Too expensive for businesses that want to bring information and natural resources to us from space, and to settle the moon and the solar system, and to create new high speed transportation systems and custom communications networks. Too expensive for a small prospecting company with a bright idea for finding gold or oil or wild ginseng.

Rockets aren't getting much cheaper, but they continue to get bigger. Did the Space Shuttle become a taxi for taking all of us into space? Is Pegasus any cheaper than Scout? Is the smallest LMLV (aka Agena) at $24M something a college class can ride into orbit?

I work with a lot of people who have clever new uses for space and who enlist my help to realize them. Among the top ten questions, including "How was your trip over here?" (these people tend to live 2 to 20 airline hours from wherever I am) is "how do you get those things up there?" As if I had anything to do with launching satellites other than the subtle art of not paying retail. Over the last 20 years, which is how long ago I stopped studying too much, relearned how to drive a car, and actually saw the outside world occasionally, the cost of space transportation has gone up. The Shuttle no longer will launch your 150 lb satellite for $50,000 (thank the lawyers and Ronald Reagan for that one), Pegasus costs more and is less reliable than the Scout it replaced, and the availability of Ariane ASAP (piggyback) accommodation has not increased.

As a rocket scientist, I don't condemn the rocket manufacturers. Launching into space is a difficult and unforgiving challenge technically. It's capital intensive, there are myriad regulatory, safety, political and environmental issues to deal with, and it's never been particularly profitable. Add to that the propaganda of those who believe transportation to orbit is in fact quite easy and prices are just artificially high owing to conspiracies and laziness, and you have a marketplace inhospitable to innovation and investment, coupled with a basically tough problem.

If we can't look forward to rockets getting cheap enough to drive space costs down, what alternatives do we have other than making satellites smaller, thereby lowering their launch cost? The logic seems pretty airtight on that one.

4.1 Smaller Size, Smaller Cost

It turns out that when you pick that road—the "I'll make it smaller until even I can afford to launch it" road—a lot of good things happen. For instance, reliability goes up. Small satellites contain fewer parts than big ones, and 100 parts are a thousand times less likely to fail than 100,000 parts, all things being equal. And they're not equal. They're mostly stacked in favor of microspacecraft. For example, small satellites are built quickly. Hence their parts can be selected from newer generations of components. And in case you haven't driven a '66 Bug lately (in their time they were hailed for their reliability), newer really is better in terms of MTBF (mean time between flat tires). New parts are more highly integrated, meaning each one does more, reducing the number of parts you need and the number of solder joints, leads, and wire connections. And all of this increases reliability. (For more on this, see section two of this volume.)

Small satellites are built by small organizations. As you may know from playing soccer or living in a family, small organizations lack certain classic organizational features, like bureaucracy, paperwork, politics, constant meetings, formalized roles and rulesets, poor intergroup communications, a high ratio of managers to doers, and obligatory office parties attended by people you don't know but ought to, that used to be scheduled during the Seinfeld season opener.

The more efficient, smaller organization helps in big and small ways. For instance, you can buy better parts, because without the top-down rules on parts, and with each engineer charged with understanding the program environments and requirements in depth, each engineer can actually be an engineer, and can take responsibility for solving her or his particular design problem the best possible way—meaning the most cost effective and reliable way.

Engineers like to work this way, and that, coupled with the short development times of small satellites and the high level of personal involvement and responsibility, means that they (the engineers) tend not to quit mid-program. Sure, we know how to transfer knowledge via paperwork. This is why Homer wrote the *Iliad*. But most engineers, talented as they may be with Matrix-X or a logic analyzer, are not Homer. The transfer of knowledge when a key person leaves a program is never perfect.

Not just the satellites are simpler when they're smaller. Their missions are simpler as well. Granted, a scientist may know that the space telescope is a lot more difficult to build than a wide field-of-view all-sky monitor. But in the one-bit world where the public wants to know "did it work or not," we're on the sunny side of the street. No taxpayers revolt has ever, in the history of humankind and taxes (I'm not sure which came first), been waged over satellite pointing accuracy and downlink data rate.

4.2 A Brief Discussion of Large vs. Small

Better yet, as individual large satellites are replaced by constellations of small ones, the impact of a failure is minor. When a reporter asked a Motorola rep about a recent Iridium satellite failure in orbit, the response was along the lines of "Why should we care?" You don't hear that from Loral or Hughes when $250M worth of geosynch commsat fails to deploy a solar panel.

Even though we could drive more safely by driving slower, we don't go 25 mph all the time. Nor do we carry umbrellas every day, even though it could rain. At the risk of whatever shred of technical faith you may have once invested in me, I have a confession to make: I haven't yet thrown out all my kitchen appliances that lack a third "ground" plug. We compensate for the intrinsic increases in satellite reliability that come with reduced size, in part by using less skilled labor (students), by using parts without space heritage, and by employing more advanced software (for auton-

omy and to do more with less hardware) and more interesting designs (for instance, leveraging new deployable arrays). Small satellites are injecting new people, new technologies, and new thinking into aerospace engineering, a field that may seem to the newcomer surprisingly staid and risk averse.

You can't do everything with a small satellite that you can with bigger ones. And this at first glance would seem to be a negative. You can't do everything with a laptop you could do with a Cray, either. But which company would you rather own? Cray or Compaq? Or Dell or even recently-embattled Apple? NASA argued that the building-filling computers of the '60s were neither too large nor too expensive since they had three or four of them, gave grants to scientists to use time on them, and didn't manage to keep those mainframes too busy. (They were mostly used on activities of only the highest relevance to Joe and Joan American, e.g., finding prime numbers with > 70 digits.) Certainly if these highly capable computers aren't that popular, smaller, less capable ones are not going to be useful or desirable at all, was the logic back then.

Small computer and small satellite people manage to find new applications that propel their existing applications, as much as they try to cram the capabilities of major systems into small, cheap packages. Amateur astronomers rival the pros, despite their limited time, apertures and experience, in finding asteroids, though certainly not in more complex research. But asteroids and other amateur targets are potentially more important to our continued life on earth than determining of the age of the universe or whether it will expand without limit, however scientifically significant that might be.

Microspace is redefining the uses of space. Students do not build Milstar to get some grasp of aerospace engineering principles. They build 50 kg piggybacked satellites. If a solar sail is ever flown to the moon or another planet, it will happen first on a microsatellite. AeroAstro is building the first satellite not designed for space research, education, communications, military, or remote sensing applications. It is a microsatellite mission combining public participation and entertainment, bound on an infinite mission across the galaxy.

Not that we have to apologize for small satellite performance. Constellations of them can provide faster communications over better links than geosynchronous satellites, which are suffering market losses to transoceanic cables for telephony and internet access where even a half-second delay is a nuisance. Small satellites can also do reconnaissance with faster revisit times than less numerous larger platforms. They can get new technologies into space faster and cheaper, and answer research questions while they still matter.

And, just as my ’92 Miata with its canvas top down and little 1.6 litre in-line four humming is a lot more fun negotiating a country road on a Sunday afternoon than my neighbor’s Four Wheel Drive bruiser, up to the bumper of which my leather wrapped steering wheel doesn’t reach, so too I’ve found that small satellites are simply lots of fun to build and use.

Problem is, fun won’t sell a mission in today’s one-bit world. Neither will fine, logical arguments about getting new technologies into space faster and cheaper than anyone else. As the Wizard of Oz told the Tin Woodsman, “What you need is a testimonial.” Good advice. Although in the 1990s, the Wizard might say, “What you need is a good one-bit sound bite.”

Chapter 5

The One-Bit Sound Bite

The world is full of information. And fortunately for us, humans are nature's most sophisticated information processors. We've written terabytes (thousands of billions of bytes) of literature, we print thousands of tons of newspapers per day, and we fill the airwaves with billions of bits-per-second of news, entertainment, and data. In fact, it's said there's more written information in a single Sunday *New York Times* than the average person living a few centuries ago was able to process in a lifetime. But lately, it seems that our ability to access this ever-increasing amount of data is overwhelming our ability to parse, let alone process, all but the tiniest fraction of it.

Small wonder then that the word "byte" is pervasive as it is. We can only afford a "sound byte" of the president's radio address, or a news byte of a disaster in some far away place. We live in a society flush with great food—and we're all on diets. Surrounded by plenty of nourishing food, what small segment of it do we select? Fat-free Entenmann's cakes of course—loaded with empty calories and devoid of nutrition.

And we live in a society flush with information. Similarly, with the richness of all the world's information resources at our fingertips, our personal information diets consist of plenty of bits—color video and stereo sound—of empty infocalories, "Bay Watch," the incessant drone of call-in talk radio and a daily mailbox-load of catalogues. (Note however absolutely no dogalogues—remember that next time you're out pet-shopping.)

What has all this got to do with your microspace mission? More than all those little ICs and shiny mil-spec bolts you've got in that little satellite. Read on, innocent engineer...

Since you are not sending a man, or even Prince Charles for that matter, to Mars, your mission will not be occupying many of the sparse bits left over in the attention span of your sponsor, boss, potential customer, and/or department chair. These important people feast daily on their infomeals of:

- *Space Fax Daily*
- Fifty or so e-mails about how to expense the purchase of an umbrella for the day that someone way out on the other end of the org chart spent on the antenna range in Arizona in August when the freak tropical storm blew in
- The n+1th rerun of the "Junior Mints" Seinfeld
- A multimedia presentation from a company that wants to custom tailor a corporate video
- The latest news clips of an apartment building fire
- More news clips: some astronauts floating around in some cramped space station module speaking either Russian or English
- A few pages of their kids' math and history text books
- A brief recounting of each of his or her kid's day (number of bits equal to 10^6*e^[- age in decades]).

Yes, the product of you and your team's fifty person-years of labor will have to be distilled down to one bit. Just how much information is one bit? Not enough to spell "C-A-T" or "You Have Already Won $1,000,000 AND MAYBE MORE!!!" No, one bit is just barely enough to answer one yes-or-no question. And that yes-or-no question is, always was, and always will be: "Did it work?"

Hence the single most valuable lesson in spacecraft design you're likely to learn reading this book (and certainly the most valuable if you're just skimming this book due to your info-overload management diet) is to make sure that the answer to this question is and can only be Y-E-S.

Dan Goldin, head of NASA at this writing, is responsible for $14 billion per year, more or less. Your microspace mission might be $1.4 million, or about 0.01% of Dan's budget. If it takes you, say, three years to move from start to completion on your mission, your project is already down to 0.0033% of Mr. G.'s attention span. If he reads ten pages of condensed briefs a day (10% of which involve program status) every day for all three years of your program, that means you're going to get a total coverage of about two lines of text, most of which will be contextual. Your sound bite might playback like this: "The PIPSQUEAK microsatellite, built by the students at El Piquante State U., was launched piggyback on board the 89th Shuttle SS-Alpha girder delivery mission, and apparently worked."

The word "apparently" got thrown in there by an aide preparing Mr. Goldin's daily briefing package as a CYA maneuver in case it turns out your press release was premature, or for some reason Mr. G. has a beef with EPSU. So, unfortunately, that brief packet of partially accurate information is your full allowance. Subsequent releases about the spacecraft performing its mission flawlessly for six years on orbit and completely revolutionizing our understanding of some subtle element of ionospheric transport, along with your team's heroic efforts in overcoming the loss of your ground station when a Mexican hurricane lofted it into the Pacific, by deftly employing paper clips, an old Gibson amp, and a crushed Sony Walkman, aren't going to make it past the EPSU *Gazette* (the weekly school rag).

If you don't believe me, ask yourself how much you know about the Helios geosynch mission—it costed out at $500 million. How much do you know about Magellan—a $2 billion mission, or Clementine at around $100 million? So proportionately, what's the world going to know about your $1 million modern equivalent of a spark-gap radio with solar panels? Zilch plus one bit.

Here's how to get that one bit right.

First off, don't claim you're going to discover the 11th planet, even if that's what you hope. In the unfortunate case that your aspirations should be publicly disclosed, the news bite will read "so far has failed to find evidence of the 9th planet." The press, of course, will have gotten even this one vital fact wrong (making you, not them, look like a total idiot), and the bite will have to go to press before you've even had time to boot the spacecraft processor and receive a test message. Instead, just say you're going to demonstrate critical technologies for an eventual robotic presence on the nearest extrasolar planet, or something equally obtuse. The report will then read (if you spin it right) that "everything is working perfectly on the EPSU P-1 spacecraft toward demonstration of a robotic assault on MIR." That's worst-case reportage, of course, but it will get somebody's attention and pass the one-bit rule. Congratulations! You're a success. Best of all, to garner that crucial bit on your scoreboard all you have to be is on your timeline, which, if you haven't even had time to boot the processor, can't help but be a true statement.

Having framed the mission objectives conservatively, the next step is to build around a bulletproof core. All the interesting stuff that motivated you to bend your entire life around the spacecraft's development is immaterial to the one-bit world out there. The 100 MIP, 100 microwatt processor, the 10,000 s Isp electric thruster, the ion velocimeter on a chip—they don't matter to that one crucial bit. Fly a VHF beacon—100 mW of RF carrier with a lithium battery that you turn on before the launch sequence starts. That will certainly work, and you can report immediate success in a "radio science experiment." Regarding that processor—why not back it up with a 68HC11 left over from your LDTV remote control (which you've by now traded in for HDTV anyway). So long as it responds to commands, turns on a radio, and says "Hi Guys," you've got mission success. Everything else is gravy.

Let's close this profoundly cynical exposé on what really buys brownie points among your sponsors and tenure committees with a great joke from the Internet, which I'll paraphrase for you in the form of a question: If you and your boss go for a walk in the woods and he or she asks about your mission, and, due to the loud CRRRAAAAACK!!! of a tree falling nearby he or she doesn't hear one word you say, do you still get a healthy bonus? Your job is to get that one bit right.

Now that we've got this concept cleared up, let's take a not-so-random walk into the briar patch of cost optimization. And let's see if we can make it out without a scratch. Or at least, alive.

Chapter 6

Nothing Is Cost-Optimized

Elusive oasis of engineering efficiency.

Spending five or ten thousand days of your life as an engineer eventually causes certain mirages to materialize within your brain. These oases of auto-fabricated reason propel you through a landscape strewn with the countless grains of fine details that form the world of engineering. The lure of spending a career upon these sands is that you will eventually irrigate them into a cool green orchard, an epiphany to gaze upon, a veritable monument to your ingenuity, that yields its sweet fruit to all earth's people in perpetuity. And whilst awaiting this uncertain Nirvana, the chance to create the next Hoover Dam or cellular telephone, these mirages prove to be a welcome refreshment for a weary mind.

In the midst of the deepest jungles of green coconut palms and cool blue ponds shimmers the elusive oasis of engineering efficiency: the maximum miles per gallon of gasoline, the fastest human-powered vehicle. But even efficiency on a more modest scale—the lightest propellant tank or the minimum amount of energy needed to perform one processor operation—creates for the engineer a way station and a satisfying illusion of progress.

6.1 Absolutely Nothing Is Cost Optimized

The minimum cost space mission—now THERE's an optimization. Who wouldn't argue that space is too expensive? Cost minimization is obviously a desirable goal. In fact, in our capitalist society it's an implicit assumption. An 89¢ quart of milk is obviously better than a 99¢ one. Southwest Airline's $59 fare to Cleveland is a better deal than $279 on United. Sounds simple but it isn't. What if the US government is doing the shopping and the cheaper milk is imported from Cuba? What if the cheaper airfare is on a new airliner that bellyflopped on the tarmac last week? Buying and making engineering decisions, particularly when government money and facilities are involved, are not so simple as they may initially appear.

Despite my reputation as one of the gurus of cheap space, I have to admit that I don't know of even one single cost-optimized space mission, even these days when people won't propose anything without appending the obligatory "better-faster-cheaper" mantra to it.

6.2 Other Kinds of Optimization

By far the preferred and most successful optimization strategy is political optimization. When the Jet Propulsion Lab (JPL) unveiled its New Millennium program to radically downsize and economize uncrewed spacecraft, it was prepared with the antidote to the program's potentially fatal flaw. Politicians do not appropriate money for the pursuit of space science, but for the pursuit of votes at reelection time. This is not the workings of evil or corruption. It's just the logic and reality of survival in a tough job. But to run a cheap and efficient program means you have to exploit a tight, highly cohesive team—a skunk works.

Did you know that despite being canceled after only a few of its budgeted tens of billions of dollars had been spent, the Superconducting Super Collider (SSC, what would have been the world's largest and most potent atom smasher built to discover the fundamental building blocks of matter) revealed one of nature's most subtle secrets? The feature of SSC that was most visible and significant to the public and to our leaders was not its physics but its real estate. Virtually every state in the union vied for the privilege of hosting the biggest scientific instrument ever built by humans. Each of the proposals—themselves the products of multimillion dollar efforts by the states, teamed with partners in colleges, universities, businesses, and municipalities—trumpeted the vision of a noble research effort funded by our sage Federal government. All agreed not on real estate, since each bidder asserted its own state was best, but on America's destiny to lead the world to a new understanding of the Physics of Matter.

After such heartrending universal endorsements of the values and virtues of this multibillion dollar investment in pure science, it was a bit disconcerting that the first down-select, eliminating all but the leading dozen contenders, resulted in mass defections from the chorus of praise singers. Apparently only twelve states still shared the scientific zeal of the Feds. The failure of the thirty-eight out-selectees to grasp the overarching wisdom behind what they now scorned as a $100 billion boondoggle gave testimony to the vision's fragile beauty and truth. The volume and pitch of the remaining twelve states' chorus overwhelmed that of the nonbelieving thirty-eight. Victory for science and humanity was within our—or at least their—grasp. Arriving at the climax of this crescendo a sudden hush fell over the country's great institutions of learning as the envelope with the winner's name was opened and read.

6.3 Texas

The answer to science's biggest question in decades was Texas. And while horns tooted and champagne bottles uncorked in Houston, Dallas, and Austin one question rose to the lips of scientists and educators and civic leaders across the barren forty-nine: Texas? Shakespeare knew well that heaven has no wrath like a state scorned by federal agencies. So, as the biggest Physics project in history broke ground in the parched Texas prairie, the world's most highly educated mob convened for a lynching party.

Had the balance of forces been only forty-nine to one, a massacre would have been certain. But even in Texas, plenty of astrophysicists and solid state physicists and plasma physicists saw their research and career options clouded if not foreclosed as the Super Collider consumed the nation's science budget for decades to come, and they too massed for the attack. Suddenly the SSC was a $100 billion sandbox for a few Texas academics, a windfall for a few Texas land owners, employment security for a few thousand Texas contractors, and a disaster for 99.9% of the Physics community.

Within a year the *New York Times* published the Super Collider's obituary. The full-page article spoofed the project, offering the unfinished components of the program, slaughtered in its infancy, at fire sale prices. An empty underground circular tunnel in west Texas could be yours for just the $500 million spent to dig it out. Few buyers emerged.

6.4 Another Take on Optimization

Advertising is repetition. Behavioral research, mainly sponsored by ad agencies, tells us that we need to be exposed to a new concept seventeen times before it is remembered and understood. Count how many times a thirty-second radio spot repeats the sponsor's name—and how many times per day that spot runs.

Learning is repetition. After your 17th visit to the supermarket in the ground floor of the Marui building near Japan's Shin Yokohama train station—wherein, during summer, you will encounter a green, nearly spherical squash filled with red meat and labeled Suika—you will definitely know the Japanese word for watermelon.

The colossal failure of the Super Collider drove home to scientists what they had heard before but never learned. Why, after all, does the US continue to launch hundreds of weather balloons every day from all over the country when they could all be replaced by one satellite saving $100 million per year? Because they are launched from all over the country—from almost every congressional district. The secret of nature, discovered by physicists via the supercollider, but which was never a secret from politicians is simple: constituency.

Shuttle and Space Station live by that sword. NASA can show you contractors in all fifty states employing ordinary Americans just like you and me building our high tech pyramids. We can't cancel the Space Station—after all we have commitments to Europe, Japan, and Russia, to the dreams of America's youth, the world's most influential constituency. And the Shuttle and Space Station projects are not planted in Texas, Michigan, or California, but are orbiting more or less equally over more or less all of us.

A true minimum-cost small satellite program is, by the inviolable law of constituency, not politically viable. To minimize cost you need a small, tightly focused development team. Ten people or maybe twenty, max. And the nature of soldering and bolting things together being what it is, most of that group will organize itself into mini-groups of three or four locations at most. Not only can't you carry Congress by gaining the votes of only four states, your budget is too small even to get on the radar of those measly four.

Cost optimization thus bows out graciously, and one of its many surrogates, in this case political reality, takes over. Tailoring for political constituency is not criminally punishable, or even ethically questionable. But there is this nagging Talmudic dictate: Know before whom you stand. Probably forged to remind Jews to worship their one God, this wise caveat has meaning for supporters of microspacecraft as well.

Trappings of a political theology include the Integrated Product Development Team, or IPDT, the unsung savior of the New Millennium. These are clusters of technical teams that are complementary in their specialties. All the Team members work together to carry off this year's Minimum Mission Impossible. Not coincidentally the four, six, or ten team members are spread among four, six, or ten states, preferably ones with powerful lawmakers sitting on critical committees.

At this point, the best way to get eight, say, reputable and respected organizations interested in the project and to get their sixteen senators and eight congressional district representatives to help in the struggle for government dollars, is to demonstrate that the program involves enough money. Since working on space stuff has a certain cachet, just $10 million to $20 million per player may be enough. After all, each will reason, we can grow our ration once the program is underway. And we'll grow from this small role in sort of a training-wheels space program to a bigger role on a subsequent "real" mission. In the meantime, the governor or the Congressperson can happily and casually mention at receptions and fund raisers the space project he or she got going down at Parchedfield State U.

Thus are all the diners' rice bowls filled, both monetarily and egocentrically. They will tell you, "all for the sake of giving you technologists the opportunity to do this revolutionary small, low-cost stuff that is going to bring space to the disenfranchised masses by cutting conventional aerospace program size, complexity, and cost."

Since you don't understand nor really care about all these politics, you'd would be well advised at this point to tuck your long, minimum mission hair under your white woven polyester clean room cap and shuffle semi-happily back to the shop.

6.5 Underground Heroes: The Real Optimizers

NASA is government. DoD is government. ESA and NAZDA are government. Telecommunications, the so often cited example of private sector space, is tightly regulated and in most countries controlled and operated by government. When we enter the smoky private club of space projects, we need to check our shock and outrage at the crass floor show of political optimization at the door. If you don't want political optimization, don't request funding from governments. The problem is, is there any other way to develop spacecraft?

Miraculously, other ways exist. Though often cited as a hallmark of low-cost space achievement, amateur organizations optimize all the time, not for minimum cost but for maximum fun. They fly lots of stuff because it's interesting or new. Amsat spacecraft are cheap for two simple reasons: Nobody gets paid to build them and Amsat doesn't have any money. Whatever they do must be cheap or it won't happen. If you ask an amateur group how to do a particular mission for minimum cost you will also find out how to do it for maximum fun.

Amsats have lots of neat stuff on board. Heat pipes, composite structures, lots of modulation schemes and a rollicking collection of frequency usage plans, voice encoders, radiation monitors. Do they really need all that stuff? Yes—because if they didn't have it, the people who build 'em would spend their evenings and weekends out in the yard or at the neighbor's barbecue instead of hunched over a soldering iron or a CAD package busily designing and building these satellites.

Strange as this may sound to homework haters and academic anemics everywhere, the education optimal is closely connected to the fun optimal. Satellites are not built by students in order to minimize their cost. Students spend a lot of time (aka virtual money) building stuff they could buy for much less than their labor is worth. But students want to learn—or at least their professors want them to learn. A student's job is NOT to work at Tower Records to save up and buy a $2,000 sun sensor. Her/his job is to figure out how to build said sensor. So students go all the way around the block rather than taking the quick shortcut over the picket fence to flight success. Often students pick missions that have never been done (successfully) before—just to go where no one has ever gone before.

Thanks to the two redeeming factors of the American educational process—the thirty-four week "year" and the four "year" career path from rookie to retiree, several students will often walk around the same block. Doing the same job two or three times is anathema to cost minimization, but in the educational setting, it's the norm.

This explains all the graffiti in high school textbooks—you are the 119th reader of the same book and the 119th doer of the same exercises.

6.6 Buy a Ferrari

Politics, fun, education...what else is there? Most of AeroAstro's satellites are mission-optimized. The scientists building a payload tell us what the nonnegotiable performance requirements are for the spacecraft, then tell us to build it damned cheap. Trouble is, if you want a Ferrari engine, Ferrari aluminum body marinated in Ferrari Red paint, Ferrari suspension, Ferrari leather, and Ferrari instrumentation, you tend to end up with a Ferrari price tag. Sure, you can save a few bucks by having your auditors disallow a few of my travel expenses and by squeezing extra hours out of our already overworked engineers and techs. But this is coupon clipping—you'll save one or two percent. Maybe.

But if you spec, instead of a specific engine, a certain amount of horsepower and torque, maybe I can get that out of a junk-yarded Ford Taurus SHO V-6. And if handling isn't really all that critical for this particular mission (the aerospace drag racer), we can dump the dual trailing unequal A-frame link arms and just bolt the wheel axles to the frame, soapbox derby style. Now you're talking major savings, by a factor of ten or a hundred.

When a client comes in and his requirements list is long, detailed, and nonnegotiable, he may think he wants a minimum-cost mission. But what he has is a requirements-optimized mission. Cost optimal actually means the requirements can be negotiated or traded off to cut cost.

6.7 Building Satellites for a Dictatorship

Here's another creature that masquerades as minimum cost. A high-ranking official of a very small country, Country Z, calls me. I pay attention, because I know that even a very small country with a one letter name is bigger than Hughes and NASA put together (a fact I find reassuring when assessing my remaining career options after making one of my characteristic statements to the press regarding Hughes or NASA). Country Z has never built a satellite before—and they want to put the beloved Z-flag in LEO. Mission is completely negotiable. Education is incidental. Politics and constituency matter less, much less, in a dictatorship: El Hombre wants. El Hombre gets.

My blood heats up—Aha! A cost-optimized mission. The Treasure of the Sierra Madre at last! Weeks later an emissary of Country Z shows up at AeroAstro, usually in a black limo complete with driver, briefcase carrier, and personal secretary/translator. In the course of the day's 300-baud conversation, it becomes clear that if the satellite happens to not work, my highly ranked visitor may never turn up in public again, and perhaps neither might I. What we have here is not a cost-optimized mission, but rather a reliability-optimized mission. If somebody holds a gun to your head and says "get these dozen eggs to Coshacton without a crack," you don't squeeze the cardboard carton under your arm and hop on the Schwinn. You end up in a convoy of a dozen Hummers with special gyro-stabilized platforms, eleven of which are decoys, accompanied by attack helicopters and aircraft. You commission an unbreakable egg crate, and you hermetically seal it within a block of titanium. Better a big VISA card balance than a dance with Mr. Death.

You say the incidence of armed attacks on program managers is minimal? Well, you haven't hung around AeroAstro. How many Lt. Colonels do you know who presided over spacecraft programs that resulted in $millions being spent, rockets being launched, and non-functional hunks of aluminum being stranded lifeless, cold and black in LEO, who then went on to brilliant careers as four-star generals? (Zero is the safe answer.)

From the previous chapter you already know that the entire state of a complex spacecraft can be described by just one bit of information. Nobody cares if the damned thing has 26 Gbytes of EDACed RAM and an 80 Mflop processor that draws 20 mW on a rough day. No, there is a fully adequate single-bit description of your entire satellite, the one you and your team labored over night and day for the past four years: Did it work (1), or didn't it (0).

Other than risking termination at the hands of El Hombre, there are other motivations for the success-optimized mission. For instance, if your 10kg satellite is being launched on a dedicated $1B Titan IV launch in order to get to Alpha Centauri while the Democrats are still in the White House to take credit for it, success is everything. All crewed missions are thus by definition not minimum cost. The reality is that so long as the cheapest launch vehicle costs as much as the total lifetime productive output of twenty union assemblers working for the GM plant in Lorain, Ohio, all satellites are reliability, not cost, optimized. Hype to the contrary, the LEO comsats like Iridium's and Global Star's are no more cost-optimized than their geosynchronous counterparts. The cost of a failure, in delays to initiation of service (and hence revenues), in investor, customer, and insurer confidence, as well as corporate reputation, are too great.

Too bad, really. Students don't like it when their satellites fail, but they still graduate. They can accept failure. Amateurs are disappointed if there's a failure, but hey, they do it for fun anyway, so they'll build another one and double the fun. They tolerate failure—and Amsat's reliability track record is possibly the best of any spacecraft developer in the world. But as we can see, the people who ought to want minimum cost can't afford it.

6.8 Another Bunch of Optimizations

Political optimization. Fun optimization. Educational optimization. Mission Optimization. Reliability optimization. On the road we follow in our noble quest for the Holy Grail of low cost, we encounter many irresistible sirens and alluring songs. A wise professor of planetary science once told me that back in the '60s, getting a few numbers and a few squiggly lines on an O-graph trace back from a satellite impressed the socks off the public and the sponsor. In the '70s, it had to be pictures—preferably color pictures. In the '80s, video (despite the incredible boredom of watching a satellite take eight weeks to "zoom" up to the vicinity of Saturn). In the '90s it's live two-way video and in the '00s, I suppose, nobody will be interested in space if they can't just go there themselves. Another optimization emerges—entertainment.

Sure, your client wants to know what comets are made out of. But what your client needs are Hubble-class images of the comet, up close and personal, preferably with a brightly colored gas nebula in the background. Or another example: astronauts are supposed to be seeing what exercise does for maintaining health during long periods in space. But don't forget those live interviews of the perspiring space traveler. We drag a bunch of audiovisual gear into orbit with us, and we adjust trajectories to get striking photo ops, and despite the fact that each astronaut-hour on orbit is worth something like $2 million, we'll force them to wave to the kids down there in Akron and send e-mail to Belvoir Elementary School—via a gigabit-per-second geosynch satellite link, of course.

Brought to you by the one guy who actually does those exercises.

And let's not forget schedule optimal. We want a cheap satellite, the customer says—to make use of a "free" launch slot coming up in eight months. When you end up shipping two engineers halfway around the world carrying just one small metal box to get a minor glitch fixed overnight to get it back on the satellite before the first shift (of three) gets back to work in the morning, when you pay three times the standard pricing to get your circuit board built RIGHT NOW, when you get your first five-figure FedEx invoice, you'll know the difference between cost and schedule optimals.

All of which is why I have the ideal job. I'm an expert in doing minimum cost missions—something everyone thinks they want, although they really don't. I'm sort of the Jack La Lanne of space. Everybody wanted a body like Jack's so they watched his show—and Jack's sponsors didn't care whether his audience watched from aboard the couch or abeam the ironing board, rather than the exercise mat. Jack got old and creased, but the fans still loved him. The guy who actually did all those calisthenics. Amazing.

Books are entertainment, and entertainment is all about emotion. The emotion you might gather I'm trying to evoke you might guess is cynicism. Maybe, but cynicism is negativity and ergo not PC. Let's call it skepticism, which has a ring of robust intellect about it. That brain of yours has a 10 W power budget that must be fueled, explaining why great ideas often end up on napkins and placemats (but some unfortunately don't stay there). Turn the page and join me for a final repast of Mission Cuisine, fresh off the poultry farm...

Chapter 7

What Came First, Chickens or Eggs—Really? Some Recipes

Missions, cost optimization, politics—it all gets pretty bewildering sometimes, kind of like attempting to answer the perennial question of life: What Came First, the Chicken or the Egg? For all the talk on the Machiavellian trials and tribulations of building and launching microspacecraft, you'd think the community of would-be low- cost launch vehicle builders, among whose august company I sometimes include myself, worked for Col. Sanders or studied under Frank Purdue.

Stop whining already. Name two businesses started up by the sheer force of arguing for the existence of a market. You think the local donut shop didn't have chicken/egg problems of its own? The owner knew there was a market for donuts—but that didn't create a donut shop out of nothing. Most likely the aspiring donut fryer invested his or hers and the family's life savings to open that shop—and they all may have worked there for years before any eggs were provided by that chicken. If they ever were. Failure rates in donuts are enormous—we all know that 95% of startups fail in the first five years—and that statistic is not at all dominated by startup rocket companies.

The ad hoc group I call the SELVs (Small, Entrepreneurial Launch Vehicle companies), meets mostly at certain airports that have proved to be our magnets—Huntsville, Albuquerque, Dulles, and Denver. We all sense that there might be some Air Force or NASA chickens nearby those airports where some fertile eggs might be available for hatching our own prize chickens—our low-cost dream machine—which would in turn lay lots of eggs—money for our investors, low cost transportation for satellite builders. And those eggs would hatch into more capital for microspace, and more satellite projects, spurred by the

Are these the guys we're working for?

low cost of doing them with our cheap launch vehicles... When we meet, we complain about the lack of vision of the people we are visiting—can't they see the breakthrough we'd achieve if we only had an egg to nurture into the flying chicken? We never, ever question our own vision—but if these guys are so sight-impaired, how come we spend our precious time and money trying to describe our vision to those who can't see? Because they have, we think, some eggs...

Some of our cadre have abandoned NASA and the Air Force. They are getting money from wealthy individuals who believe in the chicken-egg scenario. Or they are spending their own eggs, earned in some other, sensible (aka profitable) business. But the sad history of the past ten years—the amount of time the microsatellite world has sorely needed a low-cost launch vehicle—is that these formulae are not working. Of course HAL doesn't really work, neither does the Dilithium crystal (or improbability) drive. But lots of us still think about what things would be like if we could build artificial intelligence, go WARP-six or raise a trunkload of gold bullion to build a microspace, microcost launch vehicle. Amazingly, despite the clear desirability of all these things, none of them actually come to be.

If we could transport ourselves in an automobile-like thing at WARP-6 to anywhere in the universe, obviously life would be quite different. But as it turns out, just going from the ten mph horse to the forty mph automobile, a modest four times improvement in one key parameter—speed—revolutionized the world quite com-

pletely. So why do we assume we need a laser-powered rocket that puts a thousand pounds in space for the cost of a few megawatt minutes of electric power off the grid and a large block of ice for the laser to ablate (estimates for that chicken are of order $10/pound to orbit, but the egg needed to hatch that bird costs billions of bucks, if it will work at all) to change the space world?

Let's assume a scenario closer to the horse/car breakthrough than the Dilithium crystal WARP-six thing. Say we can put 100 pounds in LEO for $3M. That's a five times reduction in cost compared with the only commercially viable small launch vehicle in the world—Pegasus. But hey, you'll complain, right now on Pegasus, that's already do-able—but scaled up. Five hundred pounds for $15M. Is there any point to that modest achievement? If you are looking for a chicken (or goose) laying golden eggs, you'll search long and hard, you'll do an infinity of studies, and you'll never find the hardware. But what if you're hungry enough that just the thought of a three-minute egg sounds pretty good. Then the real world becomes your friend, not an insurmountable obstacle to goal achievement.

Do we need Pegasus, given that Delta can put five thousand pounds up for $75M? Pound for pound, Delta is twice as good a deal as Pegasus. And gallon for gallon, a school bus moves two to ten times the people that your car can. So why don't we all drive school buses? Maybe they don't fit in the driveway, maybe we don't have forty-nine friends to share our commute to work with. In transportation, if you can build a one-fifth scale vehicle for the same price per seat or per pound as the big boy, it's a breakthrough. More examples? How much real estate does it take to house ten thousand college students in dormitories—versus twenty-five hundred families of four each in single family homes? How much does a ticket on a 747 cost to Europe—vs. one-eighth of a chartered Gulfstream IV? Heck, how much does a single candy bar cost in an airport ($1) vs. one in a bag of two hundred at the grocery store's bulk foods department (5¢). We all know big is cheaper, per pound, than small.

So, one reason we don't get off the launch pad is that we're trying to launch tiny payloads at less cost per pound than big payloads. We're aiming, so to speak, too high. Rockets are more like cars, airplanes and boats, than they are like satellites, which are more like computers. While computing power per dollar has increased by a factor of more than a million in the last 40 years, how much has the cost of a Cessna, a Pontiac or a forty-six foot sailing sloop come down? Not one iota, that's how much. The marginal gains achieved in combustion efficiency have been more than nullified by the cost of four-wheel drive transmissions and airbags. The savings in materials costs by using plastics and composites are more than offset by our demand for higher-speed radial tires that last longer than our cars used to and air conditioning, not to mention the obligatory CD player. Same thing with rockets—lots of us SELVs have very cheap engines—but we also aren't living in Robert Goddard's era. We've got triply redundant navigation systems and doubly redundant, flight-qualified flight termination avionics and pyrotechnics to deal with.

A school bus gets about eight mpg. With fifty kids on board, that's four hundred seat miles per gallon. If you could get three friends to commute to work with you, not even the most efficient automobile in America could do half as well. Yet we're told—and we buy into the notion—that we need to build a rocket 5% the size of Delta and do it for less cash per kilo. Ain't gonna happen. You read it here first. Just like you are never going to find that golden egg-laying fowl.

This doesn't mean there is a law against looking. Nor a law against raising money from unsuspecting zealots and inexperienced technocrats to try to build such an animal. But you don't hunt down too many eggs in the grocery store if you're out scavenging the Serengeti plain looking for golden eggs and the birds that lay them.

Grocery store? That's what I said. The eggs are just lying there, waiting for us to use them. There's no enormous technical barrier to a five times breakthrough in cost per launch. Take the technologies that exist right now, and do your own design of a total-reality (no new technology) vehicle. You can launch a hundred pounds for $3M. The problem is not the investment, nor the technology, it's the particular egg we're fixated on. It's not golden. It's no big per-pound discount, no factor of ten or a million decrease in costs and no investors running around getting rich on rockets and pumping capital into the industry. It's the egg thing—our vision—that needs to change. In fact, it's so damned easy to build the $3M launcher that the SELVs themselves don't talk about it—anybody can do it, so each of us tries to sell something better, harder to do, that only we can do via a magic technology, like rocket helicoptering to LEO, towing a rocket behind a 747, strapping together tens or hundreds of cheap rocket engines, catapulting the first stage from a silo, floating to forty thousand feet beneath a helium balloon, or launching from a recoverable platform that drives the rocket up above the atmosphere. One day historians will look back on us idiots the way we look back on the alchemists, or the early would-be flyers with mechanically flapping wings and stepladders. When alchemists finally abandoned the lead-to-gold trick and all the attendant and misleading hocus pocus, and got to work on the art of the doable, chemistry happened. Of course we had to first blow off the phlogiston too...

Long before cars, we had trains, and ocean going liners, and even Zeppelins. Why was the car such a revolution—and why does it remain so popular a hundred years after its invention? Did a magic technology come along—a black box—and the car just all of the sudden became practical and desirable? Doubtful. The early cars used steam, gasoline, electricity—all sorts of off-the-shelf technology. The magic wasn't really so magical. The car is just the right tool for the job—for many jobs. You need to get to work 10 miles away on a rainy morning: you gonna take a Zeppelin or the Queen Mary? You need, for mental health reasons, to drop the kids by mom's house for a few hours—on the train? Get real. For all its pollution, for its capital cost, the insurance, the breakdowns and maintenance, the gas it needs, the congestion it causes and the dangers of being injured in a car accident (ten times more dangerous than an airline flight), your chances of being killed in a car are one in twenty-five hundred), the car is the vehicle of choice. It fits the job of transporting you where you need to go when your legs are not up to it and a 747 is overkill.

A hundred-pound payload, $3M launch vehicle. Would anybody buy one? It's like the market research to figure out if a bridge over a cold and turbulent river is worth the investment. Counting the crossings by hardy swimmers is not a great metric. You have to picture the commerce that could take place, the trading, tourism, commuting and improved transportation to other places beyond the vicinity of the bridge, that might happen were the bridge built...

Celestis is a viable commercial business launching people's cremated remains (ashes) into space for a burial with a heckuva view. Ashes don't weigh much—something less than an ounce all ready to fly, but families don't like to wait months or years for enough people to join the ranks of the previously alive for their loved ones' burial. Ariane piggybacks one or two hundred-pound class satellites a year, not when they want to go, not to the orbit they want to go to, and under the constant threat of demanifesting if the primary payload goes overweight. And they charge $1M or more for the ride. If you could get a $3M ticket into orbit, for $18M, you can build your own miniature Iridium or Orbcomm—plus the satellites, of course. I know of at least one satellite company (ours) that'll build the forty satellites for less than the cost of the six launches. And six or seven of those satellites will fit into the hundred pounds. And that's now, when nobody is pushing the envelope to make ten-pound satellites so that a hundred-pound rocket can populate an orbital plane.

Keep going. When a big satellite dies in LEO, how much is it worth to try to fix it? If the ailing satellite and it's launch might have costed $100M, it's not worth a $15M Pegasus and a $15M robot satellite, given that probably you can't fix it, and you'll be lucky to diagnose the problem. But how about a $3M launch and a $1M satellite? Four million dollars can get spent figuring out what went wrong via ground simulation and testing—why not go find out? For $3M why argue about whether a huge solar array, radiator or antenna will or won't deploy —just send it up and do it. Do it twice. In fact, for the price of one Pegasus flight, do it five times. Five times is trial and error.

Problem is, for most people today, one flight is life or death—and most sponsors won't take that chance on a one time roll of the dice. It's a sad fact that, Star Wars controversy notwithstanding, we've never actually intercepted an object in space. Of course, at $60M per try, we couldn't afford a lengthy education. Nor can we afford, with today's technologies, to let students build stuff and try it out in space. We say we do, but in fact our "cheap satellite" programs aren't cheap—they're worth more than the budgets most of us forty-somethings aren't even allowed to manage. And hence more money than we give the average college kid to do some playing around with.

These are not all the answers. Neither did somebody think up the fried egg and bologna sandwich, the fritata, the souffle, huevos rancheros and angel food cake before the first egg was cracked. Let's get over our chicken/egg hangup, heat up the griddle and start cooking real food. They may not be golden eggs, but they sure beat starving to death.

Part II

Missions and Reliability

Section 2
Reliability

Chapter 1

The Mantra of Reliability

The word "inordinate" well describes the amount of reading you're about to do about making missions work and especially about the parts we build satellites out of. Reliability and robustness are complicated, but the focus you need to achieve them isn't. The common denominator of all highly robust systems is their designers' fixation on reliability.

By far the #1 cause of program failures is failures to be funded—that accounts for probably 98% of failed missions—good missions that failed to deliver their promise. That's why I've been working on little satellites for twenty years, written books and book chapters about microspace, teach courses about minimum cost missions, and founded AeroAstro. As the American Lung Association might have said, if your program isn't funded, nothing else matters.

Gold plating, anyone?

If you increase cost to buy "reliability," you are risking loss of the program from this overwhelmingly devastating killer—an inflated budget. Regardless of whether you put money into better parts, more engineering oversight, longer and more thorough testing, you will only know what the development program costs many months after launch, after the bills are paid and the audits completed. Budgets kill satellites even then—there are plenty of fully capable spacecraft in orbit with no money to staff their earth stations and operate them. Another budget gotcha. Keeping cost down isn't everything, but it's hard to say what's #2. (I'll say in a minute—I'm supposed to know some hard stuff.)

Assuming you do have money to spend on enhancing reliability, would you spend it all on better parts? Only if (a) you believe parts failures are the next biggest cause of satellite failures, which they aren't, and (b) somebody phrases the question the way I did. In fact we rarely think of a parts budget as a threat to reliability, but as it rises, something is being sacrificed, and usually that something is in fact the #2 (after budget) cause of mission failures. There are only two major consumers of a program budget: parts and engineering labor. And not surprisingly, engineering is the second biggest cause of satellite failures.

There. I've said it. With all our fancy design tools and program documentation and reviews, with all our Ph.D.s and M.Sc.s and consultants and red team reviews, engineering is the weak link in the reliability chain. Small satellites trace their extraordinary reliability to two facts—they have less parts to fail, and they have less engineering.

These chapters are about reliability—but they are really about putting program resources where it matters, and for virtually all modern microspace programs, that means beefing up engineering. I don't necessarily mean by doubling the number of minimum wage engineers on the program—it might mean doing what Hollywood does and buying truly excellent talent, providing that talent with an excellent work environment and rewarding great work. We focus on whether parts have heritage—but what about the staff's heritage? Sure 747s have flown before, but would you fly New York to Tokyo with a crew that had never been in a 747 or seen Narita before, comfortable that the parts bolted onto your 747 are highly reliable?

It's ironic that a book on microspace needs a major focus on the issue of reliability. My own records indicate that small satellites are more reliable than big ones. This is particularly significant since many of the big ones are geosynchronous comsats which are quite often identical to one another, greatly reducing engineering risk. Whereas historically most small satellites are one-of-a-kind development items. It's not that small satellites are unreliable that makes reliability so important. It's that reliability is achieved differently in microspace than in macrospace. Reliability is not a feature you can add any more than you can optimize your life around happiness or contentment. These are worthwhile goals, but they are achieved via what we do, the design/build/test process in the case of satellites, or what we do every day of our lives, in the case of happiness.

It's not just that reliability is achieved differently in microspace. It is often achieved in exactly the opposite way from the norms of big aerospace systems. Thus imposing inappropriate habits, whether or not they are valid in macrospace, hurts your chances for a success in microspace. Nobody will take your word for that. This section of the book provides you with some defensive munitions against well meaning supervisors, reviewers of the "That's not how we won the Big One" school and MIL-Spec slinging bureaucrats. Read it and...be careful out there.

Chapter 2

Fun with Parts

If you have one part with a reliability of 99%, the probability it will not fail (it will work) is 0.99 to the first power, namely 0.99—i.e., it's 99% reliable. If you have two parts, both of which must work for a system to function, and each has 99% reliability, the probability of the whole system working is $0.99 \times 0.99 = 0.99^2 = 0.98$. So you now have a 98% reliable system. If you need all of ten parts to work, the reliability drops, since any one of the ten parts could fail. The reliability is now

$$0.99 \times 0.99 \times 0.99 \times 0.99 \times 0.99 \times 0.99 \times 0.99 \times 0.99 \times 0.99 \times 0.99 = 0.99^{10} = 0.90$$

Thus you have a 90% reliable system.

The first day the weather went below freezing last fall, I jumped on my bicycle heading off to an early morning meeting, oblivious as usual, bopping away to the dulcet sounds of Smashing Pumpkins, and on the first turn I hit the icy pavement with a dull thud/crunch. The thud was me hitting the pavement. The crunch was, luckily, not my hip bone but my yellow plastic "Shock Wave" (built for jocks) Walkman which had come between said hip bone and the blacktop. I was merely, albeit painfully, bruised, but the personal stereo's injuries were fatal. As any self-respecting engineer would, I saved the carcass and that night took it apart, furtively hoping I could pull off a miracle and fix it; or, failing that, at least learn something about Walkman engineering. What I learned was there is a grand total of one active component in that Walkman. So if that component is 99.99% reliable, that's the reliability of the Walkman, more or less.

The reliability of ensembles of relatively large numbers of parts is shown graphically below. If you want an ensemble of parts to be 99% reliable you have two equally valid choices. Use a very small number of 99.99% reliable parts (less than a hundred). Or use a very large number of incredibly reliable, say 99.99999% reliable, and hence hugely expensive, parts. The only difference between the two approaches is that the first approach uses one hundred parts which might cost a dollar each—$100 in parts—while the second might use a hundred thousand parts, each of which cost $250, with a parts bill of $25 million.

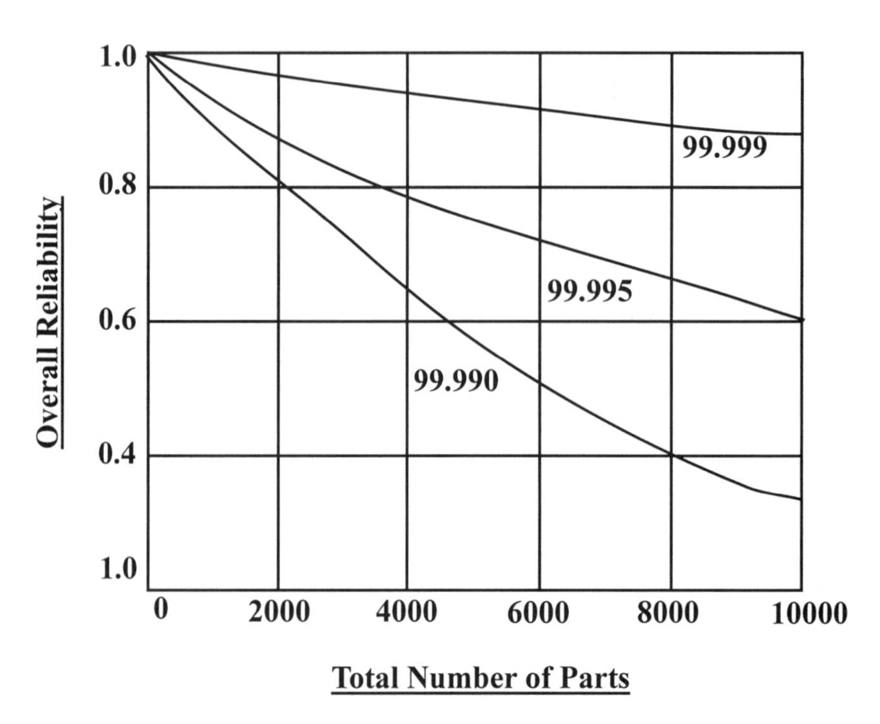

Myth: Small Satellites are less reliable than conventional satellites.

Reality: Experience shows us that they are more reliable, and simple analysis tells us why they should be.

While these two approaches may both yield the same systems reliabilities, the systems themselves are totally different. One is the size of a microwave oven, while the other is the size of a schoolbus. One is built by ten people in a year or two, the other by thousands of engineers over the span of a decade or two. One performs one or two tightly defined missions very well. The other services a large number of users and applications. The small one is built with parts you can buy in a few hours' shopping in Akihabira (Tokyo's electronics shopping district). The large one will require you to develop close and long-lasting relationships with custom parts vendors all over the world—relationships that will require you to build and pay for an organization to write specs for parts and another to procure them, and yet another to track them through the life of the development program.

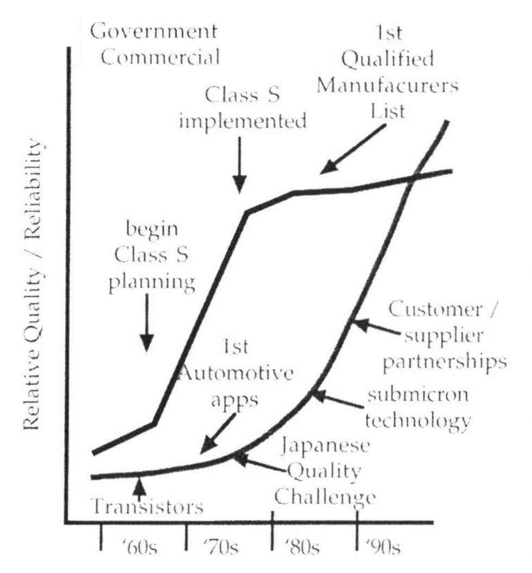

Parts reliability from the 1960s to the 1990s.

Remembering that we know about as much about component reliability as we do about tomorrow's Dow Jones trajectory, does it make sense to create and pay for S-Class parts? Are they more reliable? The chart tells part of the story. (And thanks to Jan King for his thoughts, upon which it is based.)

Way back when real men wore bell bottoms and long sideburns, and (some) business people wore leisure suits, what was the suite of parts we worked with? Discrete transistors and vacuum tubes for higher power applications, big stuff that could be built by hand, and often was. How reliable were the parts? In those days, a breakdown of your '65 Ford Fairlane every few thousand miles was normal or even healthy. After all, it strengthens the spirit to fix a leaking radiator on a frigid December evening somewhere on the Massachusetts Turnpike using a torn up T-shirt (which used to be, in unripped configuration, worn around your body to keep warm despite the chronically obstructed heater core) to plug it up so you could limp into the next service station. A flat tire every fifteen thousand miles or so was also normal. Cars had bias ply tires, and to get twenty-five thousand miles out of them, they and your stars had to be very carefully aligned.

2.1 Reliability: A Retrospective

What has changed in thirty-something years? Cars have a lot more parts: antilock brakes, air bags, myriad complex pollution controls, four-speed automatic transmissions—often connected with some sort of four wheel drive, CD players, air conditioners, rear window defoggers, fold-down rear seats, emergency flashers. My old Ford didn't even have an outside rear view mirror. Now cars have them on both sides, with electric servos to line 'em up just right. All those extra parts.

But contrary to the law of large numbers, that says the more parts, the less reliable the system, new cars break a lot less than the old ones did. Today, my wife buys a car and expects it to go fifty thousand miles, and never, ever have a single hiccup. And by and large, it performs in precisely that way. Why? How did cars get something like ten or twenty times more reliable despite their ever rising complexity? Since this isn't a book about cars, maybe I should ask what *didn't* they do to make cars more reliable,

Solid gold frisbees. 100% reliable. 100% useless.

since that's equivalent to asking what *did* the aerospace industry does to enhance reliability.

What the aerospace world did reflected the politics of aerospace in the late '60s and early '70s. We weren't overly bent out of shape about cost back then. We were busy beating the Soviets to the moon. Never mind that McDonalds and MTV might ultimately play a larger role in nulling the Soviet Threat than the moon ever did. We believed in what we were doing. Nobody thought about cheap rockets, cheap space, or space tourism much. Our role in space was to beat one enemy and show others what a threat we were, building technology and exploring human destiny. Those were good enough reasons to pry billions of dollars from US taxpayers. And, excepting geosynchronous spacecraft, no revenue source was going to be even a close second to our crewed and uncrewed space adventures.

Reliability was a big issue—particularly when you're launching humans to the moon and it's all on live TV for the world to see. So we bought reliability at any price: the birth of the S (space) classified part. Skilled technical people and engineers knew that by painstaking hand assembly, by inspecting and testing components one transistor at a time, and by tracing the materials from which stock structures and components were made all the way from the foundry to the final spacecraft part, by keeping records of every process, every step, every discrepancy, and every instance of possible mishandling, we could buy huge reliability gains. And after this, there was more. Technicians severely de-rated every part. A one-watt resistor was never used for more than a 0.1 watt application. As microprocessors became available, one rated at 20 MHz would be run at 4 MHz. Assembly standards and inspection procedures were written and compliance was mandatory. Certain types of construction, including the now prevalent encasement of microchips in plastic, were banned for space because they could not be opened and inspected.

A typical automobile has about a hundred thousand parts. Imagine if even one labor hour were spent to track each part, including keeping paperwork on all the tests and environments that part ever saw, which factory it came from, which shift built it, and what QA person ran the inspection. If only an hour were spent on each part, by a person earning just $24,000 dollars per year with a modest benefits package, just the parts for your car and the cost of quality assurance for the parts in your car, would run $2M. Needless to say, the automotive industry didn't go that way.

But satellites did. Even if your average satellite has a million parts, and even if we did add a one hour per part overhead for quality assurance, that's only $20M. A lot of money, but not for a billion-dollar mission to the moon. In fact, $20M is a bargain, since it ensures the reliability of a billion dollar mission. It's like a mere 2% tax. S-classification in that context makes sense. It caught on big time, and, at least initially, it appeared to function well. A lot of big, complex missions flew, and they mostly worked. Pioneer, Apollo, Voyager. You don't argue with success like that. Well, maybe you don't. I do.

At about the same time, the Japanese began to sense opportunities in the US car market, serviced as it was by three big players, all of whom were resting on the technological accomplishments made in their youth while spending their middle age rearranging hood ornaments and fins. The Japanese attack was in part based on quality—the technology to preclude failures. People bought Japanese cars in large part because they discovered that, unlike their US counterparts of that day, Japanese vehicles didn't rattle, their parts didn't fall off, they always started, and they didn't go out of style in twelve months. And Japanese quality was not achieved through massive applications of paperwork production and rigid assembly methods which tend to be outdated by the time they're written and approved.

Now quality is a complex question, and I'm not going to trivialize the work of thousands, maybe millions of engineers dedicated to defining the many steps one can take to stop things from breaking. But here are a few general concepts.

The Japanese benefited from a major change that swept across the electronics industry—large scale production of highly integrated parts. Enormous investments were made to automate the production process and stamp out literally millions of microprocessors, logic arrays, and other highly integrated devices—entire RF transceivers, for example. Integration, and automated assembly processes actually cut parts counts by accommodating thousands of discrete parts into a single chip, and yielded more reliable parts because their production process was not subject to human-induced variations. Once such a process was established, at least in theory, millions of identical, virtually flawless parts could be produced. And if they weren't flawless, robotic inspectors would weed out the variants.

Entire circuit boards are made today by mechanized processes, from photolithography of the traces, to picking up the minute components and placing them on the board, and then soldering the leads and final cleanup. The completed board is visually inspected by a camera and image processing computer, then tested automatically. Whereas it may take an engineer several months to hand-assemble a similar circuit board, inspect, test, and turn it on, a production run with total automation can turn out flawless boards by the hundreds per hour.

The spacecraft world didn't adapt to this algorithm for several reasons. First of all, there was no paradigm for it. We knew hand inspection was good, but robotic assembly and testing were unproven. Furthermore, when a failure is going to mean a phone call from the President of the United States, you want to go with what you know will work. And of course we don't build millions of spacecraft per month, so many of the benefits enjoyed by the manufacturers of cars and TVs are just not applicable.

2.2 The Road Not Taken

The two technologies went their own separate ways. Aerospace bought reliability with parts costs, and the mass markets achieved reliability via automation, process control, and reliance on the most sophisticated, highly integrated parts available. Unfortunately, aerospace parts became a sort of a burl on the evolutionary tree of technology. Intel's board of directors does not labor late into the night figuring out how to bring out the next generation processor for NASA satellites. The huge market just for those cute processors you get in musical greeting cards makes NASA look like a pretty dismissible segment of the market. Aerospace standards didn't allow manufacturers to work in the ways they did for consumer products. No plastic encapsulation, for instance, was allowed. For many years as surface-mount revolutionized miniature electronics, surface-mount was not allowable in aerospace as standards, originally devised with big, clunky, discrete components in mind, required each part be mechanically bound to its circuit board by more than just solder. The innovation of surface-mount—that if the parts are light enough even "just solder" is overkill mechanically—was ignored by aerospace technology, which hung on to through-holes and pins.

There are also two views of parts reliability. Why is it, for example, that experience in constructing low-cost space devices tells us that S-class parts are less reliable than cheap consumer parts? Aerospace parts, firstly, are now made by secondary manufacturers who specialize in the arcane practices this kind of manufacture dictates—practices long ago abandoned for more cost-effective ones by commercial, nongovernment manufacturers. The quality of hand-produced yields of a handful of parts can't begin to rival the near-perfection of the highest quality commercial grade parts. Plus, it takes years to qualify a part as S-class. During those years, the basic component technology in the commercial sector has already moved on and improved. If you need a fast processor, do you lash together a boatload of V-40 processors (now approved for space), each wrapped in a heavy ceramic casing for ease of inspection, thus packing up a huge electronics box with boards, connectors, and mechanical supports to keep it all together? Or do you just buy an off-the-shelf G3 Power PC processor mounted on a board slightly larger in size than a playing card? Even if the individual V-40 parts were more reliable than the PowerPC chips—which they won't be—by the time you put hundreds of them together on all their discrete circuits, write the software to make them work as a parallel processor, and package the whole thing into a large, heavy, power-hogging box of gear, your reliability is lousy compared to that one Power PC chip, a few outboard devices, and software that is being used by millions of people every day: the ultimate testing environment.

There is a whole world of technological dead-ends out there. Automotive steam engines, for example, used to be competitive with internal combustion engines back in the Stanley Steamer days. But with the hundreds of billions of dollars we've now invested in IC engines, steam would require a major external upheaval in the market in order for it to make a comeback. Static RAM chips used to be the memory of choice for satellites, and before that, the magnetic bubble was popular. But the mainstream of computer users doesn't care much for expensive chips that keep their memory when power goes down. They just reboot off the hard disc when necessary, or stay plugged in all the time, or use a lithium-ion battery. And as more and more money was poured into improvements to the dynamic RAM chips used by virtually every PC in the world, SRAM and bubble just couldn't keep up. The densest SRAM chips held 512 kbits of data—and cost a few hundred dollars each. Today's densest DRAM hold 100 times more data, costs 0.1% as much per bit, and draw so little power to read and write, and so little idle power, that overall it draws considerably less power than the SRAM. The DRAM is also about 10 times faster and the chip yields (reliability) are roughly a thousand times better. SRAM is essentially dead. Nobody improves 8mm film movie cameras any more. Not much money is invested in improving 8-track tape players, or canvas for airplane wings, or diamond needles for playing LP records.

2.3 The Fable of the Evil Twin

My wife's a twin. Luckily, she's not evil, so far as I know, and neither apparently is Jean, the person sharing her DNA. But let's say Nancy and Jean each set out to build a satellite. (Since Jean is taller, darker and more inscrutable, we'll flip over all the cards and make her an evil twin for purposes of this fable.)

Jean is a bit more conventional than Nancy, so she wants to be absolutely certain that the satellite we're building actually works. Nothing sinister about that. "Reliability is paramount," Jean says. "After all, I don't care how fancy it is, if it fails on orbit, none of that cool stuff is going to matter. We've got the security of the free world to attend to! No messing around! " (I guess she's building spy satellites. Jean's always telling me she's glad I'm out there protecting our country building satellites for DoD, and I haven't yet disabused her of that notion.) Nobody argues with any of that, so Jean specs all S-class parts, all Mil-Spec assembly procedures, comprehensive, cradle-to-grave parts and materials tracking, and so on.

Nancy, who wears a different color contact lens every week just to make sure everyone notices her eyes, drives an old Mercedes except when she's on a bicycle, and has never in her life owned a hard shell suitcase, figures that a satellite is sort of a space-bound laptop, and uses the utilitarian Epson technological approach.

Each sister's engineering team goes to work. Nancy's team has access to all the very latest parts, vendors are on virtually every street corner, price competition and performance competition are rampant. Nancy's people buy COTS (commercial off-the-shelf) processors, radios, rechargeable batteries, and a COTS operating system. Custom components like the antennas and power controller are built up from commercial parts. The whole job has minimal parts count, minimal cost, and no flight heritage at all. Will it work? Nobody knows, so they test it extensively over a range of environments. The engineers scrutinize the materials and manufacturing processes used to make it to figure out if there's any risk of serious outgassing or radiation problems, and out the door it goes. If it does work on orbit, it will work great considering its small size and low cost. If it fails to work, due to lack of traceability of the parts—Nancy's team isn't into paperwork—we might never learn why.

Jean's team, meanwhile, is still on the phone. There are only a few vendors of S-Class parts, and most of the parts have six- to twelve-month lead times. Also, because they are manufactured and tested to order, usually you have to buy an entire batch. Typically, there's a fifty part minimum with a batch cost of $10,000. Plus, none of the parts these vendors carry are in any of the data books from this year or last year or even the year before that. They're all from ten years ago. But nobody's got the software tools to program those things, and none of Jean's young staff has a clue where to get tools or how to use them if they did find them. But the design process plods ahead. Jean's satellite needs about ten times as many parts—older parts don't do as much as their newer counterparts. There are no programmable gate arrays, so the logic is all executed in discrete devices—lots of 'em. No lithium-ion rechargeable batteries for Jean. She gets a single $400,000 NiCad battery stack. No, better make it two. And there's no dense DRAM in S-class, so Jean goes with an $800,000 S-class tape recorder. No, make it two of those. Can't gamble the free world on one thin strip of mylar tape.

The parts orders finally go out, but there's plenty to do during the year's wait for the parts to be custom manufactured, tested in batches and sent to Jean along with thick stacks of paper tracing each part back to the foundry where its materials were made. Because those tape recorders and NiCad batteries are big and heavy, so are the boxes they use to package up all the circuit boards they need to build the satellite out of ten year old technologies. Plus, they all need so much power that Jean's going to need to deploy a couple of large solar panels that track the sun continuously. The MEs (mechanical engineers) go to work, designing structures, designing deployment and tracking mechanisms, figuring out how to heat-sink the massive power-conditioning equipment, and creating complex finite element thermal and mechanical models to predict if the whole complex will fall apart when they try to launch it, or whether it will overheat in vacuum.

Finally the Jean Team starts construction. There's a lot of new team members, since for every engineer there's either an inspector or QA/QC engineer making sure all the procedures and standards are adhered to, and monitoring for anomalies, each of which is documented and analyzed to see if it could recur on orbit, and if it could, how to prevent or cope with it. To manage the larger team, more managers are brought in, and the team is divided into numerous subteams, each working to its own interface control document, each written by a specialist on ICDs.

Well, at least with all those reliable parts, testing is not an issue. But the engineers did the FMECA (Failure Modes Effects and Criticality Analysis) and guess what? All the additional parts made the system fairly unreliable. Hence you need the tests. Plus the tests are required by the Mil-Specs, and anyway, you need the tests to verify the finite element models Jean's Team invested so heavily in, and for good reason.

Jean's program took a while to get to that testing phase. What with the cost of those parts, only two flight versions of each board were built, plus two protoflight versions. When time came to start turning the system on, there was only one system. And twenty engineers, nineteen of whom at any given moment are waiting in line for their chance to work on the system. Nancy's crew, using mass-produced parts and fabrication techniques, built ten or twelve of everything, and had one complete system for every engineer on the program. Plus one batch of Jean's specially made parts were destroyed by an accident at the factory. Well, not really an accident. Turns out the hepafilter in the clean room hadn't been replaced per the approved schedule and was beyond its service lifetime when the custom packaged ICs were built. Their QA manager made this discovery and dutifully wrote it up. All the parts made there for the past three months were recalled and destroyed, including Jean's. Insert another six-month wait.

Which was really no big deal, since the software development was bogged due to the lack of tools and documentation on how to fire up the ancient hardware available in S-class versions. Not that there aren't experts on that old stuff. But they exist only in the minute aerospace eco-niches. The mainstream of software expertise has moved on.

A couple years later, a few tens of millions of dollars later, Jean's Team finally got to the launch site. Nancy's machine, built with cutting edge technology, miniaturized down to 50 kg had been piggybacked into orbit on Ariane years ago. Jean's machine, with the redundant S-class batteries and tape recorders, and the attitude control system beefy enough to move things (and not move things) with all that mass on board, plus the mass of all those discrete components and the boards, boxes and harnesses necessary to house them, and the spacecraft structure necessary to hold all that stuff together, was half a Delta launch. The other half of the Delta payload, unfortunately, had a funding glitch which itself was precipitated by a critical performance test failure on part of the payload. Another year's delay. Nancy's satellite was designed, built and launched in two years, and the mission was done for a few million dollars. Jean's was designed built and not launched in 5 years, and the cost was tens of millions of dollars.

While the credits roll, which one is more reliable? That's the point, according to Jean. Nancy is risking a lot of nonspace qualified parts, but they were tested, and do parts really know if they're in space or not? Mostly not. Carl Meinzer, head of Amsat in Germany advises satellite builders to "know your parts." If you know what you're flying, he says, you'll know whether anything you've selected might cause a glitch—a seal that could leak or a problem surviving small doses of radiation. Nancy's small, highly interactive team understood all the spacecraft environments, and selected and tested their parts to ensure they would survive a long time.

Jean's risk is not the parts. The parts cost tens of millions of dollars to ensure that they can't fail. Unfortunately, in the process of virtually eliminating parts failure risks, all the other odds stacked up against Jean. She has a system with a bloated parts count, and statistics tell us that lowers reliability. The satellite is physically complex. It has deployables, two active steerable arrays, a lot more power to switch and to dissipate, requiring an active thermal control system, a lot more current to possibly form a current loop and bias the attitude control system. The redundancies she built into the system require arbitration, and what arbitrates among the arbiters? The few engineers, skilled in the arcane field of designing circuits and software for components of a bygone era, had to build a much larger software and hardware package and attempt to test it all.

So the question isn't "Will any parts fail?" The question(s) is (are), how can we possibly understand this entire system and make sure all of the interfaces play, all the deployables deploy, all the software runs without glitches and without unplanned states which can destroy the mission? How confident is Jean in the team that, over the six or seven development years, has almost none of the original members still present? And what about the quality of software and hardware developers content not to learn about the technology developments of the last ten years, and to live in the technical backwater of S-class parts? And does a human technician soldering parts by hand on a bad day make more or less reliable parts than a billion dollars worth of automated fab and test equipment that is producing a million parts per day with virtually perfect yields? S-Class parts are severely de-rated. Are things only working because, though they are out of spec (poor design), they just barely get by due to the overcapacity of the part? Eventually those walking wounded will fail.

These are the big questions. And they are fundamentally unknowable. What is known is that parts selection standards drive the design process, drive the design team, drive the team size, drive the schedule, drive the management architecture and design process, drive the architecture, drive the software development effort, drive the budget, and even drive the selection of the launch vehicle. As I said a while ago, when faccd with concepts we can't really understand, like random parts failures or solar eclipses a couple thousand years ago in what is now Europe, we deal with it through sort of an occult belief—in the present case a belief that using hand made, vintage, expensive parts makes a reliable system.

Both satellites, assuming they don't run out of money, eventually get launched. Jean's satellite, made with such supreme care, and Nancy's, built with the latest of everything, are each launched on launch vehicles far less reliable than even the least reliable spacecraft. They are deployed from the rocket with mechanisms less reliable than the least reliable rocket. Jean has taken on the lesser of one risk—the risk of part failure. But she's also taken on more of the more insidious and harder-to-predict failure opportunities. Yet neither twin can control the most unreliable part of their respective missions: the launch and separation from the launch vehicle. Jean has spent tens of millions of dollars strengthening the strongest link in the chain. But that's another subject. So let's drive on to the next chapter....

Chapter 3

The Logic of Auto Parts

Not that you're superficial or overly goal-oriented, but you might have totally ignored Joplin, Missouri. It's one freeway exit on the route from St. Louis to Albuquerque, New Mexico. As you walk down that exit ramp, your rental truck lazing a mile behind you in the sticky midwestern summer sun, you wonder—just why did I volunteer to drive that satellite to Phillips Lab? Oh yeah, the romance of seeing Heartland America from end to end. That was it, wasn't it? Next time, you conclude, you'll leave exploring the soul of our country to college kids and to the likes of Forrest Gump. People with time on their hands and an appetite for discomfort. Two assets you've lacked for twenty-five years, if you ever had them.

Luck is with you. Sort of. I mean, if luck had really been with you, that fuel pump in the rental truck would've hung in there another minute or two. If luck were really with you, you would have talked yourself out of this adventure before you started. The best luck of all would have been to draw the winning Super Lotto ticket, enabling your comfortable early retirement to Cortina with a million tax free dollars a year for life to support your penchant for skiing the South Face of the Alps. In this particular instance, luck has amounted to the stunning vista of a car parts store just down the embankment from this off ramp.

After a brief, semi-controlled slide down the dried grass and mud that such embankments tend to harbor, and a quick jog across the marginal road, dodging a pickup and a kid on a big motorcycle, you walk in and find a native Joplinian handling the counter and the register. "Help ye?"

"I think I need a fuel pump for a GMC truck."

"Rental crap out on ya?"

Perceptive guy. You nod, casting your gaze in the direction of the freeway.

"All them fuel pumps is about the same fer those trucks. Lessee now..."

And Les, as his shirt says he is named, starts rummaging around a cardboard box which has been slit in half at its equator to make a parts bin. Les snorts a monosyllable not found in any dictionary, but which definitely does not signal good fortune whether you are in your doctor's office or in an auto parts store in Joplin, MO.

"Hmmmmmm."

Les' face is serious. An auto parts guy without the part. Then he brightens just a bit.

"Lemme check out back."

An idyllic setting for your lucky day.

And with that he leaves you standing in the storefront. You hear him in back, rustling through several similarly-truncated cardboard boxes. Bored and tired, you seek entertainment by perusing the dusty racks of self-serve stuff. Check this out. You can buy a 60 amp-hr, 12-volt battery for $29.95. Let's see, that's 720 W-hrs. About 100 W-hr at 15% depth of discharge for about the cost of Xeroxing the viewgraphs describing the battery on the satellite in the back of the stalled truck. The satellite battery has a single Mil-Spec battery stack with a bit less capacity, but it cost about $300,000. A whopping ten thousand times more. Yeh, but the battery in the satellite is guaranteed for ten years. But heck, the $30 battery is guaranteed for seven years, and they'll prorate your new battery if you need one. But of course, the car battery has liquid acid in it. You'd never flight qualify it, even though you could possibly fly one of those gelled electrolyte batteries they sell for off road motorcycles and skidoos.

Your train of thought is derailed by Les' sudden reappearance.

"This's yer lucky day, fella."

Funny he should say that.

"I found just whatcha need. It's a might dirty, but it works. Ten bucks and it's yours."

You think of the thousand plus miles yet to go on this misconceived journey and the hours of delay and hard work you got stuck with thanks to this ten-dollar part.

"You don't have a little better one? I don't want it to break again."

"You work fer NASA or somethin?" Les quips back. "Ain't too many customers 'round here beggin' to spend more money. We're not buildin' rockets here, ya know?"

What choice do you really have at this point? You add a few miscellaneous "just in case" parts—some gaskets and sealant, some degreaser/cleaner (Les is by now sure you're a bit too fastidious for his tastes), and two hose clamps. For under $20, you're back on the road, or on foot, at least.

Another hour or so later, the truck starts up. You stuff the old fuel pump and the receipts in the glove box, figuring you'll make a big deal out of it when you return the truck. And, except for the fact that you've lost a couple hours, you're out twenty bucks, and you're a ball of sweat and grease, it's been a non-event.

Do I work for NASA, you think, mulling over Les' jibe. What a joke. We build the cheap tiny satellites NASA doesn't even care about. We scrounge for parts. We keep the project small and self-managing, and we don't normally go Mil-Spec. The batteries in the bird in the back were customer-mandated. Normally we just buy commercial grade NiCad C-cells—$6 each.

Here you were on a freeway off-ramp in Joplin, Missouri, and what's the one fact the shopkeeper knows about NASA? That they spend a lot of money on parts. You realize that's one thing even a kid knows about aerospace gear: very expensive parts. $600 hammers. $900 toilet seats. But what about those $400 memory modules? Commercial ones are $20, if you can even buy that outmoded 512k stuff anyplace anymore. A Rad-Hard, Mil-Spec 10 MIP processor is over $1M. But you can buy a 500 MHz Pentium with a keyboard, monitor, power supply, and a gigabyte hard drive for less than 0.1% of that price.

Maybe it's the lack of air conditioning, maybe your allergies are causing your brain to swell against the inside of your skull, maybe you need a fresh supply of Power Aide or some sleep, but you're suddenly bothered by a flood of parts questions, the nagging, niggling questions you usually wake up with but quickly drown out with a flood of hot water, punk music, and telephone calls to marketing leads. Today they won't go away.

3.1 Dueling Brains

Your Left Brain asks:

Where do failures come from anyway? Are parts what fail on satellites? How important are parts costs? I would rather spend $20 on a fuel pump and not have to fix it again.

Isn't most of a satellite's cost in the labor needed to design, build, integrate, and test it? So shouldn't we buy the best possible parts? Why not spec S-class "just to be safe?" How about using cheap terrestrial parts? They're not just cheap: they're made in huge quantities and they're pretty reliable. Aren't there people using nonspace parts successfully? But haven't there been some disasters that way too?

But your right brain knows what's really on your mind:

Was Les right? Am I really just a fussbudget?

How much longer 'till dinner?

Anything interesting on the radio around here?

I bet—no, I know—my kids think I'm dorky. Well, one day they'll thank me. Well, I'm pretty sure at least Howard will thank me... I don't know about those other two...

Leaving the right brain's chaos to your AAA Triptik and your analyst, Franz, who gets paid for that sort of thing, let's see what answers we've got for the analytical you, who's just trying to build cheap, reliable microsatellites, ego and self doubt more or less notwithstanding.

3.2 Where Do Failures Come From?

It's unfair. The world doesn't respect engineers. Want proof? Just how important to your day-to-day existence is the Heisenberg uncertainty principle, $E=MC^2$, or the charge on the electron? The latter is just an arbitrary number. I've never gotten my bike up to anything near C (it's not even on the speedometer). And not simultaneously knowing the momentum and position of any given subatomic particle has never adequately explained why I can't find my Oakleys, or prevented the police from discerning it was me doing 100+ mph, testing my tune up job on the old triple SU dual down drafts. This irrelevance to daily life has not stopped the likes of Milikin or Einstein or Heisenberg from winning tenured positions at Caltech, getting extensive treatment in serious textbooks, or even bagging the occasional Nobel Prize.

By contrast, engineering is not just important: it's life or death. Do you want to try getting by without a house to live in, food to eat, or at least a telephone to make reservations? Nobody got much help out of Millikin, Einstein, or Heisenberg in producing any of those necessities. Or in printing books or *People* magazine for that matter. We needed engineers.

Getting back to the question at hand, let's explore why things break, and whether good parts are a solution to this problem. Let's talk about a, if not the, guiding principle of applied engineering: Murphy's Law. Succinctly, Murphy said anything that can go wrong, will go wrong, and in the worst possible way and at the least convenient possible time. Like your fuel pump giving out. Like your car deciding not to start when you've finally landed at an airport seventy miles from your house, on a dark, cold, rainy night, where the last shuttle just dropped you off in a distant long term parking lot. Or like the day of that big job interview.

Cars don't refuse to start when there are lots of alternative cars in the garage or driveway, when you're just going for a Sunday drive for the hell of it, or when you've got all the time in the world. Murphy discovered the principle that affects every one of us every day of our lives. Did Murphy pick up a cool $100k in Stockholm? Does his name appear in pop science books, in college text books, or even on a plaque outside an engineering building at Stanford? Regarding the contempt society has for engineering compared with the exaltation of Physics, I rest my case.

Anyway, who are we mere mortals to fathom why things break, when even one of the century's engineering geniuses, Mr. Murphy, could only state the most general principle while postulating that we will never be able to predict where and when failures will strike? Without trying to establish general principles, we do know about the Amsat experience in building ham radio satellites over the last thirty-five years.

3.3 Poor Design

The number one observed reason for satellite failures, the subject toward which we should aim most of our attention, is poor design, not poor parts. Satellites fail because somewhere within their thousands of lines of code, there is a software state that allows the satellite to commit suicide. The Air Force's Clementine satellite managed to accept a command to turn on a thruster without a command to turn the thruster off. It spun itself up faster and faster until all of its propellant, needed to take it to its asteroid target, was exhausted. Or how about the French amateur satellite that, having erroneously concluded that radio #1 was broken, switched to the backup radio, which actually *was* broken, and could only switch back to radio #1 via a command received from radio #2?

Parts sometimes fail not because they are faulty, but because we failed to protect them from startup surges, or from excessive cycling of relays while dithering between two almost equal states, or from interference from other parts, like radio transmitter energy destroying a sensitive receiver front end. The first Swedish Maxus rocket failed because a subcontractor-supplied thrust vector control system was not adequately insulated from the rocket exhaust, and it burned up. Remember hearing about a major rocket failure that cost insurers over $300M? It was traced to a missing comma in the guidance system's C-code. Mars Observer exploded due to overpressure in the hydrazine propellant delivery system, a design error that had already been observed on the ground in tests.

None of these are failures we can prevent with better parts. They are caused by faulty design. The space business is (in)famous for the dollars it devotes to the exquisite perfection of parts production and screening. But are we famous for the dollars we devote to paying for the best engineering talent money can buy? No. Hollywood is famous for spending huge sums to get the best talent. So is major league baseball. When Boris Yeltsin needs a quadruple bypass, he doesn't go to the lowest bidder. On the contrary, aerospace is notable for the government's relentless selection of the contractors paying the lowest salaries and providing the least expensive and least capable tools that these low paid engineers need to do their jobs. Yes, your satellite is built by the lowest bidder. Thank goodness neither my car's brakes nor my uncle's pacemaker can make that claim! They have to work.

3.4 Misjudgment of the Environment

The number two killer of satellites and rockets is also not parts failure. It is misjudgment of the environment in which they operate. Our ALEXIS satellite almost succumbed to that fate. Manifested on an experimental Pegasus flight, the vibration environment was as yet unmeasured, only estimated from models and ground tests. We discovered about two thirds of the way up to orbit that the coupled structure of ALEXIS and Pegasus had a vibration mode resonant with the first vibration mode of one of the solar panels. It took six weeks of nailbiting and hard work by a few genius engineers (none of whom were even nominated for the Nobel or any other prize) to recover the satellite, which then went on to exceed all requirements and mission expectations and to live and function for over ten times its design-lifetime on orbit.

One reason the LEO communications satellites have taken so very long to get going (at this writing, the clock is still ticking) is the very noisy radio environments they have to overcome at their allocated radio frequencies. Environment is more than just vibration and temperature. Satellite programs run late and overbudget in part because people exist in the satellite's environment, and occasionally drop heavy instruments on the device's delicate parts. Or the roof leaks onto them. Or nobody built a diode protection into the power supply for when a tired, harried, and underpaid engineer puts the plus wire on the minus, and vice versa. That's an environment—the environment of the real world where people, no matter how many procedures we write or checklists we provide, make mistakes.

The modern jet aircraft is a model of design for its environment. It can handle fog, rain, snow, ice, and darkness. It can avoid thunderstorms and other airplanes, those solid blocks of aluminum dotting the otherwise low density atmosphere they prefer to travel in. But aircraft are also designed to resist errors by pilots. Pilots have years of training and they're highly motivated. If they make a mistake, they very well might die. (Such is thankfully not the case for spacecraft engineers!) Nonetheless, pilots do make mistakes, and their airplanes complain, with stall warning indicators, low altitude warnings (including a synthesized human voice urging "pull up, pull up!"), and myriad warning lights and stick shakers and fastidious copilots and flight engineers. All there because human screwups are part of the airplane's environment. How many features have you seen on a satellite designed to protect it from the human environment? Maybe we cover the solar panels to protect their thin glass from downwardly accelerating screwdrivers. Other than that, the assumed environment is faultless engineers and technicians.

3.5 Human Error

Is the third most frequently observed cause of satellite and rocket failures parts? Unfortunately for the value of space grade parts and the peace of mind they might impart, the answer is "No." Human error, already mentioned as an "environmental" factor, is clearly next. A whole Ariane IV crashes over the Caribbean, bringing three or four hundred million dollars in shiny, fully documented aerospace grade components down to a watery orbit. A technician had left a rag in a fuel line. Stuff happens, right? Ultimately, humans built the faulty O-ring that downed Challenger, and humans made the decision to launch the ice-laden rocket. Much less dramatically, humans send faulty commands to working satellites and sometimes turn them off or otherwise render them useless.

What do we conclude from the big three reasons for satellite and rocket failures? Something like what the NRA has been trying to tell the American public for decades: rockets don't kill satellites, satellites don't kill satellites, ground stations don't kill satellites, parts don't kill satellites. People kill satellites. In fact, satellites are usually killed by the people who know and love them best. There ought to be a society for the prevention of human cruelty to satellites. Let's take a look at this.

A few of our social habits say it all. We don't eat food from farms—we eat food from factories. Health food stores are stocked with chemicals. We work long hours behind a desk to afford a car and a health club membership, so that when we finally do get off work, we can drive to the health club, fight over close parking spaces, ride the escalator to the check-in desk, and then plug away at a stair stepper machine. We teach kids to be good citizens by a) packing them like lemmings into boring little classrooms, and then b) setting them in front of Terminator II and Bay Watch, while we c) work long hours to afford to buy them $130 sneakers and pay for day

care and after care. We have extended our success covering snafus with euphemistic language to the new realm of euphemistic behavior—acting like we are solving a problem as a substitute for actually dealing with the problem.

We do things that don't necessarily address problems. The cure for overweight isn't a bigger sweater or a redefinition of the government weight charts.The way to reduce traffic congestion isn't always to build more roads and parking lots (over half the land area of LA is now devoted to those two applications). The best way to commercially exploit space and to enhance space technology is possibly not to build a space station worth more than some Midwestern states, and the cure for failing satellites is not mindlessly upgrading their parts, since it's rarely the parts that fail.

In polite aerospace company, just in case you find yourself at an AIAA conference or a big meeting at NASA headquarters, people don't talk about mistakes killing astronauts. Challenger is rarely, if ever, used as an example of how things can break. But you want to know something? Engineers kill people every day. Not on purpose usually. But engineers build cars that kill their occupants in certain kinds of crashes, as well as the airbags and seatbelts that protect those occupants. Engineers build bridges that sometimes collapse. They build bicycles that kids fall off of, sometimes onto a curb with an unhelmeted head. They build airplanes that sometimes fall to earth—my dearest flight instructor, good old Al, was lost just that way—by a poorly engineered propeller controller. Engineers build artificial heart valves that sometimes fail, high tension wires that sometimes electrocute people, air conditioning systems that can breed Legionnaires' Disease, space heaters that sometimes ignite the rooms they are heating, and all kinds of purposely lethal stuff—bombs, missiles, neutral particle beams. You name it—engineers somewhere made it.

Lawyers only rarely make a life or death difference. Accountants almost never do. Ditto a lot of MDs. I'd rather go to a poor ophthalmologist than get in a car with an antilock brake system that goes haywire on the San Diego Freeway at rush hour. The fact is, what we do, what engineers do, is not just pretty important—it's often life or death.

For all that gravity, for all the importance of reliability—what colleges have majors in reliability and robustness engineering? What very expensive engineering software tools are used to find faults, to estimate reliability, and to rationally plan comprehensive and affordable test programs? Do the top engineering salaries go to reliability people?

Most programs don't have "Reliability Assurance" in them at all. Rather they have "Quality Assurance" and "Quality Control." I have several very low quality gadgets that are more reliable than most satellites—they are engineered to work despite the low quality their low price necessitates. Who really cares about quality? Maybe if you are buying rich Corinthian leather, but in a satellite what matters is reliability, not necessarily quality. Our singular focus on quality misses the point.

How reliable are our reliability predictions? They are rather unreliable. The Shuttle was supposedly capable of twenty-thousand flights without a fatal accident. Until the twenty-sixth flight failed catastrophically, whereupon the reliability was lowered. Not by a 1% or 10% margin, but by a factor of four hundred.

Is this just too hard a problem, or stated another way, what can the lone program manager do about reliability, given the lack of available expertise and understanding? The lesson of the TCP-IP protocol that has enabled the Internet's great rise to prominence is one model.TCP-IP had a singular focus in its development: it had to be bulletproof. This became an obsessive focus of its engineering team. They were willing to sacrifice all of our holiest grails—performance for instance. There was little question: if performance and reliability had to be traded, performance was sacrificed. Complexity. We labor under the misconception of KISS (Keep It Simple and Salient, or the more common Keep It Simple, Stupid)—that that which is simple is reliable. If you think about today's automobile vs. one of thirty years ago, modern reliability is much higher in virtually any measure. So is complexity. For instance, occupant fatalities, a form of "unreliability" in the vehicle's ability to protect occupants, which is in fact what the car body is there to do, have been cut dramatically via increased complexity. We have seatbelts, airbags, crush zones, specially placed structural blocks to prevent the passenger's body going under the dashboard, improved (and more complex) glass and interior materials, breakaway knobs, collapsible steering columns—a lot more complicated than rigid steering columns. We also have complex interlocks to keep us from running ourselves over while, say, putting down the garage door—also adding complexity but increasing reliability.

Focus on reliability means that requirements can be sacrificed to increase reliability. It means that more engineering dollars might be needed to get the best talent and make it productive, as well as to handle the additional complexity reliability often requires. Until a practical formalism is developed, which could well not happen in our lifetime, the program manager will be force fed irrelevant quality solutions—like QA inspectors, program reviews, and parts quality specifications. I can offer only the following table to help decide which measures you might want to take on, and which ones to ignore. Good luck and remember, if it fails, you can't claim innocence by virtue of following Rick Fleeter's advice regardless of its grounding in real-world experience. So if coverage of one's rear end is paramount, then I suggest the Mil-Specs. They don't address microspace reliability too well, but they do provide a handy alibi.

Piece part reliability.It is easy to see how it could kill a satellite, but few, if any, modern missions, regardless of parts quality used, fail because of it. Thirty years ago, when our industry learned its obsession with parts, this wasn't the case. But thirty years ago, not only were parts highly unreliable (driven a thirty-year old car lately?), but we had a large number of parts, since highly integrated parts (ICs) were many years away from space application. Remember that highly reliable parts not only sap money, but add complexity because they tend to be older and less highly integrated.

Assembly, like parts quality, is amenable to standards and procedures. Like parts, we can't ignore it and solder the satellite haphazardly. But the standards are there, well known, widely practiced and easy to implement. Unlike parts, going with accepted standards does not preclude best engineering practices.

Human errors are hard to prevent. But rushing, understaffing, emphasis on cost and schedule instead of careful work done methodically, and all the bad things we do when we are overbudget due to spending too much on parts, and late due to the long lead times of highly reliable parts, all contribute to human errors. Note that there are no MIL-Specs on preventing them.

Misjudging the environment is itself misunderstood. Satellites can be hurt by testing over too wide a temperature range. We misjudge the space environment, and weaken some links in our chain through overtesting. Humans are the most dangerous element in the satellite's environment. Software in the ground station and satellite must be built to recognize dangerous command combinations and question, highlight or even ignore them. We all know about vibration, temperature, cleanliness and electrostatic discharge. Let's not forget about screwdrivers falling onto solar panels, and time. Most spacecraft sit around a long time awaiting integration onto launch vehicles, or being driven around to tests and integration. Is there an effect of months of sitting, and low level vibration? Is the spacecraft designed to be easy to lift and manipulate into position on the launch vehicle? Handling is definitely an environment.

Modern microsatellite failures are caused mainly by the engineers who build them. Their errors may be rooted in understaffing, insufficient support and tools, too short schedules, compromised test facilities and scheduling, and just plain inexperience of the development team. The MIL-Specs say nothing about this subject. In fact the Federal Acquisition Regulations and government contracting officers won't allow payment of salaries to staff and consultants sufficient to bring the most experienced people to a program. Our parts have heritage—but for our engineers, they are often if not building their first spacecraft or microspacecraft, than filling a role they've never had before. Design reviews are only going to catch the major guffaws that can be spotted by a casual observer. Deeper rooted problems—the 1 W resistor that should have been 10 W but is getting by because it hasn't yet been on long enough to burn through, the timing conflicts from the CPU clock that, after a few months' degradation in space, will start causing resets.

These design flaws, buried deep within the spacecraft, are not going to be found in a viewgraph presentation on the digital design. They aren't going to be Mil-Spec'ed out of existence or prevented via certified soldering technicians. They may or may not crop up in exhaustive testing. They are subtle and insidious. They are only

Ever had a really bad connection?

going to be prevented by a highly motivated crew of talented and experienced engineers provided with sufficient staff, tools and time, and managed crisply, to do the job right. What a concept.

3.6 Connections

Reason number four, the fourth most notorious killer of aerospace devices has gotta be parts, right? Wrong. But we're getting close. Connections are number four. Not connectors — connections. Connectors are fantastically engineered devices. Probably the award for the highest ratio of engineering hours to component complexity should go to the lowly D connector. Some of them have only ten pins, something like one millionth the number of connections on a $10 microprocessor. D connectors don't fail. But connections fail. If you were born in or near the first half of the 20th century and became an engineer, chances are you "programmed" an analog computer. You patched together amplifiers, integrators, switches, adders, subtractors and mixers—today we'd call them "objects"—using simple pieces of wire with banana connectors on each end. Lots of them in a great tangle, analogous to the classic telephone operators' switchboards. What they did have in common with programming in BASIC or C is that the programmed-up computer never worked on the first try. Students always assumed they had messed up on the circuit design—the brainwork, the programming. But inevitably, at least one of the reasons it didn't work was that one of the connections looked real—but in fact the wire inside its bright red or yellow or green vinyl cable, was in fact two wires—separated somewhere by an invisible (under the insulation) little break. Back in those days people used to build Heath Kits, which were collections of all the components needed to build an FM stereo receiver, an amateur radio set or an electric organ (now referred to as synthesizers to improve our self-image when we play them). These included all switches, cords, wires (often precut) connectors and even labels for the plethora of knobs sticking out of the chassis. As a general rule, the components in Heath Kits were of decent quality and worked well, and the designs were solid. The instructions were clear, and most Heath builders knew a few things about electronics, soldering, and careful craftspersonship. And the product virtually never worked the first time you turned it on. If you were lucky, the failure would not be the kind accompanied by tendrils of smoke rising mystically from a power transformer. The culprit, inevitably, was connections.

Most micro satellites are basically boxes of electronics. At least that's how they look in the models we build to show customers what the satellite will one day look like. They also look like boxes of electronics in most reliability analyses. We estimate (guess) the reliability of the various boxes, and multiply those numbers together, as if that will tell us the reliability of the ensemble. But both physically and in terms of reliability the real things look different for two main reasons. One is they're covered in thermal blankets. Usually added to simplify the thermal design calculations—though thermodynamicists will give you myriad other reasons (ignore them)—these blankets don't often cause reliability problems. But they illustrate the problem with reliability based on the box-eye-view of a satellite. Where in that calculation do you figure in the errant blanket whose adhesive or velcro lets go and flaps into the way of a deploying solar panel or opening telescope door? And what is the probability of that happening? Is your velcro 99.999% reliable, or maybe 99.99999? It matters, and frankly, you have no idea. And my guess is this lack of information has caused no one to walk into a CDR and tell the assembled experts that they have no idea what the system reliability might be.

The other reason satellites and their models look differently, and act differently is that the real thing is strangled under a macrame of cables. Hundreds, thousands, maybe tens of thousands of wires run around the satellite, bundled into cables wrapped up in plastic tape and fitted with connectors on either end. To get from the last part on board A to the first part on board B, our precious electrons have a lot of wickets to pass through. Our coddled component is soldered to a circuit board. The solder is a connection. The circuit board trace goes to a sliver of copper onto which a connector grips. Another connection. The connector plugs into its mate—another connection. At the back of the mate, a wire is soldered to the connector pin. Another connection. The wire snakes from wherever part A is housed to wherever part B is—nominally another board inside another box in another part of the spacecraft. The wire, if it is broken anywhere, is a lost connection.

Let's talk about these wires. They are, of course, inside an opaque insulating jacket, so you can not tell if it's fine, fraying, or severed in there anyplace. Even in clear insulation like Hi-Fi speaker wires, you could never visually find a break. At a graduation forum at my Alma Mater, one of the professors said, to a packed lecture hall, "Coronary arteries are very persnickety, and their failure will account for the deaths of half the people in this room." Wires are the coronary arteries of your satellite, and usually don't figure into anyone's physical or reliability model. Technicians lay out these wires, one by one, on large wooden boards. It is sobering to see how long these wooden structures (and hence the actual wires) have to be, and how many wires are laid out on them in making just one of the tens or hundreds of cables in a satellite. People are lucky—they have just a few coronary arteries, and they are not very long. Also unlike wires, coronary arteries are alive and can to some extent repair themselves. They often give warning signs when they are in distress, and they degrade gracefully in the sense that there is some opportunity for collateral circulation to supplement them. None of which is true of wires—once fractured, they are fractured forever. You may not even know that right away if the insulation is holding the two ends of the wire together (temporarily). Most wires are not redundant, and collateral circulation's only analogue might be for ground wires. Otherwise, a broken wire is a serious failure. I remember as a kid reading that all the veins and arteries in the body, if strung together, would go all the way around the equator a few times. I don't know how that calculation was done, but the average microsatellite probably has several kilometers of wires, and those wires are wrapped into all kinds of angles and twists, scantily insulated from their intimate neighbors, often crushed and manhandled in development, shaken to death in test and in the launch, and then thermally cycled thousands of times in orbit. Despite this, wires rarely figure into reliability calculations, which says more about those calculations than it does about the wires.

Remember part B where our electrons are headed? We're just about there. Assuming all these wires happen to all be intact, each wire connects to a connector pin, which connects to its mate's pin, which connects to another circuit board which connects, via a circuit board lead, to another solder connection—to our coddled Part B. That's nine connections, possibly several meters of wire and at least two connectors, to get from A to B. If you've got a thousand wires, that would be nine thousand connections. This is not counting the thousands of solder connections within each circuit board. I doubt there's a satellite out there without twenty-five thousand connections. Many small satellites, you know, the simple ones that don't have too many parts or circuit boards, might have a hundred thousand. Not counting the ones inside the components themselves. But that's another chapter.

The spacecraft world is certainly aware of this issue. We have standards for soldering, standards for cabling, standards for mechanical support of parts so their solder joints aren't stressed. We have expensive, stress relieved, gold-plated connectors. With all those standards and all that attention, the problem is under control, right? Ahem... Ever notice the label on those obnoxiously loud air compressors they drag around cities on trailers when they're going to rip up the street? They're usually called silent-flow or something equally reassuring, giving you false hope that you might still get some useful work done within a half mile radius of the thing. Peaceful names notwithstanding, they're as loud as hell.

How about standard shaving razors? Their names all tout smoothness and safety. I cut myself with my razor a lot more often than I do on my Mac keyboard, which doesn't even have the words "smooth" or "safety" in its name. How about the ever-ready alkaline batteries you bought in preparation for that earthquake? Are they ever ready? You'll find out after three or four years without an earthquake when the big one comes and you don't have power for a week. Ever is the one kind of readiness they ain't got. Candles, which really are ever ready, don't say so.

So it is with spacecraft. Where there are myriads of standards, there's a reason for myriad standards. The reason in the present case is that the damn connections fail with utterly frustrating regularity. And not just in dramatic ways during launch or on orbit, when the mission comes crashing down around you and the only sound you hear is a whimper from the primary insurance carrier. Connections fail in test while you're still checking out boards on the bench, sometimes creating those miserable day and night crises where you think you'll never squash the bug while the rest of the program waits and the schedule slips. Connections fail in vibration and thermal vacuum testing. After all, that's the point of the testing, to find those kinds of latent defects. But then you've got to find which of the thousands and thousands of connections have died, never knowing if perhaps it's not a connection but a component, or a line of code, or a malfunction in the ground support equipment reading data out during the test. Those failures don't kill the satellite or rocket, at least not immediately. They kill the schedule and the budget. They test the durability, the patience, and the ingenuity of the engineers. And sometimes programs are killed because they are too late, too overbudget, and face too many bugs. That's connections.

3.7 Piece Parts

Number Five. Piece parts. Remember parts—components? Those are the things that everybody, even that auto parts guy in Joplin, MO, knows we spend a fortune on. If the Space Station had a gadget the size and complexity of your rental truck's fuel pump, would it cost $10? $10,000? More? All I can tell you is if you go to a fairly upmarket hardware store, you can spend $10 on a door hinge. But if you buy a hinge for a solar panel made by one of the industry's younger and more aggressively efficient vendors, you can spend $10,000 on each hinge. How often do satellites stop working and rockets crash because of part failures—part failures not caused by poor design, other human errors or failures of connections between parts? It happens. But rarely.

But piece parts quality is what we're famous for, and we pay big bucks to get and document the use of space quality parts. Why? The factual historical reason is that S-classification started back in the '60s, and life was different then for three good reasons.

One reason was that parts were a lot less reliable back then. This was before the microprocessor, hence before automated and robotic part assembly and inspection. Aerospace was pioneering new parts and there was no assembly line cranking them out by the millions with 99.999% yields. There were no significantly complex electronics in automobiles or other mass products for consumers, and there were no Japanese competitors forcing US and European manufacturers to upgrade quality or die. So DoD, the U.S. Department of Defense, pioneered space qualification to ensure that a higher standard of reliability could be met in what were basically hand assembled parts, over what the rest of the parts consuming market was willing to tolerate.

And despite the dramatic increase in parts quality forced upon industry by the increasing complexity of consumer devices, particularly automotive electronics, and despite huge reliability gains from computer controlled manufacturing, inspection and lot testing, and from increasingly higher levels of integration—essentially folding many parts into a single component—the arcane procedures that served us then, and the people and the jobs those procedures keep alive, die hard (if at all). If your dog scores a hamburger or steak bone by bringing you your slippers a few times, he's liable to bring you every shoe in the house whether you need 'em or not. Each program manager is motivated to follow procedures. If you follow all the rules and the satellite fails, you've at least followed the rules. The rules don't get hanged at sunrise, but you could if you bend or break them and a mission is lost.

Reason number two is that parts failures is a heck of an easier blame to place than fingering individual engineers and proclaiming that they screwed up. It's very hard, in a large program, to even find the appropriate engineer to finger. So many people get involved in design. Is it your fault, or your supervisor who signed the drawing, or the program review team that critically attended all the design reviews? Engineering is ultimately the science of the possible. We solve that which we can solve. Parts selection and specification is controllable, so control we do. Ad nauseam.

Reason number three is fear. Why add even an incremental risk? We're doing the best we can with the team we've got. Let's give them every advantage by ensuring that at least the parts won't fail.

3.8 Piece Parts, Space Qualification, Blame, Fear, and the Psychology Thereof

So we have three great reasons to make an engineering decision: (1) protection of the collective ass of program management and all the jobs of people who spec, build and inspect S-class parts, and maintain their paper trails; (2) protection of the engineers and their managers from exposure of incompetence; and (3) fear. Another reason for the continuing use of S-class parts is that they are so well documented. In theory, if there's a failure on orbit, we'll know all about the part that failed and we'll be able to improve it and not repeat the disaster. But the same is true of commercial parts. If one is flawed—say the rear door latch on a minivan—the market is liable to eat you alive, resulting in the recall of perhaps millions of vehicles for repairs. And if that doesn't do it, there are also a few lawyers willing to help society ensure that the guilty parties compensate those who suffered losses due to the flaw. Talk about effective feedback!

Large programs rely on parts lists and space qualification for another reason. Not every engineer in every nook and cranny of a program knows what all the requirements are: all the environments, thermal, radiation, vibration, and so on, and all the test schedules and lifetime requirements. So the engineer falls back on approved parts and parts lists, shifting the responsibility to choose appropriate parts from the engineer to the people who make the approved parts lists—people who presumably specialize in the environments and requirements imposed on the entire spacecraft. Those parts selecting people do not have to live with the difficulties the limited number of approved parts impose on the engineering team.They are accountable in case one of the approved parts proves unsuitable. Given that motivational mix you can surmise the conservativism in those lists.

By contrast, the microspacecraft development team typically consists of six to twenty people, and each of them is responsible for knowing those environments and using their, and their buddies' and managers' and reviewers' collective engineering judgment to select the right parts.They can trade off the risk of a new part vs. the added risk due to complexity of using older, less capable parts.And they can assess whether their time is better spent investigating a new part vs. designing with older parts.

All of which addresses the least significant problem in preventing failures—building the design around highly qualified piece parts.

Unfortunately, the limited number of qualified parts (which are usually quite ancient, due to the fact that it takes so long for them to be approved), their cost, and their long lead times contribute to the other four types of failures. Companies with burgeoning businesses selling, say, microprocessors, don't take their best people off the Motorola or Apple account to custom build 30 parts for the Air Force. Space qualification is often handled by secondary suppliers who buy and rebuild complex parts. By the time demand for the part is anticipated, and the long process of learning how to manually package, inspect, and test it is completed, and then lifetime and environmental testing is complete, years pass, and the original manufacturer is a generation or two ahead of the space qual part.

Thus the designer is forced to use parts nowhere near as capable as those currently available commercially. Less powerful parts means more outboard components, hence a higher parts count and more connections. The hottest circuit designers in the industry tend not to specialize in arcane, outdated components' applications, nor are they likely to work for government auditor-approved salary scales. The result? Poorer designs that rely on poorer designers using poorer parts.

Of course more circuit and software complexity, the result of using less capable parts, fosters human errors. While they don't contribute to misjudging the environment, space qual parts result in larger, heavier designs that are harder to engineer for their environments. And a few tests, even a few hundred tests, applied to a few hundred parts, is nowhere near the testing given to mainstream parts built by the millions and put in daily use around the world in myriad applications.

So why do things fail? Aren't you sorry you asked? Here are the five big killers ranked according to their lethality:

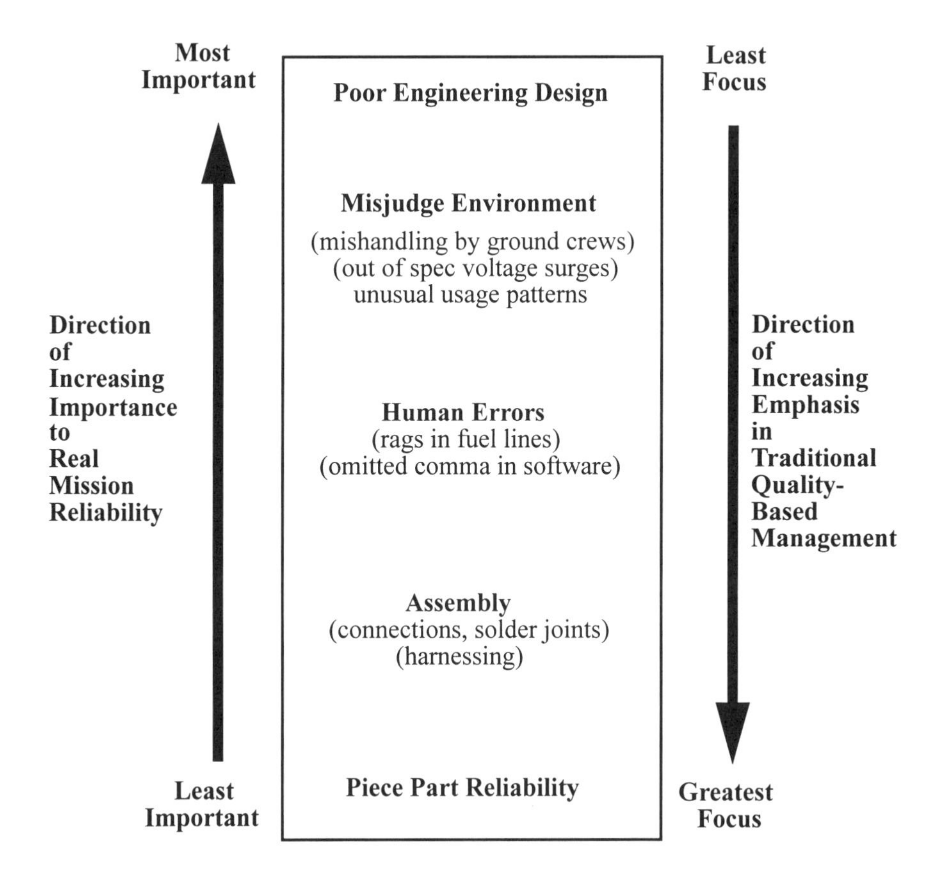

If you want to know the order in which the conventional space business expends resources to prevent failures, just reverse the list. Totally logical.

3.9 How Important Are Parts Costs?

Isn't most of the cost of a satellite in labor and launches? So why not spend a little more on parts? After all, if they do fail, it's really expensive to break open the whole system and fix 'em.

If you've ever torn open a truck to replace the fuel pump, or pulled every V-belt in the world off your Ford Fairlane to replace the water pump, you know you'd spend an extra ten bucks to get a better part. On the other hand, if you purchased all the new parts necessary to build a $20,000 Chevy out of them, you'd spend $150,000 or so just on the parts. For that price, the vehicle you build ought to be really reliable! But that's an expensive way to buy spacecraft reliability, since it isn't a law of physics, or even a principle of engineering, that space parts are necessarily more reliable. Space qual parts carry that label because they meet certain standards of fabrication, inspection, testing and documentation. Modern parts are likely not designed for the mostly manual steps prescribed by the space qualification requirements, which are often manually assembled and inspected, which lowers reliability. The documentation doesn't add anything to your product's reliability. And the additional people needed to handle all this documentation cut down on team cohesion, create new management layers, and hence lower the focus of the team effort. Amateur satellite experience has demonstrated that cheap commercial parts are more reliable than military standard parts, and those in turn are more reliable than the most expensive space qualified parts. Moral of the story: if you want a reliable satellite, spend money on top notch engineers, not on space parts.

When we built the ALEXIS 100 kg astrophysics satellite, parts were about 30% of the total cost. As a rule of thumb, parts are 25% to 35% of a custom development project budget. Experience across a range of programs shows that that ratio holds no matter what the parts cost. So for every additional dollar you spend to upgrade parts, you'll spend an additional $2.30 in labor, overhead, G&A, and so on to use that part. Why? More expensive parts mean more paperwork. The paperwork is part of what you're getting for more money. The parts are fully traceable. To get any value from that, you too have to track them through the assembly, test, integration, and retest steps. Do you think over the course of a two-year program an engineer could spend an hour documenting the whereabouts of a $10 part? That engineer's hour costs the program at least $50.

Besides that, more expensive parts are, unfortunately, older parts, many requiring more outboard parts to support them. That adds to cost. You can't buy just one S-class part. They are built and tested in lots of typically thirty. If you need, say, two, you'd probably like to buy five. Thus you'll have twenty five spares you don't need, which adds to cost. More expensive parts have longer lead times. Often a commercial part is off the shelf, but the space qual equivalent is a six to twelve month lead item. The delays they cause in your program have real costs which probably dwarf all the others. What would your fuel pump have cost if you had to wait a few days in a hotel in Joplin to get it? Probably a couple hundred bucks in hotels and meals (the $300 fuel pump), and maybe a year in therapy (the $6,000 fuel pump).

So the short answer: Parts counts drive all the other program costs, and the assumption that more expensive parts will give you less trouble is not always valid.

3.10 What About Using Terrestrial Parts?

Luckily for those of us who make a living pondering the big issues, there's no single correct answer to this question. Otherwise you wouldn't have bought this book, nobody would hire me as an overpriced consultant, and my younger sister would probably be building little satellites just to keep her hands busy while the kids are doing their homework. Let's put the long story off for a minute and jump to the answers. You have three choices:

#1: Shell out for the space qualified stuff. You'll spend plenty of money, but nobody will criticize you for doing anything too radical or for threatening Mission Success. In fact though, S-class parts can decrease reliability by complicating the design, by requiring additional management and tracking, and by diverting resources from more efficient means of ensuring flight success, like hiring the most experienced engineers and spending more time testing. Good engineering judgment aside though, if something does go wrong, you can say you followed the book. Before you decide on this option, make sure you have the schedule and the budget to burn.

#2. Use a terrestrial device: Many parts don't know they're in space. So like the cartoon character who overruns the edge of a cliff and just continues running horizontally until he looks down, these parts carry on aboard the spacecraft without incident. Typical among this group are discrete components (resistors, etc.) excepting some electrolytic capacitors, ICs and FPGAs (Field Programmable Gate Arrays), NiCad batteries, nuts, bolts and wires. Then there are parts that require a little special handling, but with care can be used reliably in space. These include lead acid gel cell batteries, some optics, solder, deployable mechanisms, RTV, connectors, and terrestrial instruments like magnetometers and cameras, which generally need to be potted and have other modifications to withstand the shock, vibration, and outgassing specifications for spaceflight.

#3. Build your own part from scratch. The arguments in favor of home brewing are intrinsic to engineering. We wouldn't be engineers if we didn't think we could design or build it better. But let's look at the cons. Even the simplest device, a PC's mouse, for instance, is not so simple. It looks pretty simple—maybe three moving parts. But look at the questions you have to answer before designing just a lowly mouse:

- What kind of wire to use in that cable that so easily flexes and never takes a "set" so that the mouse moves freely?
- How much stiffness to add as the cable merges into the mouse, so the wire won't fatigue?
- One microswitch or several? And how many clicks of the mouse should the thing withstand?
- How much depth of teflon on the skids on the bottom? Too much and the motion's vague, too little, and it sticks to the table.
- How do you open it for cleaning, so that when you click it back, the geometry remains perfect and the ball on the bottom sticks out just the right amount?
- What is the right amount anyway? What size hand are you designing it for?
- Does it have its own antibounce circuit, or does the computer handle that?
- Are you building one or one million? If you just need one, you can't afford the manufacturing techniques used to build the custom plastic body.
- Is there any electrical hazard, for instance if my hand is wet or I glide the mouse through a Pepsi spill?
- Will it withstand a drop off the table, a long trip scrunched in a suitcase?
- How heavy should it be? How will I trim weight if it is too heavy?
- What kind of bearing do I use to support the little ball—and how do I keep it from getting gunked?
- Would an optical mouse work better?

Even if we get by all this, there's testing. Real mice have millions of hours in real use in offices and homes and airplanes. They have failed in ways designers never dreamed of, and the failures resulted in redesigns with clever little "tricks of the trade" to avoid those subtle failure modes. How many years have you got to rigorously test that home-brew mouse, and how many millions will you build to test simultaneously?

We live in a world of $15,000 automobiles and $1000 laptop computers. We equate those prices with the value of the product. If a car costs $15,000, it must take the equivalent of one person working a few months to build it. Thus a $1000 laptop must be a pretty simple thing to build. It costs as if you could build it in a week. Wrong! The first laptop cost its developers tens of millions of dollars. And it was only that cheap because they, and many others, had already done hundreds of other laptops and many of the product's subtleties were now well known. Otherwise the price of the first laptop would be in the billions of dollars. The first new car off the line, looked at this way, cost $2B.The second one and subsequent ones cost $15,000. Same for the mouse. The fact that they cost $69 at PC Warehouse doesn't say anything about the engineering effort to make one. Again it's in the tens of millions of dollars. The latest new shaving razor, a simple device you hold in your hand, with no software, no power source, no wires, no harnesses, no electronics—cost $1B to develop and test before being brought to application in the market. Keep that in mind next time you read an article about a $900 hammer. Serial #0000001 of that $9 one you just picked up at Ace probably cost a lot more.

Given those grim figures, you are now realizing that in a $1M spacecraft program you won't have the R&D budget to home-brew even one fairly complex widget, let alone a whole satellite full of them. That may be a bit pessimistic, but skepticism is a good vantage point for taking the measure of the heady enthusiasm of the early days of a program.

The moral of one obscure Zen parable is "When the student is ready, the teacher shall appear." Having signed up to build some incredibly complex (compared with a razor blade) spacecraft before you realized that $1B doesn't buy much these days in a development program, let alone some small fraction of $1M, you may find yourself in need of a teacher. Here another obscure Zen parable will come in quite handy, this one about a student who is given the opportunity to ask one question of the greatest of Zen masters. To ask any one question of his choice, the student travels hundreds of miles on foot to the Temple of the old Master. The young student, exhausted from a journey of a full year, arrives at the Temple, passes by a haggard old man sweeping the front steps, enters, but does not find the Master anywhere. He sleeps on the sidewalk outside the Temple, which has no great gardens as he had expected, but rather is squeezed up against a busy thoroughfare. He resumes his search in the morning, with no greater success. In fact, there is no one at the temple at all, except the old, apparently hapless peasant incessantly sweeping the two or three front steps. He decides to await the Master's return. He finds a room in the Temple to stay in, sleeps there, wondering how he can find the great Teacher, even dreaming of his meeting—the anticipation consumes him that night. The third morning, alone in the Temple, he interrupts the man, who is still sweeping the steps."Do you know," he asks, "where is the great Master?" To which the old man, expending the student's allocation of one question, replies: "I am he."

You know those hard core shoppers, the ones who barge right into suburban shopping malls on Christmas Eve and then turn around and face daunting crowds the following week to get those last-minute and post-holiday bargains? These are people who dig through all those shoes piled up in big wooden bins at the entrance to the department store, and actually find their color, style, and size at 60% off. Their refrigerators are mere props used to magnetically levitate coupons for cents off on a thousand products, and they don't throw away certain catalogues, ever. You need an oak bar stool with hemp seats? They know who's got 'em—on sale. Need a gadget to use your Cuisinart to grind coffee? They hand you the catalogue that's got it. These people are your teachers when it comes to selecting parts and components.

We can't build everything. In fact, we can scarcely afford to build anything. Spacecraft engineering is in part the art of shopping. Knowing what's out there, where to get it, what you can use it for, and what you can't use it for. And even knowing if somebody else ever used it the way you want to use it. This requires an immersion in the arcane culture of space. And there are plenty of Masters out there steeped in that culture. Unfortunately, very few of them have ever built a $1M satellite or $10M rocket, and they will not have the answer to even your first question.

Hardcore shoppers on Main Street. Looking for reliable parts, no doubt.

Chapter 4

Darwin Predated Satellites but Engineering Obsoletes Evolution

It pays to remember that we weren't originally created out of completely redundant parts with virtually indestructible titanium-ceramic composites. Prehistoric humanity was based upon an organic chemical system which evolved as sort of a kludging together of primordial parts that, in turn, evolved from things like sponges and lesser creatures. The whole human body was totally unworkable. As you might predict, it failed regularly, necessitating the messy and contentious process of mating and reproduction. Worse yet, our ancestors' mortality was the entire focus of their attention. They worried about nutrition and exercise, the cleanliness of the environment, the safety of home appliances. What anxiety for them to realize that they were eminently destructible units in the rugged world we inhabit, with its sun, moon, oceans, storms, and non-UL rated extension cords.

The biggest institutions in that ancient society were hospitals, places where one would go to slowly and painstakingly grow, cell by cell, replacement parts of the organic body—skin, bones—stuff like that. And not everyone succeeded. They actually died, and the landscape was littered with cemeteries in which were laid to rest perfectly good beings hopelessly trapped in failed mechanisms. In ghoulish fact, none of their bodies lasted more than 80 or 90 years, and millions died every year simply due to the physical limitations of the organically engineered mechanisms that supported them.

Clearly the species had to migrate to a different platform. This was first demonstrated in a now-legendary tree felling competition between a person wielding a hand axe and a person employing a motorized device. The motorized device, despite its crude design, won handily. This proof of the inadequacy of the organic platform was repeated countless times. Computers beat human chess players. Trains beat runners and bicyclists. In retrospect, this doomed species' support of its inadequate technology was as gallant as it was pathetic.

This continuing dilemma finally resulted in a human biological systems design review organized by The Aerospace Engineering Corporation. We aerospace engineers, knowing how to make really reliable stuff, chaired this massive review. The human body, as it was then somewhat chauvinistically called, when weighed in the balance, lost big time.

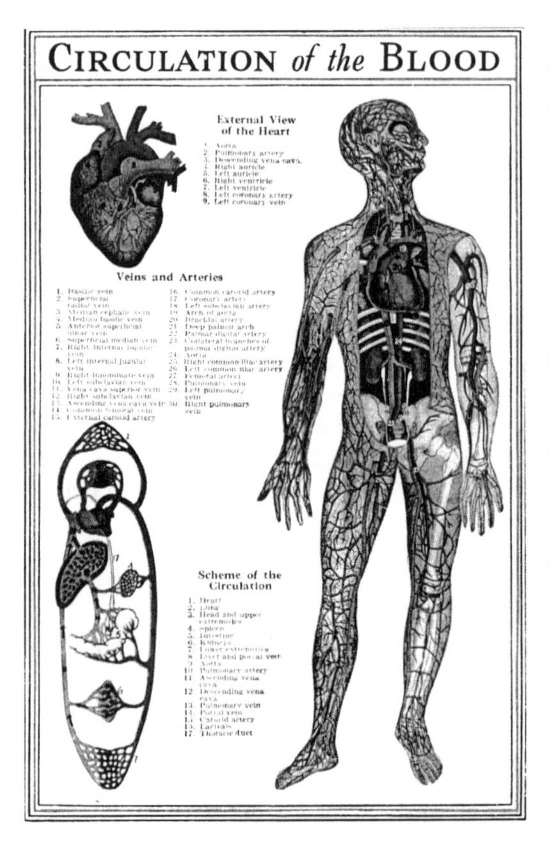

Example of a highly-flawed system design. Where were the engineers on this project?

The number one problem of the body was that it was not engineered—it just evolved. Imagine building a spacecraft from a bunch of randomly selected parts that just happened to work long enough for the machine to arrange other parts in the same way and get them going before it conked out. When it came to the body, no thinking at all had gone into engineering simple concepts like redundancy. The entire body was controlled by a single processor called the brain. One hard whack to the head—which housed the brain—and the game was over. There was no onboard state saving either. Most times when a hard reset of the brain occurred, the memory was lost. New beings, despite being derived directly from the parental unit, required reprogramming from scratch. Since the interfaces and organic memory both operated at an incredibly low bandwidth, this reprogramming took 25% of the lifespan of the body, not a very economical way to do business. And worse yet, the reprogramming was fraught with packet errors, causing major discontinuities in learning and culture from generation to generation. Version control? Forget it—nobody ever thought about it.

Worse than the effort of software loading, the teams that developed it, usually called "parents," had no system, no training, no qualification for this kind of work. As a result, the software they wrote was totally unstructured, and adults of the species often spent years with professionals, known as psychiatrists, trying to work through their tangled web of software incongruencies and glitches. As you would expect, saddled with unstructured systems like that, software upgrades were impossible. Hence, for tens of thousands of years, the same basic features were locked in, with new features impossible to create.

Ironically, people considered this state of affairs a good thing. They would read books thousands or even ten thousand years old and marvel that "human nature" hadn't changed in all that time. How could it! It hadn't changed because it was mired in its own lack of structure, seriously deficient quality control, and a flagrant disregard for modularity. The machines, millions of years after their evolution, were still running Rev.1.0 of their operating systems and applications—and they were proud of it. Go figure.

Both fuel and oxidizer were circulated around the body by means of a single, completely nonredundant pump known as the heart. In America alone, millions of people died every year from the failure of that one part. But did anybody build a body with a backup? Crazy. And reactants (remember, this is organic chemistry, not electronics) were carried not by electrons, but by a viscous red liquid—blood—which had a nasty propensity to leak and hence must be fortified with Vitamin K to plug the leaks. Nonetheless, lots of people died from these leaks, via wounds or failure of the arteries carrying blood, or from circulatory failures or strokes when the blood clotting mechanism got too aggressive, or when it wasn't aggressive enough, along with countless other indescribable discomforts.

What really stumped us was that lots of structures were redundant, though didn't really need to be. You can live without toes, but the body had ten of them. And ten toenails, all of them totally useless. Such is the illogic of evolution.

A severe limitation was the body's power system. Simply running at idle—in front of a word processing program at a keyboard, for example—after 12 or 14 hours the body would require 8 or 10 hours of sleep mode. Why? Nobody knew, or even seemed concerned about it. They just dropped off to sleep all the time, sometimes in dangerous situations like driving a car. Under heavy loads, like running fast or taking the Graduate Record Exam, the duty cycle was much lower. At stressful times like these, they could run at peak efficiency about 1 hour per day—a less than 5% duty cycle. After a three hour brain-intensive session, they all needed to retire to bars to drink beer and attempt recuperation, which seldom actually worked. It was truly sad. Here was a species trying to make progress and grow into a technological society, but it spent most of its time resting for lack of a decent power budget.

And when it wasn't resting, it was spending a heck of a lot of time and precious energy mating and reproducing, then raising new beings, before each body would give out, and then working in hospitals trying to repair and regrow broken ones. What a rat race that must have been.

It is at times hard to believe this species evolved on Earth, where we encounter a pretty heavy dose of radiation every day. The body was not at all rad hard, neither to ionizing radiation commonly found in earth ores like Uranium, nor to ultraviolet from the sun. Nuclear energy was never developed because of the susceptibility of these organic devices to the radiation, and its horrific effects. And people still died of cancers (unregulated cell growth) caused by UV and ionizing radiation. A non-rad-hard animal in a radiation environment would seem like formula for disaster, but that's how God built those poor creatures.

We could have gone along with all that, but for one fatal flaw. I mentioned that human bodies had no decent data port, and of course there was no external power plug that might have provided a simple solution to the low duty cycle issue. But there was also virtually no on-board instrumentation. To monitor anything required really obtuse, expensive instruments. Where a single 9 pin D connector would have sufficed, this species built multimillion dollar machines like MRIs to obtain a low-resolution look into their own bodies. People bought thermometers and held them in their mouths to get some second order metric of metabolic rate, and wore blood pressure cuffs around the arm to try to figure out how the blood circulation system was doing. It would have been laughable had it not been so bothersome. Well, the lack of instrumentation kept a lot of high tech industries going, but at what expense to society. A lot of these poor people died from overeating, lacking even a fuel gauge!

Well, that pretty much settled the necessity for the phaseout of those organically driven platforms. Besides the miserable and unproductive lives their inhabitants lived, it just wasn't safe for the rest of us. Here we had this incredibly buggy software, and no instrumentation to figure how it was installed or what was the state of the hardware it was running on, attached to a body with a lot of mobility, dexterity and strength. Some of the body's muscles were strong enough to really harm things—break windows, fire weapons, that sort of thing. And there was absolutely no safety interlock. If the body decided to take a swing at you, it would—and it quite often did. Not that they were going to damage any of us. And even if they did, we keep backup copies of our software, and platform mechanical modules can be reinstalled easily enough. But we can't live in a world with a bunch of hair triggers about to go off, breaking people and their property just because some chemical synapse failed or because of some faulty programming from 30 years ago.

That old organic system got us out of the primordial soup, but there is such a thing as taking an idea a bit too far. Organic engineering was a chapter in history that nearly destroyed us. But thanks to aerospace engineering, it's been banished to prehistory along with the dinosaurs and 8 track tapes. We know how to build reliability, redundancy, tough materials, and reprogrammability.

And one day one of our satellites may last 20% as long as 3 billion people routinely do right now on earth.

Chapter 5

Baby Boomer Risk Reduction: Revisiting the Clean Room

5.1 Progression of Value

Even for people like me involved in product (spacecraft) development, most of our experience is spent using, not building, things—cars, computers, clothes. But even the IRS admits that things depreciate with time. That car you used to park in a far corner of the parking lot all by itself, the one you washed and vacuumed religiously every weekend inevitably devolves into the wreck suitable mainly for parking at the airport while you charge off to Huntsville or Albuquerque.

A developing satellite follows a different trajectory—one that's a bit similar to that of the aging Baby Boomers. As kids in high school and college, the boomers took risk, and even occasional damage, in stride. They marched and protested just an arm's length from angry police in full riot gear. They smoked marijuana, dropped acid, ate magic mushrooms. Sexual freedom blossomed twenty-five years before "protected sex" even entered their vocabulary. All that when they each had on average fifty-plus good years ahead of them to put at risk. Now baby boomers are careful. They join health clubs to keep their aging hearts healthy and build stronger bones. They subscribe to magazines like *Health* and *Self.* They drive Volvos and two-and-a-half-ton suburban trucks with dual dashboard and dual side impact air bags, six sets of three-point safety belts, antilock brakes, and full-time four-wheel drive. All that just for a quick trip to GNC to refill that bottle of Vitamin E. Better to protect those final few years allotted to them, they've outlawed leaded gas, asbestos, and soon (perhaps) tobacco. Even a fine bottle of vintage Barolo warns them in eight point type of the hazards of demon alcohol.

It is thus with a twinge of embarrassment that I make this admission—the maturation path of cheap satellites is just about the same. In the optimistic days of their youth they grow atop a lab bench, desk top or kitchen table. Engineers and techs work on the satellite dressed in the uniforms rigidly prescribed by our profession—jeans and T-shirts. We do, of course, stand the fledgling spacecraft on a blue antistatic pad, a perfunctory nod at reliability assurance.

5.2 The Logic of the Aging Boomer

But by launch day, a satellite has no better chance escaping the confines of the clean room and the company of humans under dust free caps masks and gowns than a six-year-old can avoid being stuffed behind a Windsor knot and enveloped in a sport jacket on encountering one of life's little potholes—Easter, or Christmas, or Yom Kippur, or Aunt Phoebe's third wedding. Did we lose our small satellite pioneer spirit when we turned forty and sell out to please The Order? Do all of us, however anarchistic, have a deep-seated yearning for ritual motivated by desire to reunite with our youth when we were stuffed behind that Windsor knot and enveloped in a sport jacket and mom kissed us stickily on the forehead and extolled our cuteness?

Who do I look like, Rabbi Dorfman? As a satellite engineer these questions are way out of my pay grade, but if you want my opinion, which is probably why you bought this book, yeah, that's definitely part of it. But most of the answer lies in the logic of the aging boomer's risk aversity. Satellites and adults reach a state of maturity where the investments made in us and the responsibilities we have make it logical to take some precautions.

A screwdriver dropped onto a solar panel the first time it's been tentatively fitted onto the satellite does damage about equal to the labor it takes to get it back to the vendor, fixed, and returned. Not much money, maybe the equivalent of a Camry plus or minus a Geo Prism. That same event occurring on launch day minus one can cost those shipping and repair costs (amplified by expediting to get it done overnight) plus missing the launch date, meaning possibly loss of a sizable deposit, shuttling a crew back to Virginia from Plesetsk and doing it all over again months later. That'll run you, for a "bare bones, commercial, small, cheap, efficient" program*, probably no more than one Testarosa and a Silver Spur, plus or minus a one bedroom pied-au-terre in Manhattan or London.

---

* A random sampling of the adjectives our clients use to seduce from me the lowest possible price quote. Far from being forgotten, they are repeated throughout the program both by the client and by us. In the latter case usually prefixed by "I thought you said you wanted a ----- program" and in the former case couched more as "Just because this is a ----- program doesn't mean we should be forced to live without..." where "..." means things like massive program reviews in front of consultants intent on justifying the need for additional reviews, an independent quality assurance oversight team reporting directly to the client, or a fully redundant complement of MIL-Spec parts built by companies normally servicing some of our nation's largest and inevitably least efficient bureaucracies.

5.3 Clean Rooms and Polypropylene

What has any of this got to do with clean rooms and white polypropylene bunny suits? If a surgeon were to sneeze or perspire into your opened abdomen, you could suffer a nasty infection and the physician, possibly, a proportionately nasty law suit. And, if you get a really tough lawyer, maybe even a disproportionately large settlement. If a lash falls from an Intel technician's eyelid onto a new silicon wafer, it ends up in the trash heap. Clean environments and lint free clothing to protect your work from you are clearly good ideas.

To the extent that satellites are really just electronics, structures, and mechanisms (and many cheap satellites have no mechanisms), why do we work on them in clean rooms? Occasionally, rarely, they carry instruments which are hard or impossible to clean, and which are sensitive to dirt. UV and X-ray cameras are the canonical examples, though the organic molecules which obscure UV lenses are not necessarily any less prevalent in clean rooms. But at least there's less dust to accumulate. If you do have exquisitely fine mechanisms which have to work (deployable arrays and booms, for example) and which could be jammed by dust and dirt or an eyelash, a clean room is a good idea, though your mechanisms are a disaster waiting to happen if they're that sensitive. If you are screwing together propellant lines, particularly if you are building up hydrazine and other contaminant-sensitive propellant systems, you need ultra-cleanliness. This level of clean is not even achieved in the clean room, but it's a start. Final cleaning of these systems is done just before propellant loading, but you definitely need to start with a debris-free system before you can even attempt to clean it.

Now let's talk about the real reasons for most clean room operations. Lots of stuff gets built outside of clean rooms. Your MD doesn't move you into a clean room to give you a flu shot, or even to lance a boil. Your car radio, VCR, computer and television/monitor were not built in clean rooms. Similarly, most small satellites aren't built in clean rooms. They just end up there later on. Clean rooms keep more than dust out. They keep people out. Dust doesn't kill too many satellites, but people do. One program I worked on was seeing about a failure per week in the integration lab. This did not bode well for even a modest 6 month (26 week) mission. Miraculously, the satellite built by this program worked for many years on orbit without a glitch. In space it is highly unlikely a person will happen by and connect the wrong wire or drop the occasional screwdriver on a connector or solar panel.

What if the president of the US showed up at his inauguration or addressed the United Nations dressed as I usually am in shorts and a T-shirt? People would disapprove. They might even not elect him next term, whether he otherwise did a good job or not. Why do airline pilots wear suits, ties and black leather shoes? Most pilots say they fly best in jeans and running shoes. We try to match people's expectations. Customers expect to see satellites in clean rooms, and the customer is always right.

But isn't a clean room better? It does reduce the risk, however small, of a bit of foreign material jamming a mechanism or shorting a tiny circuit board lead or connector.

Yes, those failures might be minimized in a clean room. But how many satellites die of an attack of vicious dust? Many, many more die of attacks of savage budget cutting, and clean room operations cost labor hours and hence dollars. Suiting up and changing multiple times during a day to go in and out for supplies saps time and money from engineers' and technicians' productive workday. Every tool, test equipment item and even logbook must be thoroughly cleaned before it goes in, and even just speaking with the person inside there requires use of (cleaned) headsets, intercoms and PA systems.

5.4 Dressing a Three-Year-Old

My guesstimate is that cleanroom operations double the cost of a satellite compared with building it on a benchtop in the lab. If you think this sounds like a lot, imagine performing some mundane activity, like dressing your three-year-old for preschool, in a clean room. First you've got to thoroughly clean the kid and wash all the clothes he or she might need. Coming out of the washer, the clothes should be wrapped in plastic bags, double bagged, and sealed. Of course they're damp, so you'll dry them in the clean room. The kid can't be wrapped in plastic, so you'll bring him/her in, preferably still wet from the tub, then swab down with alcohol and lint free cloth. You'll keep plenty of (expensive) lint free lab wipes on hand, and a sealed trash container for used ones, in case of sniffles. You both might as well sit in there and bond while you wait for the clothes to dry. Maybe a thoroughly cleaned hair dryer would help speed the process a little. I suggest though that you plan on enrolling in the afternoon half day session. This will take a while.

Note that you are meanwhile dressed in a white bunny suit. If you get uncomfortably hot dressed in plastic, too bad. If you decide you need water, you'll have to invite, via an intercom, a coworker to come in and mind the kid and hold the hair dryer while you go out for water. You really shouldn't bring drinks into a clean room. It would be just a bit nonproductive to ward off all that threatening dust and then spill Mountain Dew on something. But maybe some filtered water in a plastic bottle would be OK. Check the procedures manual first, and if it's not in there, write a memo to the supervisor requesting permission. You'll get an answer for sure in a few weeks.

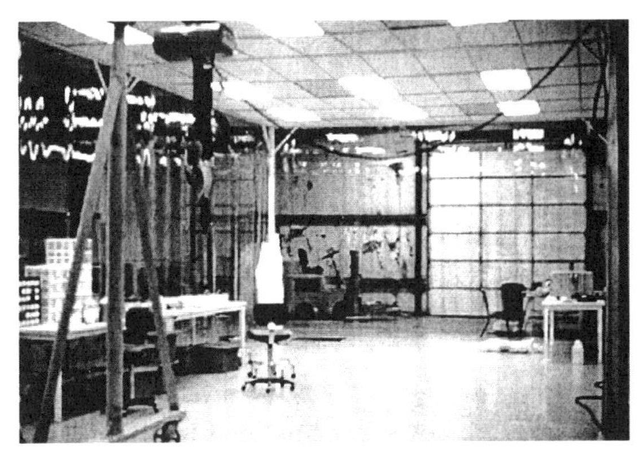

Back in the clean room, the clothes are finally dry, and you get the jeans and shirt on the kid, and you're working on socks when you realize you only have one sock (or the kid throws a tantrum that those aren't the right socks). Now what? Call that coworker and wait ten minutes while he or she gets off the phone and suits up again. The coworker takes over care of the kid, and you go out in search of the other/right sock. Good luck. You dropped it by the sink where you had washed it. You rewash it, wrap it in more plastic, and carry it back to the clean room, suit up (again) and relieve your coworker, who goes back to the phone, and spends most of the rest of the day trying to reestablish the phone meeting you interrupted.

Meanwhile, you dry the new sock(s) and, fifteen minutes later, are about to finish the dress-the-three-year-old job. When the kid announces a trip to the bathroom is an emergency.

Sure, satellites and test gear don't have bathroom breaks. But they do break, and when they do, they often have to be taken out of the clean room for repair. You don't want to contaminate the clean clothes, so you undress the kid, hang all the clothes up still inside the clean room, and leave. In the mezzanine, you put back on the kid's non-clean room clothes (the equivalent of the factory packaging) and cart the two of you off to the bathroom. Mission accomplished, you come back to the changing area, but now the kid is no longer clean. Rather than go through the bath thing again (2 baths in one morning?—that would definitely incite a riot)—you settle for another alcohol rubdown on return. So, you suit up for the 3rd time, take the kid's "factory packaging" off in the mezzanine, enter the clean room and do the alcohol cleaning thing with fresh lint free cloths again. You wish you had some talc, but that's not allowable in the clean room, so you figure skin cream would be good. But you can't use just any skin cream. You settle on vaseline, your 3rd choice, but better than nothing. Ten minutes later, a couple hours after you started, you're finished. You've dressed a three-year-old in the clean room.

What did this cost? Instead of fifteen minutes, it took hours, entering and leaving, drying clothes with a hair dryer, forgetting a key part, re-cleaning the subject, waiting for coworker relief when necessary, going out for water. Your coworker has killed several hours too, constantly being interrupted to spell you. A job one person can do is changed into one requiring two—doubling the team size. You burned up two boxes of lint free wipes, a bunch of alcohol, plastic wrap and some vaseline. And of course the rent on a clean room. Running all those blowers and lights and built with fancy wall materials and an antistatic floor, a clean room is a lot more than the typical kid's bedroom. That's overhead. And it adds to cost. Dressing a three-year-old can be exasperating, but it is not a complex operation compared with, say, debugging a gamma ray detector or aligning some UV optics. Imagine all the special tools and equipment you can forget to bring with you for those tasks—and how much time you'll spend preparing them all for the clean environment, and making do with what can't be cleaned.

Tired of all this? Imagine how the poor satellite constructors feel going through this drill every day for a couple years. The less time spent in clean room operations, the more money you save. You can save enough money to justify some covers on sensitive equipment which you remove before flight or open on orbit. You save enough to buy a few percent more photovoltaic area for solar energy collection so that you can spec thicker cover glass which is easily cleaned at the launch site. Money spent to stay out of the clean room is almost always money well spent. And moving into the clean room late in the program, only when the cost of failure gets large owing to the time constraints programs face near launch, and when the sensitivity to last minute glitches gets intolerable, turns out to be the sensible thing to do.

Chapter 6

Engineering Religion

Pop Quiz:

You are the project engineer developing a small satellite, and you have just laid out the program plan at a mammoth design review. You have presented the design, the parts selection choices, the testing plan. Your team has gone through the design and operations details. You stand up at the end of the review and present the budget. Fifty percent of the money is already spent, and something less than that fraction of the work is completed. A consultant to your customer stands up and suggests, that in the interest of mission success, you need to hold more reviews, to upgrade parts quality in certain critical systems, and add a redundant charge controller. Your response is:

a) You agree that all those steps are improvements that will enhance mission success and promise to implement them.

b) You agree they are desirable but plead for more money so that you won't truncate other parts of the planned activity.

c) You state that all those steps will decrease the probability of mission success and refuse to implement them.

(Save your answer for later.)

Psychologists tell us that our appetite for risk is a constant. Each increase in safety is well balanced by a commensurate increase in inattention to safety. Antilock brakes haven't lowered the highway accident rate not because they don't work, but because people with ABS drive faster and closer to things, relying on the ABS to stop their cars for them. Unfortunately, the physics of stopping is not altered by ABS. On the relatively rare occasion of a Washington, DC snowstorm, cars are often seen abandoned in the ditches into which they've slid. Several of us obsessive-compulsive engineers have observed that many of those cars are four-wheel drive vehicles. Possible explanations for this include the probability that more people take their FWD cars out in snowy weather, or worse, people who don't know how to drive in snow tend to buy FWDs, hoping that these vehicles will make up for their lack of driving skill.

They don't. My theory is that people in FWDs just feel safer than those of us driving around in rear wheel drive Miatas with tiny narrow tires. Hence, they drive a bit more aggressively (equalizing their risk), and again the physics of traction or the lack thereof eventually prevails.

American society has a fix for this propensity toward constant risk: risk magnification. Since our perception is that safety and health innovations are shrinking risk, and our response is to act more carelessly, the solution is to blow up the existing risk perception to a much larger proportion of the whole. For instance, you don't see Bosnians fretting much about secondhand smoke, or a strong Israeli demand for fat-free Fig Newtons.

Saturday radio talk shows are mainly infomercials on the anticancer properties of various rare and expensive pills and ointments, and magazine racks are well stocked with reading on extending life, staying healthy or getting healthier, or at least warning you which cars might tip over in violent maneuvering. At the hardware store, you have to decide if your family deserves a 1" deadbolt, a kitchen rewired with ground fault sensors, or lights that pseudorandomly switch on and off throughout the evening, which are purported to confound bandits. How many airbags are enough in your car—one, two, three? Is it irresponsible to leave your kid standing at the bus stop? There are kidnappings, you know.

Our bodies are constantly fighting a battle against invasion by all the bugs that want to set up housekeeping at our expense. Around the world, nations are constantly preparing for war. Our environment is stressed to maintain its thermal and chemical equilibrium despite our massive CO_2 production and clear cutting of forests and jungles. Meanwhile, in a virtual simulation of the outside world's real threats, our minds are locked in a constant war against risk. None of us have the slightest notion if $700 spent on an airbag buys you more safety than $7 spent on vitamins, or 70¢ spent on a safety yellow reflector decal for your bicycle. Consumers don't have any quantitative data, but we sure have opinions and we make resource allocation decisions every day, for some reason believing we're always right. In short, everybody's an expert in probability theory.

Little wonder that opinions fly fast and furious when a bunch of engineers review the reliability of a satellite or rocket. Everybody wants to be on the side of safety and flight success. Do you know anybody who wants a flight failure? And though we have absolutely no numbers, we think of ourselves as paragons of logic and quantitative reason, well-equipped to speak out on just how to make satellites reliable.

Engineers and scientists practice rationality. The scientific method, quantitative analysis and documentation of technical progress do not apparently leave much room for whimsy. Let alone mysticism and religion. Americans tell themselves from childhood that we are a country of laws, and the phrasing of our constitution appeals to cool rationality—eschewing, for example, Lords, Kings and Queens with powers inherited by birth.

Or so I thought until I got my nose out of the books, so to speak, and started attending spacecraft program reviews. The speakers and the audiences at these august ceremonies of worship to the evil gods of the glitch present themselves as the paragons of rationality and of the quantitative methods of modern technology. While the typically subdued lighting of a Design Revue venue might be reminiscent of a seance, the viewgraphs are crisp products of digital PowerPoint binary logic, filled with numerical calculations, decision trees and circuit diagrams. All of which is fine so long as the topic is precisely quantifiable—delay lines, Ohm's law or maybe a link margin.

But let the topic turn to reliability and the laptops slam shut, the viewgraph machine falls into disuse and more tribal instincts tend to prevail.

Not, as Jerry Seinfeld said, that there's anything the matter with that. If you have no idea of societal norms, but you have an instinct not to die, it's not too tough to come up with a tenet along the lines of "thou shalt not kill." What parent would not have composed rules along the lines of "honor thy father and mother." These commandments are not derivable—they are unquestioned postulates upon which we build our civilization. But I for one don't need to believe they were handed to us out of the sky. Or at least if they weren't, we would have invented them anyway.

Ok, those are the easy ones. Let's say you notice that out at sea, whenever the sun sets in a deep red sky, the next morning brings gorgeous weather, but when the sun rises red, within a few hours it's rain and rough seas. Despite having no clue that your weather typically is convected from West to East, nor that sunlight is composed of a spectrum of colors which can each be refracted differently by particles which tend to be suspended in a dry atmosphere, you might come up with a line like "red sky at night, sailors delight, red sky at morning, sailors take warning." Observing that some people die after eating raw foods might induce adoption of societal norms for cooking.

Human survival depended on our ability not to necessarily understand mechanisms, but rather to recognize cause and effect patterns. In fact a lot of science happens this way. Nobody understands cancer completely, but we notice that smoking a lot of tobacco correlates with various cancers, so we do the only logical thing and continue producing cigarettes by the millions so long as the companies that make them pay the government and lots of plaintiffs' lawyers billions of dollars.

What we now know though is that cause and effect aren't infallible. Do Jehovah's Witnesses live longer because they are Jehovah's Witnesses, or because they are vegetarians? Are blacks and women financially less successful than white males because they are inferior, or because society historically treated them that way? A lot of people are confused about this, as evidenced by sales of shark cartilage based on the observation that sharks don't have knee problems—ignoring the fact that sharks not only don't have knees, but don't really have a taste for basketball or marathon

running. Are people really weirder when there's a full moon, or do we just expect them to be? Why not toss a little salt over your shoulder, or knock on wood, or call the 13th floor of the shimmering AeroAstro tower the fourteenth floor—you never know...

Back to our design review. With half the money spent, and something less than that fraction of the work completed, the project manager gets told that in the interest of mission success, we need more reviews, we need to upgrade parts quality and adding a redundant charge controller wouldn't be a bad idea either. What responsible denizen of aerospace engineering is going to come out against our equivalent of apple pie and mom—redundancy, better parts, more eyes focused on the design and procedures and the Parent of God herself, mission success? Not one who intends a long career in the business. (Rational managers choose answer (a) from the Pop Quiz). Or, looked at another way—nobody stands up in a design review, defending a spacecraft that is a bit late and over budget, and says we should not have high quality parts, we should have less redundancy, and we should have fewer reviews and reviewers. Right?

But what is the rational basis for these recommendations? Is there any analysis to support them? What there is are things like common sense and experience. Any first year engineering student can tell you that the failure probability of a system is dependent on the failure probabilities of its components. Doesn't this mean better components make a better satellite? Who can argue that on the third or tenth or thirtieth look a reviewer might find a heretofore undiscovered flaw in a design? If a radio fails, isn't it better to have a backup? Better parts, more reviews, redundancy. Common Sense.

6.1 What's the Matter with Spending Money to Reduce Risk?

Who could be against airbags, or forcing the French to put guardrails around their tunnel structures to protect drunk drivers and their passengers? One missing fact is that whether we realize it or not, everything is cost-constrained. The French highway folks only have so many francs, and they elected to spend them on beautiful and thorough lighting of their tunnels instead of building guardrails. Not long ago, legislation was introduced in the U.S. to prevent parents from carrying children on their laps in airliners. In a crash, a child is much safer in its own seat with a proper child seat and seatbelt. Unfortunately, the result was that more parents started driving their kids to their destinations, unable to afford all those airline tickets, and mile for mile, driving is ten times riskier than airline travel. Now that this rule has been reversed, we are saving children's lives by allowing them to fly without seatbelts.

Spacecraft programs appear bent on paving a road to failure with good intentions. As soon as the risk card is played, real engineering stops. Everybody has to focus on risk. Who is checking and rechecking documentation for errors? How do we know a redundant system won't fail just as the primary might? What's our confidence level in component W? What program manager has the guts to tell the askers of those questions to crawl back into their holes and let the engineers work in peace?

On the other hand, surgeons are seldom asked mid-procedure if they're sure that the artery they're working on is really the right one, whether they really sewed it securely, and do they have confidence the anesthesiologist really knows what he's doing. Talking about risk is great, but interrupting minds at work causes errors. I was packing for my last trip when, in the middle of putting clothes into the suitcase, one of my coworkers called to remind me not to forget a particular document. I didn't, but I did forget to pack socks, the drawer of which I was just opening when the phone rang.

Way down deep in the design process, or even later during system test, it's easy to forget why things are as they are, and changing them to make them more reliable tends to erode reliability. It's easy to add a line of code to make sure that Radio A definitely switches over to Radio B if it doesn't hear from the ground within seventy-two hours. But if Radio B doesn't know to switch to Radio A, then a seventy-two-hour outage at the ground station can kill a whole mission (and did exactly that to one expensive satellite). A solar panel hinge is binding on a wire, and a small amount of metal gets machined off the hinge—the wire is safe, but the hinge later fails.

Every dollar spent on risk reduction improves mission success? Then why pay engineers at all? Let one person build the computer, software, radios, structure and thermal systems, and pay a hundred people to ensure it all works. Do the risk reduction bureaucrats engineer better than the engineers? If so, throw out the engineers and put the risk reduction forces to work. The reality is that every dollar spent on risk reduction is a dollar not spent on hiring the most capable, experienced engineering talent, and giving those crucial activities the time and capital resources needed to do the best job possible. There isn't a spacecraft program in existence that isn't understaffed and rushed. Spending more money on quality assurance people who add to the engineering burden isn't necessarily helping ensure a great job gets done.

The blind faith we put in ostensible risk reduction activities, like the faith that sacrificing enough virgins into an active volcano, preferably at the full moon, will eventually bring an end to your latest drought, are poor substitutes for understanding what causes failures (or droughts). To wit: If a second or third read on a design is better, what about a fourth, fifth, tenth, thirtieth or thousandth? Clearly at some point most people would conclude we are overdoing a good thing. So why is three better than two, but one thousand is not better than nine hundred and ninety-nine?? If highly reliable parts are better, how much more highly reliable? We could not just buy great parts, we could buy hundreds of each great part, have them each tested extensively, X-rayed, microscopically and ultrasonically inspected. Why is a certain grade of parts better, but spending ten or one hundred times more than even the best parts cost, not better still? Even redundancy—if it's better, then why aren't the human heart, brain and esophagus redundant? Are we an inferior design that should have evolved differently? Your car has redundant brakes—but no redundant steering, seat belts or airbag. Should we call Ralph Nader?

There are two missing ingredients in avoiding cancer with copper bracelets and making better spacecraft with parts, redundancy and reviews: cost and focus. A dollar spent on better parts is a dollar not available someplace else—for testing the system, or maybe for training the ground ops crew, or for scrutinizing something other than the spacecraft that could fail. Faith in a copper bracelet might lead to complacency about diet, exercise or seeing a physician for a possibly uncomfortable physical exam. Redundancy not only costs money, it adds complexity, creating new branches of the tree of possible spacecraft states, already so vast that we haven't got the resources to travel down each of them. When you have only one radio, you are focused on making sure it works. When you have two or three, you're a bit less focused. Another review of the design could uncover an error, or it could just exhaust precious resources, particularly the program's most precious resource—the engineers who are supposed to be building and troubleshooting the satellite. Or, it could uncover an apparent "error." Months ago when the designer inserted it, it played an important role and in fact for some subtle reason doesn't actually have any serious deleterious effect. Without that "error" the satellite won't work—removing it while the rushed engineering team is too busy to remember why it got in there in the first place could also be the outcome of another review. Those infamous last minute fixes we are all counseled to avoid.

Judgment, experience, intuition—these are qualities experience teaches us to respect. And for good reason—they are the result of that experience base which matured humanity from the Serengeti to San Francisco. And when the patient isn't breathing, there really isn't time to create a decision tree and calculate the relative efficacy of all the possible causes and their remedial steps. But without quantitative reasoning, without a scientific method, we might still think seat belts a bad idea because they could slow your escape from a burning vehicle, that a shot of whiskey calms your nerves before getting in the cockpit and that since diseases travel in blood, less blood means less disease. Why else would God have created leaches?

Reliability is unfortunately not easily calculated. We don't know the true reliability of parts and we don't have a good track record predicting the inventive ways satellites die (or are killed by their operators). And parts are less and less often the failure source. When failures happen, they are more often the result of human error—design flaws, mishandling parts or the satellite, sending up the wrong commands. Less often but still a concern, is another human error—fabrication error. Each part has many leads coming off of it, and each lead may be serviced through multiple circuit board traces, solder joints, wires and connectors. Some of these are mechanically "staked," others aren't—how do we account for each connector and connection's individual circumstances?

6.2 We All Want to Add Stuff

...in case the stuff we've already got doesn't work. In flying passengers around in light aircraft, you meet many, perhaps the majority, who believe that a twin engine plane is safer than a plane with one engine. Well, it depends what you mean by safer. Statistically, pilots make fewer mistakes in single engine planes. In tests of pilots learning to fly twin engine planes, a large percentage, maybe 20 or 30%, when confronted with an engine failure, will think the working engine has failed and "safe" (shut down) the only remaining engine on the aircraft. Pilot skill notwithstanding, a plane with no engines running is heading for earth. If an engine has a 1 in a million chance to fail, two engines have a two in a million chance to fail, so the chances of an engine failure are doubled when you double the number of engines. The implicit assumption is that the airplane will fly with only one of the two engines working, which may or may not be true.

When we double the number of radios, processors, ACS sensors, or separation pyrotechnics in the name of reliability, it's easy to forget that we've nourished a sinister beast—complexity. Your PC is not requiring a reboot for the nth time this week because a transistor someplace in an IC is failing. It's locking up because its software is so complex that the company responsible for coding it hasn't been able to understand and fix all the possible failure modes. A satellite with one processor doesn't need a complex system to decide which processor to use and switches and logic to move critical functions from one processor to another. And wiring harnesses to enable either processor to control all the spacecraft functions. Processors aren't attached to satellites by one or two spokes. They have lots and lots of wires and other processors and logic circuits and memory chips attached to them. Switching all that stuff seamlessly is complicated. By adding a backup, we add lots of new ways for things to break.

Lacking solid understanding, systems like Keplerian mechanics, historically we've turned to Zodiacs, Zeus, rainmakers and the Farmer's Almanac. Maybe in their day they were better than nothing. Fortunately, we aren't in their day any more. Unfortunately, people are still relying on them. The first step to doing better is to admit all that bluster about reviews, high grade parts, and redundancy is noise. Noise that may have served us well when transistors were the rage, cars had an MTBF of about fifteen hundred miles and satellites flew with no software at all. Without our candles, Ouija board and ceremonial robes, we're then equipped to begin a search for understanding.

Here's some logic to get the ball rolling: satellites are built by engineers. If you have zero engineering, the probability of mission failure is 100%, regardless of the parts quality, redundancy, or number of oversight committees. Thus increasing the resources spent on engineering, at least in the limit near zero engineering, is the best thing you can do with precious resources. At some point, adding another hour of effective engineering might be less valuable than taking the engineers off the job to make viewgraphs, travel across the country, live out of suitcases, put on creased ties, silk blouses and/or wool trousers and stand up in front of reviewers. But how do we know when reviewing is more valuable than engineering? I don't know the answer—and the answer varies from program to program. But, using as a weather vane the moment when a person who makes her or his living doing reviewing tells you, after spending 3 hours on a program you've been laboring on for over a year, that more reviewing is better than more engineering, is probably not the answer.

As if this question weren't complex enough, it gets more complex. There is not simply a tradeoff between engineering and reviewing, both of which are necessary and need to be done right. There is also testing. Many kinds of testing. Subsystem testing, integrated system operational testing, environmental testing, acceptance testing.

Amsat's approach historically was heavy on engineering, very light on reviewing (since that involved travel no one could afford) and heavy on testing. The logic was that however it got built, if it worked for a long time in a wide variety of environments spanning more than the range of circumstances the satellite was ever going to encounter, it was working. When I buy a new car, it's a substantial investment. How much inspection do I do of a new car before accepting it? Usually a walk around to squabble with the dealer over a scratch. How much reviewing of the materials and procedures used to build and prepare it? Of course none—I entrust that to the manufacturer. But I test it—I drive it. If the new car survives a month of real world operation, that's the strongest evidence I have that it's good. Of course a satellite can't get launched for a month trial—but terrestrial testing for all but a few systems is more difficult to survive than the on-orbit environment.

But testing takes resources—away from engineering, away from reviewing. Upgrading parts absorbs resources from engineering, reviewing and testing. Building more performance into a satellite to ensure it will be working fine, with adequate power three or five years after launch takes money away from engineering, reviewing, parts quality and testing.

Testing, reviewing, engineering, parts quality, redundancy. They have one thing in common—the resources of time and money needed to do or buy them. And in a world of fixed resources, it's not obvious that doing more of one at the expense of another improves anything.

And as we spend money on quality inspection, on consultants to drive engineers to distraction proving the thing they haven't had time to design can't fail, and on redundant Kim Wipe suppliers, the grim reaper, the genuine angel of spacecraft death, smiles wryly and hobbles ever closer to our mission with his (or her) dark wispy cape and sharp sickle.

The answer must be more money so that we can do more of everything and not less of anything. Unfortunately, in addition to love, there are other things money can't buy. As the clock ticks toward launch, adding more tests, awaiting higher quality parts and attending more reviews tend to slow the schedule, even if more money is made available. A rush at the end of a program is a formula for a disaster on orbit. And can absorb even more money than the increase provided for the additional activities. Remember the hurried and disastrous launch of Conestoga. Late, out of money, and under the gun from NASA and management to do something—they did. The resulting explosion was the final punctuation mark in the saga of a doomed program.

Parts that arrive late in the program don't get tested as well as the ones we've lived with through the whole development effort. More money does often buy more people, often people who haven't been with the program since its inception. These new faces need to be brought up to speed by the in-place team. This slows the schedule, costs more money and decreases their focus. As the team grows, its organization will have to evolve. A small program with a staff that ate lunch together every day enjoys excellent communications with very little communications overhead. Double the team size and in addition to paying for twice the staff plus the additional time needed to spin up the team, you will need to budget for enhanced documentation, more travel, less efficient utilization of people, more management. Taken to extreme, you may not actually have a small program at all anymore. Hierarchical management will bring with it the approved parts lists, standardized procedures, external QA/PA teams and myriad other complexities and expenses.

After you pay for all that, and the cost increments here are not typically 10% or 25%, but rather factors of three to ten greater than the original program plan, and you're still not going to have bought love or happiness, or even reliability. The inevitable miscommunications working across a large, heterogeneous team can lead to failures. More ominously, budget growth kills programs. In fact, budgets kill a lot more programs than parts upgrades, additional reviews, QA/PA teams, masses of paperwork and redundant charge controllers can ever save. Many of these lost souls are not obvious to us because they don't exist. They are programs—good programs—that never happened or never got past a conceptual design because the money wasn't there.

Limiting ourselves to programs progressed beyond concept, to the land of the design reviews and rising anxieties and expectations, excepting the Space Station, NASA has no appetite for overruns. Even if this program survives, what are your chances of doing another one? The client, their client—everyone—will be mad at you. Even if you accurately estimate the costs of the upgrades, they won't believe you. When the final numbers are in, the costs of the changes and delays and added staff, the reengineering for the unexpected when changes in requirements and parts are made, will be shocking and you will be the scapegoat.

Only about 5% of satellites put into their correct orbits fail. But 98% or more of satellites fail. That 98% of satellites never get launched. These infant mortalities wipe out most of what we could be doing in space. It's as if our society spent all its health care dollars on exotic transplant research to extend the lives of the aged while the overwhelming majority of our children died of measles, malaria, and malnutrition—relatively simple problems we can fix for a few parts per million of the cost of those highly complex procedures. The grimmest reaper is the budget. Most satellites die for lack of resources. They may never get study money, or they get study money but no budget to move on to a system design. Or the effort to sell the mission saps the money that could have been used to prototype a key technology, and the mission dies because there's a perception that there's too much risk in relying on an untested system.

Dan Goldin, NASA's chief, has killed a few missions that were designed, built, and tested because they ran more than a few percent over budget and didn't have the dollars to get launched. I happen to agree with him. If you don't send a message that you're serious about cost, the system will eat you alive. But he has underscored the real threat: running out of money.

6.3 If Budget Isn't the Culprit, Schedule Is

Not too many satellites are killed for missing their launch, but lots of 'em are killed because their missions go out of fashion or become irrelevant. More mistakes get made because engineers only have a certain half-life. The longer a program lasts, the fewer of the original crew remain, disappearing along with them the knowledge of why we made that hinge as thick as it is, or why the software really wants that extra line of code in there.

Spending more money, spending more time, interrupting the process of engineering, adding complexity. Do these sound like ways to increase reliability? But those are the steps we take when we get nervous.

And we all get a bit tense as programs move on. We've expended years of our lives, weekends and evenings of overtime, loads of personal sacrifices, and plenty of headaches, all invested in that satellite. We'll do anything to make sure it works. But what we do, acting on intuition with no solid data, is distort resource allocation. Take money, or time, or focus away from the engineer's work and refocus it on inspection, test, backup systems, upgraded standards, and complexity. We can't prove anything is more reliable, but by spending resources on reliability, we feel better.

Or you feel better. I don't. No secondary or tertiary oversight or additional review has saved any of my missions. Good engineers and managers with ample budgets and schedules who stick with the program from start to finish have saved all my missions. I may not be a reliability expert, but I go with what works.

The chapter began with a pop quiz. Here it is again. Are you so sure of the right answer—the same one you picked five pages ago?

Pop Quiz again: You are the project engineer developing a small satellite, and you have just laid out the program plan at a mammoth design review. You have presented the design, the parts selection choices, the testing plan. Your team has gone through the design and operations details. You stand up at the end of the review and present the budget. Fifty percent of the money is already spent, and something less than that fraction of the work is completed. A consultant to your customer stands up and suggests, that in the interest of mission success, you need to hold more reviews, to upgrade parts quality in certain critical systems, and add a redundant charge controller. Your response is:

a) You agree that all those steps are improvements that will enhance mission success and promise to implement them.

b) You agree they are desirable but plead for more money so that you won't truncate other parts of the planned activity.

c) You state that all those steps will decrease the probability of mission success and refuse to implement them.

Here's a last bit of security for you. If enough people read this book, you'll at least have company while awaiting your appointment with the corporate outplacement specialist.

Chapter 7

Where to Look for Historical Underpinnings, Term Definitions, and Revolutionary Zeal Turned Up to 11

7.1 Seeing Small through Several Lenses

In 1986 I got the idea that a never-ending homework assignment aka a monthly newsletter on small space would be fun. Which it is compared with, say, being the subject of a Ken Starr investigation or a ten day CDR including sessions on documentation and quality assurance which the entire team is forced to attend on a sunny Sunday from 0800 to sunset. We people who swim at 6 AM understand the timeless wisdom of "Eat a frog every morning and nothing worse will happen to you all day." Other than an exercise in self discipline, my 'Zine, presciently founded years before the Web or 'Zines or even *Wired*, was a countercultural voice in the space wilderness. Its definition of small:

> That which the satellite establishment, including the U.S. Government's FCC and WARC, have to date not taken into account in their planning. Those that are out of the realm of the previous legislation, regulation, and policy formulation of almost everyone. That's a small satellite. That nihilistic definition would leave virtually nothing in the small satellite category now. Satellites the size of your finger nail are now unabashedly thought and written about and even funded by the world's big government space bureaucracies, and maybe some day soon even by some serious business people.

The 'Zine's resident wacky professor, Dr. R. Zlitz opined: "Small—that's not the question. Space doesn't care how big your satellite is—space is really big—plenty of room. Doesn't charge by the square foot, you know.... It's cost. If the satellite is cheap

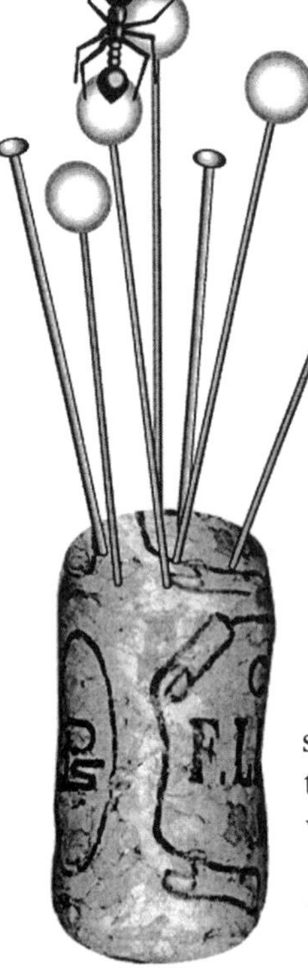

SMALL? We're talking small.

enough, and you can launch it cheaply enough (so, for that reason maybe it needs to be small), then you have a lot of advantages. You can take risk.

"Counter example: the Space Shuttle off-loaded its last ferrous memory beads in 1991, when 4 Mbit chips were already in lots of salespersons' sample kits and 1Mbit memories were at Radio Shack. The Shuttle is not cheap, cannot take risk. Small satellites are where new technologies are welcomed into space.

"Redundancy. Maybe you don't need it. A small satellite failure won't sink the treasury of a mid-sized country (refer to Risk), but more importantly, reliability is intrinsically high in simple systems. How often does your Sony Walkman, suddenly and without notice, just croak? Probably never, not counting the time you dropped it out of your beach bag into the surf at St. Kitts. When was the last time your car, even an old car, just suddenly stopped running in the middle of the freeway, not counting running out of gas? Probably not since you traded in that '65 Ford with the tube AM radio and the Bug Eye Sprite made that final journey to Lucas in the Sky with all of your Diamonds. The point is, a system with relatively few components suffers relatively few failures. If you want redundancy, fly two.

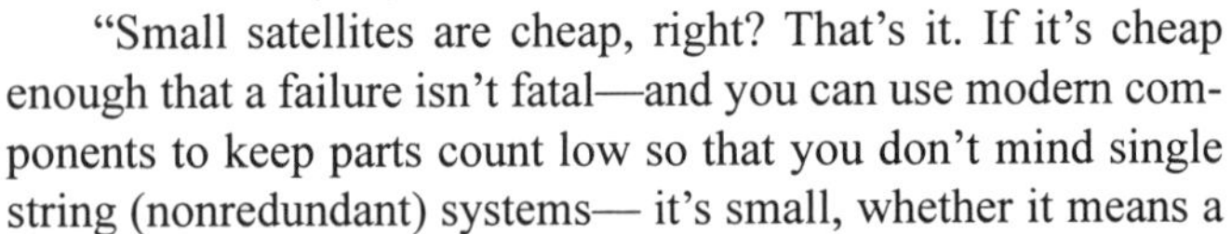

"Small satellites are cheap, right? That's it. If it's cheap enough that a failure isn't fatal—and you can use modern components to keep parts count low so that you don't mind single string (nonredundant) systems— it's small, whether it means a satellite, launch vehicle, or brewery. Speaking of which, there's a microbrewery just down the street..."

From DARPA: "Small means smaller than what people were using before we came along—to do about the same job. Doesn't matter if it's 100 g or 100 kg."

From one Military User: "Single mission. Big satellites carry multiple payloads. The mass of the interface control documents exceeds that of the spacecraft and approaches the weight of the launch vehicle. Little satellites mean you can afford a lot of them, and systems are composed of a network of small satellites instead of one or two big ones. That means hardness, so that destruction or failure of one may mean degradation of the system, but not total loss of capability. Small satellites means multiple satellites doing missions previously done by smaller numbers of larger ones."

From Scientific Users: "Something you develop in two years instead of 10. System simplicity means fast development time, which means you get research done while someone other than your grandchildren still cares about it.

"It also means the science payload is in charge, not a guest. Scientists dictate where the spacecraft is pointing, when the payload is operating, and how to optimize the use of on-board resources like power, computation, and memory space."

From the Program Manager: "It means the program organization is small enough to avoid documentation beyond the actual design drawings—schematics, layouts, and block diagrams the engineers need to build their systems. Systems engineering, quality control, test, integration are all done by the engineers building the system. They all understand the whole system, and cross-disciplinary engineering is the rule, not the exception. Built by a team where everyone, or no one, is a systems engineer, there are few design rules. Each engineer selects materials, components, design, and fabrication methods that meet the top level program objectives. If I can manage like that, it's a small satellite."

From the Client/Customer: "It means keeping the faith. Small satellites put a lot of emphasis on the individuals building them, instead of on the bureaucracy that controls them. You need to use a strong, capable, diverse team and let them perform with a minimum of bureaucracy. Burdening the program with the reporting I'd otherwise like enlarges the team and the cost. Pretty soon, the development group is too big to communicate intimately, you need paper interfaces, you need people to read and write specs, people to go to meetings, people to enforce quality and standards, and money to pay them. And you don't have a small satellite program anymore. All in all, not the most comfortable thing to purchase—you can't hide behind the way it's always been done. The small satellite is a way to get into space when the alternative, on my budget, is not to fly at all."

From the Engineer: "The design process is interactive. You've got a lot of constraints. You can't provide hundreds of watts of power, microradian pointing over 4 pi stearradians, cooling to the helium lambda point, a gigabit per second downlink, a terrabyte of memory, and a couple of Cray-XMP equivalents to handle it all. When that reality finally dawns on the customer, a conversation starts that lasts almost throughout the program. It can get unpleasant for everybody as we grapple with the various envelopes—mass, power, volume, cost—whose constrains we have to live within. The user sometimes has to downscope expectations on the number of payloads that can be used, how big they can be, and also on the quality of the components and the redundancy. It's difficult and, to keep costs low, we can't spend forever on every decision. In the end maybe nobody gets exactly when they wanted at the subsystems level, but everyone gets an efficient, low cost system—80% of what they wanted for 20% of what they would have paid, and flown in 20% of the time needed for a conventional satellite program. It's a compromise users are making more and more."

7.2 Small Spacecraft Time Capsule: The Way (We Think) We Were

Meandering among the hundreds of attendants and rooms full of industry displays both from small organizations and almost all the major aerospace companies, the sales representative men in their ties and jackets and women in dresses and makeup all crowded into the Utah small satellite conference mostly selling whatever could be sold to a sea of other sellers. Sven Grahn from Swedish Space Corporation lamented to me: "I don't think this bunch needs us any more."

Achieving a state of total redundancy might be the only sensible goal of adulthood. In the MTV world of news bytes and bungee jumping, it may be the only justification, outside of a challenge to health care technology and economics, for the existence of us over 30-somethings.

Around the time our so-called industry congealed out of its primordial soup, Utah was a bunch of engineers, most of them from Amsat, giving talks on satellites we had built or hoped to build depending on how much money we could steal out of the cookie jar at home or how many parts could be salvaged from the stores warehouse back at school. The industry displays were some Amsat models on card tables. There was no industry. That was 1986. The Air Force Chief of Staff had yet to come out with the simplifying declaration of the decade—that there was absolutely no use for these small satellites and that they were a waste of time and resources. Nonetheless, a few military men and women came to Utah in civilian clothes, just to play it safe.

Just to remind you what 1986 means—that era was also before laptop computers; many companies didn't yet have fax machines (considering them, one presumes, a waste of resources too); and it was also before Reagan's military buildup reversed its first derivative, making over 100,000 aerospace engineering and management jobs redundant. It was the end of history, at least that was what was predicted for Aerospace Engineers bailing out of all the big companies and design bureaus.

The notion that a small satellite is just a small large satellite is a sorry devolution of the initiative we all undertook to found this field. Think how far we've all come from Victorian England and the Puritans, where for the sake of some abstract ideology men and women bound their bodies in corsets and tight boots, in long jackets and puffed out dresses, and in some pretty monstrous neckware and heavy duty hats and plumage. While the monotonic trend over the past 250 years has been toward informality, Utah has gone from the land of Jeans to the land of Suits.

So Sven, I don't think we're totally redundant yet. We have yet to leave behind a few strands of small satellite DNA encased in amber that perhaps will enable some mad scientist of the 21st century to rebuild what the crush of commerce, busily saving jobs at the aerospace dinosaurs, seems so intent on ignoring. Hence and forthwith, some organic remnants of a great endeavor, building cool little satellites, often without money.

KISS. What does it really stand for? Advice given to English Barristers in training: Keep It Simple and Salient. Nobody says salient anymore. Most people think it has to do with an ionic solvent widely used by chemists, biologists, and contact lens wearers, a sorry fact when you then must experience the products of the generation of lawyers that has forgotten it.

The almost perfect success record of small satellites is not a result of our supreme talents, unfortunately, but rather of the application of KISS. When you build a satellite that weighs 50 kg, you have about 1% the parts count of a satellite weighing five tons. Ignoring redundancy considerations (and both small and large satellites have redundant systems, making the argument itself somewhat redundant), the probability of a critical component failure in a conventional satellite is, assuming we're using equally reliable parts, about 1000 times greater than in a small satellite. Experience bears this out in everyday life. What's the probability of your car collapsing on the way in to work this morning? Not too great, assuming you're not driving a '69 MGC equipped with full Lucas electrics. But what's the probability of seeing two or three cars stranded along the roadside during your 10-mile commute? Virtually certain. For large numbers of cars, representing in total a huge number of parts, several are always going wrong, but for small numbers, their reliability is remarkable, especially remarkable for us graduates of the Ford 289 V-8 school of 1960s driving technology.

The wonderful reliability of portable radios, and for that matter of laptop PCs, is again a result of their very low parts count. My walkman AM/FM has exactly one IC chip, two connectors (battery and headset) and three knobs—on/off, volume, and tuning. It refuses to quit, but my Drake TR-4cw ham radio transceiver, with over 25 vacuum tubes and multiple circuit boards loaded with discrete resisters, capacitors, and inductors, a panel full of knobs to tweak and over 100 pins of connectors on the back, breaks about bimonthly. If I don't use it too much.

We small satellite mammals have just picked a much, much easier problem. Our reward has been excellent reliability, perhaps despite our lack of extreme attention to quality control and detail. Biggest hazard? Getting underfoot of big mammals!

LSMFT. Does it really stand for Lucky Strike Means Fine Tobacco? Or is it Lighter Spacecraft Mean Faster Technology Transfer? Probably the former, but when a client drops a significant fraction of the GNP on your RFP, you can bet the war cry

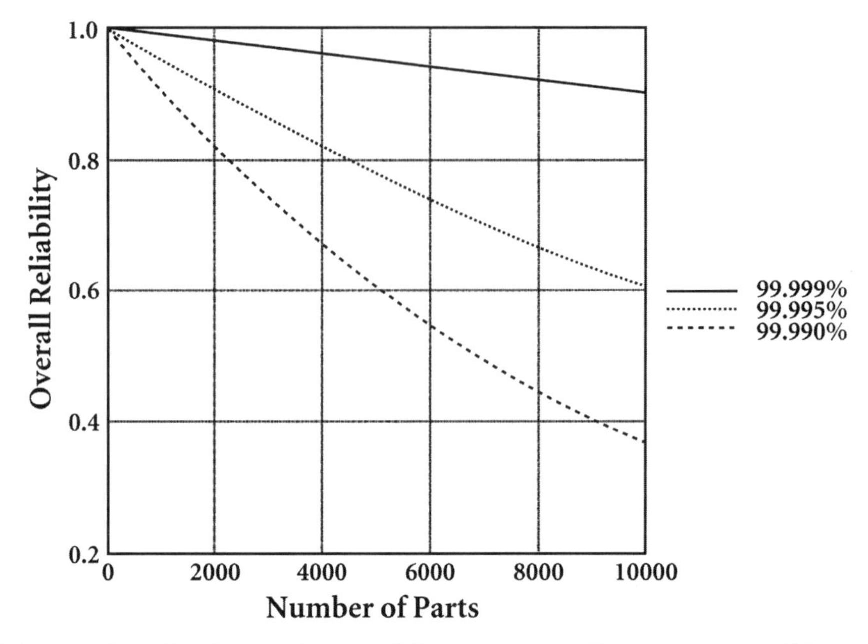

A simple system with a few parts is more reliable than systems with more parts—even if the simple system is made with inexpensive, less reliable parts.

goes up: Space Qualified! What does "space qualified" mean? Like the rest of society, the space world lives in its euphemisms, much as the Sufis live within their own contradictions rather than face those of a world of hunger and disease. In fact, space qualification is pretty easy. You test something through the range of environments it will see in space. Computer chips and software really don't care much about zero gravity, so putting them through thermal vacuum, vibration, and maybe radiation testing pretty much constitutes space qualification.

But the true Cult Of Space-qual Technology (COST) prefers that every damned part on the satellite has flown in space before. Clearly in 1957 this was tough. The only things "space qualified" according to this definition were comets, asteroids, planets, and their moons, stars, and some dark interstellar media, none of which are all that useful for telephoning your aunt in Poughkeepsie from vacation in Pago Pago. At some point, roughly when Apollo 13 limped back from the moon, NASA collected the toys of its youth and decided those would be the only toys of its adulthood. Imagine living on a diet of Mousetrap and Swinger for 60 more years? NASA's world has no Trivial Pursuits, no Bungee jumping, no Where in Time is Carmen San Diego? Just Monopoly and Scrabble.

Why else would the Space Shuttle computer system have been dwarfed by the awesome computing gazorch of an old HP-35 calculator carried into space by one of its crew? Is hiring a team of quilt makers from West Virginia to weave together ferrous beads a better way to build computer memory than integrated circuitry? The Space Shuttle made do with 2000 bytes of RAM, about 0.1% of what's built into the computer your kid uses to play Kung Fu Cat on. But that 2k was Space Qualified!

One problem with space age retrotechnology is that it drives parts count up enormously. Failure probability goes as the product of the reliability of the critical parts. A single part might be 99.99% reliable, or 0.9999 in probability theory parlance. Two such parts in series have 0.9999 x 0.9999 = 0.9998 reliability, or 99.98%. One hundred such parts have reliability 99%, and one thousand of such parts in theory have reliability 90%, a significant 10% failure risk. The figure above illustrates the overall reliability of a spacecraft vs. its part count for three classes of individual component reliability. The miracle isn't that little satellites are so reliable, but rather that big ones work at all.

The "What's the worst thing that can happen" reliability corollary.

When the space shuttle went down, the loss was measured in human lives, discussed and debated in our national news, scrutinized by Congressional inquiry. A serious blow was dealt to the US program in space, indeed to our national prestige. The Shuttle, then our only means of launching a large fraction of our space missions, was grounded for two years. This was the equivalent for space of the gas lines of the mid '70s. Like those gas lines, it has faded from memory for most of us. But then when Mars Observer went reclusive on us, the loss was measured in billions of dollars and thousands of engineers' and scientists' careers that had been dedicated to the Mars mission. Once again society questioned our leadership in space. The interesting reaction from at least some quarters at NASA was "no more major missions." Really? Then why do we need 20,000 civil servants and 80,000 (those aren't NASA employees, of course) in-house contractors under one roof at NASA?

By contrast, the failure of a $1M minisatellite is a small paint scratch— 0.08% of NASA's budget, and maybe 0.03% of our total national space budget. (Yes, Virginia, the CIA and NSA have space budgets too, as do NOAA, DoD and DoE. NASA is no longer THE space agency; it is A space agency.)

The initiative by USRA to launch two space science missions known as STEDI (Student Explorer Demonstration Initiative) and UNEX (University Explorer) for $24M is flawed in that it's too small. USRA should press NASA for $240M to do 20 missions—each year. They are built inexpensively, largely by students. Hundreds of graduate and undergraduate students design, build, test, fly, and operate satellites within the span of their scholastic careers. They undertake significant technical risks to do innovative science in new ways. They won't be tested to extremes and their designs extensively reviewed. Even if they achieve failure rates unprecedentedly high in small satellites, like 10% or 20%, the nation will succeed in launching 24 successful science missions a year on 2% of NASA's budget.

The cruel history of life on earth is that a lot of us don't make it. Mortality is the rule. Yet we survive because of the enormous number of attempts we make at life. An insect might lay 10,000 eggs, and an individual eco-niche harbors millions of similar insects. Not every one can succeed, but enough do succeed to make insects one of the most successful species on earth, both in terms of evolutionary longevity and sheer numbers. The way to do small satellites is not one at a time. It is 5 or 10 or 100 at a time. Accept failure rates of 10% or more. When one of Motorola's Iridiums (Iridia?) stopped working, the official Motorola line was straight from Alfred E. Newman himself: "What, me worry?" By flying 11 satellites to get 10 in operation on orbit, you can bypass most of what makes satellites so expensive, such as high quality parts, redundant systems, extensive testing of each unit, restrictions to use only "space-qualified" components. With these steps, satellites can be built for 50% or in some cases less than 10% the cost of a traditional design, and a 10% reliability tax is lost in the noise.

What's the difference between a drunk and an alcoholic?

Drunks don't have to go to all those classes. But one thing they teach at those remedial classes is that you can't do everything. Small satellites don't do everything. They do a few things pretty well. The way to design a small mission is not to decide to cut the mass of Space Telescope from 10 tons to 10 kg. To do so might require an investment equal to or greater than the US GNP. Low cost missions result, in part, from limiting our sights to those missions that nicely fit. The key is, fit into what? I'd say, fit into a work package for the entire mission that 15 people or fewer can accomplish. Limiting management complexity and keeping overhead low and efficiency high, the resource requirements and number of active components definitely remain modest enough to allow innovation without undue risk.

Is high technology the road to low cost?

As a swimmer, I know a lot of other swimmers. Many of my swim buddies train for swimming by lifting weights, pulling on elastic cords, training on complex machines. None of these people swim faster than I do, but they are much better at lifting weights, pulling on elastic cords, and operating complex machinery. Nothing wrong with that, but if your goal is swimming, there's nothing like practice in the pool, which is itself an interesting lesson for people who might want their own low cost satellites. You might try contacting some of the people who build them, as opposed to the people who instead build very high cost satellites. Compelling logic, don't you think?

If your spacecraft development goals are simplicity, reliability, and low cost, there is no reason not to embark on a technology development program. It won't help you achieve simplicity, reliability or low cost, but it's fun and could even pay off down the road. For instance, you could come up with a new clamp for attaching chain link fences to posts, sell it to the Department of Highways, and retire with millions. Oops, too late. Somebody already did that. Besides lots of fences still attached to their posts, this innovation allowed its inventor to open a very nice golf course too. But nowadays neither he, nor a satellite technology R&D program, turns out a cheap satellite at the end of the day, month or year, even with multiyear follow-ons.

Governments and banks are very conservative institutions. Unlike people who write amusing little baubles on space technology, leaving their typewriter keyboarded Mac only for the occasional swim or bike ride, they have responsibility for people's lives and retirement nest eggs. They can't loan ten thousand depositor dollars to small struggling dry cleaners and microsatellite factories down the street from them. They can't blow taxpayer dollars on mere functionality—hey, isn't that a job for industry or venture capital? No, they have to put hundreds of billions of dollars into Indonesia or Russia, or bailing out hedge funds. I think this is what Nixon may have termed "the big lie." Well, nothing is so big about small satellites. Our little big lie (not a cattle ranch on TV nor a description of your golf ball which has come to rest inches from the water hazard) is that to be useful, we need oodles of technology development—the techno equivalent of elastic stretchy bands, weights, and complex exercise machines. Radical is my department, since no orphans are dependent on me, no politicians vote money for me. Technology development is the biggest single detriment to development of small satellites.

What drove Epson, the computer company, and their CPM operating system, out of that business? Innovation. They were stuck with a fortune in old machines while loyal customers who loved the machines and their operating system (the Linux of the '70s) awaited the next innovation about to be released. PCs didn't suddenly become corporately purchasable because their 8-bit wide CPU's clocks finally ticked above the 4 MHz barrier. Enough people had started messing around with the technologies that existed to a) find clever new ways to put them together and package them to appeal to consumers (e.g., laptops) and b) clever new ways to use them despite their incredibly stone age technology (e.g., Wordstar and Lotus 1-2-3 spreadsheets). Almost twenty years later we have a lot more chromium, color, sound, user help, fancy print options and data about the data, but we're still looking at the A through ZZ columns and the rows that number from 1 to accountant eternity. If all you want to do is surf the web, you can buy a used machine for $250 and it will do fine. The technology is icing, but experience using the stuff was the key to figuring out what is the value proposition of a small computer, any small computer regardless of clockspeed, interface bit width, hard disc capacity or screen size, is. The Beef, to quote a hamburger commercial of that same era of enlightenment, is the way of consumer demand instead of technology push.

Small satellites will suddenly solve the world's problems as soon as the processors are even smaller, or the radios are X-band instead of S-band? I don't think so. Instead, we keep sending our users a message—the technology isn't ready for prime time yet. Luckily Alexander Graham Bell didn't await the 56k modem, and Edison made do with waxed cylinders instead of awaiting DVD.

Really, the damned satellites are already small enough. What's big is their apertures—photovoltaic cells, radio antennas, radiators. If you need 10 square meters of solar panels, you don't care if the satellite weighs 1 kg or 10. Deployables and other ways of creating the capabilities of large apertures in small packages is a key technology. Most of what is out there passing as critical microsat technology is training wheels for young engineers and companies wanting to produce revenue and earn some "flight heritage" of their own. But it's irrelevant to finding applications for microspacecraft.

I can snub our minute investments in technology because we microspacecrafters partly achieve low cost by feeding off the R&D products of huge programs that can afford to do real technology development. For example, the digital hardware systems of the most advanced small satellites exploit microprocessor, memory, and logic technologies pioneered for laptop computers, a business about four orders of magnitude (ten thousand times) larger than the low cost satellite market. Even a very generously funded technology program cannot compete with the capital resources Boeing invests in its advanced aircraft structures, or IBM and Apple in the microcircuitry of the G4, or Toshiba and Union Carbide in advanced rechargeable batteries. The surface mount technology built into your cellular phone, walkman, and car CD player is just now migrating into small satellites without benefit of much small satellite R&D support. We let Motorola & Co. pick up the check for that lunch.

Herewith, the advice of the ages (and aged): Go forth without shame, learn about the hot new technologies that are driving new product developments in industry, and apply your space expertise to choosing those that really offer promising improvements to your spacecraft design. High density static and pseudo static RAM have been a boon to microsatellites, and memories of multi-gigabit capacity have already flown on sub-50 kg busses. Miniature hard disks, so important to the laptop market, have not migrated to space in large number. They need air tight envelopes, and they demand relatively high power. In addition, their sensitivity to the launch vibration environment (no, surviving a fall off a desk to the linoleum floor isn't good enough) and the disturbance torques their motors generate, coupled with the fact that the cost of RAM is not a driver for even the lowest cost satellite, just haven't made them a logical choice. Flash RAM, which requires no idle power and is intrinsically reliable, is now being increasingly used in lap and palm tops, as well as in PDAs, Personal Digital Assistants like the PalmPilot. Other promising technologies are smaller, lighter Nickel Hydrogen batteries, already widely used in laptops. Ultimately, rechargeable Lithium batteries will be adapted to space.

Another area ripe for scavenging is test facilities. The world is chock full of shock and vibration labs, thermal vacuum chambers and RF ranges searching for a new raison d'etre. The lease/buy decision is pretty trivial here.

Fun

In the space world of the '60s, there was fun. The old guys don't like to admit it, but when they launched a rocket out of Cape Canaveral daily, they were loving it. Who wouldn't enjoy calculating, on a slide rule in real time, the trajectory to the moon for three astronauts? What would you have paid NASA to let you work in mission control during a lunar landing?

Let's face it, the space world of the '70s, '80s, and '90s is no fun. We live in fear of failure. None of the good we do can offset the negative feelings loss of a mission creates. NASA is a huge bureaucracy, not a playground for the best minds in the US. Cruel, but true. What was fun in the '70s and '80s? Silicon Valley was fun. Apple was a renegade and thousands of little companies sprang up across the landscape like the first growth after a long drought. Thanks to the digital era, most of these Silicon Valley upstarts weren't even in Silicon Valley, except maybe in spirit. They are just as easy to find in Canberra or Columbus as in Cupertino.

In the '90s, Silicon Valley has gone corporate. Bill Gates' severe case of megalomania ultimately seeks to destroy the competitive environment that has so strongly benefitted computer users. Apple, no longer the oddball utopia it was, offers some small refuge from lock step compatibility with the C:\ crowd. Software development is an industry now, bureaucratized and overcrowded with competitors. It's managed for productivity, not innovation, and the breakthrough applications that rocketed new companies to prominence are very few and far between. Big success stories like Quicken are more marketing marvels than new innovations in using synthetic thought.

Microsatellites are fun, at least for now. Maybe not as much fun as Silicon Valley was, but we aren't serving a consumer mass market of one billion people yet. Building our industry on a handful of customers limits applications and innovation. The thrill of building a whole satellite in a group small enough to share two pizzas and a watermelon for lunch is real and infectious. Just as Apple, AST, and other startups enjoyed conquering the computer giants who made the Cybers and the 360s, there is a competitive excitement to a company of three people walloping TRW, Boeing, and Ball in a competitive spacecraft procurement. Maybe the days of the giants aren't numbered, but their eco-niche is a little narrower.

What good is fun in the post-MacNamara era? Pragmatically, the best technical, managerial, and marketing minds gravitate to fun even more than to money. Rocketry made great leaps when our society's greatest engineers aspired to work in it, and when the toughest curriculum at MIT was Aerospace Engineering. In the last twenty years the attraction has been Wall Street and computers, at least one of which has revolutionized almost every aspect of our lives. But in a capitalist society, miracles take money. Progress resulted from a combination of sophistication in the financial and technical realms, and Wall Street sophistication is what financed the revolution in computers.

Today's microspace entrepreneurs are seeking a similar unity among financial, technical, and market forces. For remote sensing and communications, a new regulatory environment is also needed to effect a new order in services from space, as evident in the flurry of FCC filings, recent WARC proceedings, and new initiatives in commercial remote sensing. This new activity is the direct result of mixing a new ingredient into the morass of the old aerospace doldrums, a powerful boredom solvent known as…

Part II

Missions and Reliability

Section 3
Critical Design Review

Chapter 1

Critical Design Review: A Meditation

The divine pursuit of a meaningless month...

A line of thunderstorms creates its own threatening darkness on a midsummer's afternoon, then shatters its own blackness by hurling yellow and blue flashes of plasma at the ground. The explosive shock wave driven by the lightning hits our ears harder with only the neutral background of falling rain to rumble over. Buddhism teaches that without the canvas of silence we cannot make music, and without death we cannot live. Without age there is no youth, and without war we cannot understand peace. A bikini is more about what we can't see than what we can. Our eyes filter out the complex texture of the static forest seen from its tangled interior, until the smallest motion of a squirrel, or perhaps a snake, transfixes our attention.

Only when you understand Yin and Yang, id and ego, or the living tree towering over the crimson fallen leaf of autumn, can you understand the Critical Design Review. For it is born of paradox, and it attracts us no less than those Californians who are drawn to the narrow eclectic coastal zone between the bland, blue-green chill of the Humboldt current, and the relentless, dull heat of the Inland Empire and desert.

We engineers are the cloistered monks of aerospace. Long ago our ministry ordained us as Doctors of Philosophy and Masters of Science. Then we disappeared into carpet-walled cubicles or the invisible sanctum known as the home office to spend a lifetime ordering ones and zeros. Outside our sheetrock office building walls great congresses convene to argue budgets, violent proposal wars are waged, and managers pursue petty skirmishes over charter and IR&D budgets. Machines hum, metal is drawn and cut, solder is melted and refrozen. Sometimes, rarely, our counsel is sought, but it is seldom heeded. Our lives are, by and large, given over to the silent pursuit of order as perceived by our own compilers and ICs. If the LEDs flash, it is good.

Zero entropy is impossible, and if a tree falls in the forest, no one may hear. Dissipation: like chemicals, we can't live without it. But neither is the Universe unaffected. Thus it is that we are submitted to a Process. Our carefully arranged bits are sampled. And those Selected Bits occupy us for weeks on end, as we endeavor to display them in colorful, pseudo random patterns splashed out on transparent plastic sheets inside a glorious frame of colorful corporate, institutional. and mission logos. These are printed on paper, copied, collated and stapled even as we queue onto aircraft bound for a distant conference room. Wherein we are seated in solemn dimness, we intensely private creatures of email, simulation, and spreadsheet, all of us deaf, mute, and blind to a world of milestones, Mil Standards, and programmatics.

Shakespeare and Mozart are only vibrations to a snake. For a week we are massaged by the magnificent but meaningless bits from engineers of other specialties, specialties that, had we had any interest in them in the first place, we would have studied ourselves during training. We fall into a state of meditation. Unaccustomed to seeing our fellows standing by viewgraph machines and dressed in the British regalia of silk ties and seventeenth century riding jackets, as opposed to our spartan jeans and T-shirts, our attention wanders, focusing on the raw impression of their figures, their sounds, and their messy hair. Deprived of information carrying comprehensible meaning, we count people, chairs, stripes on the wallpaper. We memorize the names on the "attendee" list. (Is there such a thing as an "attender?") I frequently busy myself with the design of a viewgraph projector that doesn't spew hot air, leak bright white light at the audience, and emit a whoosh drowning out the speaker it is there to accompany. LCD PowerPoint projectors lead me to theories of mechanized geneticists. They share their ancestral shortcomings.

Sometime during the week, I stand at the viewgraph machine. I am a drummer in a large synchronized marching band with bright red uniforms and gold embroidery. But there is no marching band and no uniform, no music, no grandstands, no football game and no gridiron. I can't read the audience the full volume of my work, only the Readers Digest version. Should Herman Melville be judged by the Cliff Notes on *Moby Dick*? Can we judge Van Gogh from a black and white halftone of the exterior of a museum in Portugal that contains some of his paintings, itself printed in the travel section of the New York Times? My focus is simply damage control and projection of myself as an amicable enough character who would never harm a young satellite—one who also happens to know inoffensive answers to superficial questions and won't create waves.

Amazingly, this works. Or it doesn't. There is no difference, other than perhaps the length of the action-item list that was hurriedly tabulated on Friday before airplane reservations and dinner engagements finally dispersed the seance. What works just as well is showing up on time, taking "notes" with a quiet keyboard, eating lunch with the group, and staying until the end each day without leaving too often to service the cellular telephone.

Within a week, life is back to normal, the action items are being responded to, and final versions of the original displayed bits are being FedExed to the hosts and reviewers. Accumulated junk mail is recycled and hundreds of email messages are read and answered. Thermodynamically, the state appears unchanged from three weeks ago. Productive work resumes.

Miraculously, albeit painfully, the soul has been cleansed. We have visited the world of airports and aircraft, milestones and meeting rooms, shoulder to shoulder contemplation and meditation, and what managers interpret as camaraderie. Not that the nascent spacecraft perceives other than our collective absence and apparent neglect. It knows nothing of three weeks having passed that a reset of the onboard clock can't erase. But having invested 1,663,200 heartbeats abstaining from progress, we dig in with renewed vigor. Sitting silent in front of screens, we are dressed once again as ordinary campus dissidents, designing, simulating, estimating, composing, running and re-editing, communicating via email, and prairie-dogging over office partitions, consuming the accustomed diet of coke, pizza, and for some, cigarettes. Life is meaningless, but it is also good.

Chapter 2

The Dilbert Wars: The Front Lines of Program Management

Two things are found throughout the universe—hydrogen and stupidity.
—Gary Larson

One percent of people think. Four percent of people think they think. Ninety-five percent of people would rather die than think.
—Ben Franklin

It wasn't until graduate school when for some reason I started to believe I might have to actually study if I ever planned to do real engineering, that I experienced the discomfort of thinking. As an athlete I'd enjoyed the feeling of relaxation that rewards a hard workout and welcomed the mild ache of fatigued muscles. Brain exercise is different. Unsolved differential equations haunt restless sleep and unresolved software bugs consume the enjoyment of what should otherwise be a pleasant dinner. Sure, there are those exhilarating moments, but the ubiquity of homework sets, labs, and written and oral exams speak to the manipulations required to motivate even good students to work their brains.

2.1 A War Waged for the Sake of the Clueless

Blissfully, most of life lacks professors, textbooks, and tests, allowing us to join the comfortable majority that we applaud for "simplifying their lives." Kids do this by assuming adults are clueless, for instance. A convenient oversimplification. The comic strip Dilbert grew to a cult that has a similar conviction: management is clueless. Historically, management has been happily returning the compliment to engineers.

Having been a member of both management and engineering, sometimes over the course of a single phone call, I'm loathe to take the classical way out by declaring that either management or engineering is clueless. But we wage the classical Dilbertian struggle anyway, every day in each of our projects. It starts like this: One day we don't yet have the big IsenTropic Solar Coronal HEAt Program (IT'S CHEAP) and hence we fear for our jobs and mortgages. The next day, after we've won it, there we are, buried in fixed milestones, prices and specs to meet. Plus the hangover from the big win party the night before.

2.2 The Struggle Begins

The Dilbertian project manager locks into combat with the Dilbertian company manager. One is determined to bring this project in on budget and on time. It's a cause without detractors—fighting the good fight in the name of microspace and the customer. The other is trying to keep the company solvent, figuring that it's no good to engineering, management, or the customer if the IRS, the bank, or the Feds shut us down due to lack of cash or failure to comply with the federal, state, and local regs. At the first tick of the cosmic programmatic clock, harmony reigns. The second law of thermodynamics warns us where we inevitably must go from this state of perfect agreement, however. The project leader materializes in my guest chair—the one that doesn't recline forcing the visitor to hunch forward, increasing the meeting's urgency before one word is spoken.

"I need to hire twelve more engineers right now, or we'll never make PDR" (Preliminary Design Review, most programs' first big milestone). "I'll be spending $160k for headhunter fees plus $80k for computers. All overhead charges to the company, so I thought you should know." Having barely survived the years of effort and expense it took to win IT'S CHEAP, and having spent the last of our cash on last night's bash, $240,000 looks like the peak of Mt. Fuji viewed by an intrepid hiker with two sore feet thrust into poorly fitted hiking boots standing at its base in a predawn drizzle. "How about Vartan, Inga, Bob, and—what's that other kid's name?" I ask.

"Hey, this is a tough program, I need real engineers, not part-time interns, a Ukranian refugee, and that other kid is Daniel. No offense, he's your nephew and he's great at HTML, but he's only fourteen years old. Here, check out this resume: Alice Kinyama, MIT Ph.D. and seven years at Lincoln Labs. I want to move her out to the San Diego office."

"A move?"

"$10k max." Genuinely worried now, I dig a little more. "$80k for a few computers?"

"They stopped making Commodore 64s ten years ago."

Thus do the Dilbertian wars commence. Well-meaning engineers and project leaders try to bankrupt the company, while I try to sabotage their programs. As seen from program management, trying to get the right people complimented by plenty of tools to keep the team small and efficient, it's a sinister conspiracy by corporate management to derail the program in order to squeeze a few more profit pennies out of the company.

Since moving to Virginia, I've learned a little about the American Civil War. For one thing, those soldiers really knew how to walk. There are battle fields strewn throughout this vast Eastern state. Subsequent World Wars took advantage of aircraft and rockets to move the battlesite around. Dilbertian wars, as mobile as bits, are in progress everywhere you look.

Marketing is a prime battleground. We've all noticed that while technology has advanced, the people using it haven't. Unprogrammable VCRs are a silent testimony to this. I can fly to Australia in eighteen hours, but it still takes my ancient technology body ten times that long to get used to the time change. It used to take fifteen years to build a satellite—a few years to figure out what we wanted to build, another few to sell the project, five or ten years to build it. Now we build a microsatellite in eighteen months, but it still takes five or six years to figure out what we want to build and then sell it.

With program cycles so short, but the human process of developing a mission and selling it and getting it under contract hardly compressed at all, selling has to happen two or three program cycles ahead of production. Which means that engineers busy trying to build satellites are needed to crystalize ideas for new satellites, design and cost them, hypothetically, and help sell those hypotheses.

2.3 Taking Hostages

For this reason, I steal engineers off programs from time to time to help sell the next program, or even the one after that. This triggers more than the casual skirmish. Here's a professional working under a tight deadline to build some highly complex radio, or perfect a guidance algorithm, and his bozo boss calls to recruit his help on some flaky target years off in the future with a win probability of at best 5%. Meanwhile, engineering deadlines—PDRs, CDRs—are looming, suppliers aren't answering phone calls, the lab computer hard disc crashed overnight, and the last thing an engineer needs to worry about is creating a classy looking PowerPoint presentation with zoomy sound effects and fades.

2.4 Yet Another Reason Management Wants to Kill the Engineers

People like company presidents and marketing people tend to be optimists. They go around telling people about the advantages of their product, how happy their customers are, and how great their team is. While that information may help sell satellites, it doesn't help build them. The personalities of the best engineers tend toward a fixation on details, particularly those that caused problems, and the anticipation of the inevitable new problems down the road.

When I do manage to recruit an "open-minded" engineer to help sell Project X, I can anticipate that, rather than telling the prospective customer how much time and money the microsatellite will save their organization, the engineer will focus on war stories about software taking 16 years to debug and programs running factors of two or three over budget. This is a bit like having dinner with a bunch of pilots all talking about how various of their colleagues over the years met untimely deaths, and then them expecting you to get on their next flight with no particular qualms. As an ex-pilot, I can tell you why there's a locking door between the passenger cabin and the crew. It keeps passengers from hearing what pilots talk about. Actually, I used to fly passengers without the benefit (to them) of the locking door. But those passengers were special. They were already dead, cloaked in black body bags and on their way to their own funerals. They were impervious to pilot talk. This is not to imply that from the sales point of view, the only good engineer might be a dead engineer. On the other hand, dead engineers tell no tales.

2.5 Military Dress

In a healthy engineering organization, problems get discussed, documented, passed from generation to generation, and even published in journals. But the guys in the grey pinstripe uniforms naturally want the opposite—pristine records unblemished by gotchas. Similarly, the airlines want passengers to believe that their planes

are always clean, there's money aplenty, and the airline's employees couldn't be happier. The neatly-uniformed flight crews get in the clean white school buses that pick them up at the airport. It's almost like one of those Madeleine stories. But when you step into that bus, you realize the inside is bare metal walls, the seats are suitable for a go-cart and the crew is ready for mutiny at the first opportunity.

All companies do that. You visit the president, and he or she has a beautiful spacious office, usually decorated with a globe Columbus might have used, a clean desk of manicured old wood, a gentle and unrushed secretary. Serenity reigns. Just down the hall, the lab is jammed with smoked circuit boards and glitchy test gear, the engineers have been living on Coke and Snickers bars for thirty-six straight hours, and the program manager is stuffing boards into antistatic bags and packing them up to make the FedEx deadline in twenty-six minutes.

2.6 A Battlefield Littered with Bodies

Program leaders have tough lives. They pull together their own mutinous teams; deal somehow with constant setbacks and screwups, perpetually shifting requirements and renegotiations, and the irrationality of the client and of their own staff and management; and plow through a relentless schedule and a budget pinched more tightly than the waistline on most men's trousers—always just a bit too tight. The one thing that redeems this job, however, is that they get to focus on a single objective: Deliver the damned satellite on time and on budget.

Getting to this goal means subordinating a few things that might be nice to have. For instance, does a PM (project manager) care if half the team quits, so long as they quit after their part of the project is finished and the ribbons are all tied? Well, actually, he or she should. At AeroAstro, the PM was and maybe still is an engineer, and usually plans to return to full-time engineering after the stint as leader is over. There's plenty of sympathy for overwork and the difficulties of engineering. But ultimately,

program responsibility and loyalty are the ultimate drivers in decision-making, not a staff of happy campers ready for the next project.

In the Dilbertian wars, the battlefield is strewn with the lack of bodies. For instance, the bodies of the engineers who really deserved a raise that the program couldn't afford. Only their spirits remain when their part of the program is over. The rest of them went to the competitor with more money or to a fresher program. Ditto the lone gunslingers who at the start of the project wanted to do that part all by themselves. Tired of working with "colleagues" who don't carry their own weight, these hardy souls drown in the quicksand they've created themselves by underestimating the work necessary to get that part done. By the time they realize they're going under, it's too late to run for help since bringing in reinforcements isn't so easy. They need to be "spun up"—and coming into the rat's nest of software typical of gunslingers and being at all productive takes time, if it's doable at all. Which will cause the solo artist to be even more convinced he should do it all himself.

Oftentimes, the program budget kills off the engineers. Nanoseconds after the engineer finishes his or her part, the PM says soothing words like "I can't afford you anymore, go find something else to charge your time to." The exhausted engineer feels, with reason, that the reward for hard work is a layoff. It's a little like war. The reward of valor in battle is often having your head blown off, contrary to Hollywood's portrayal of heroism as an uncomfortable deed always followed by just rewards.

As the company president, I'm again pitted against the PM, but ironically this time on the side of the engineers (vs. marketing, where the PM supports his or her people by stopping me from distracting them in my effort to win the next Big One). The company needs the continuity of the engineers' expertise to guide the next program and to lead new engineers. Besides being cruel, it is irrational from the company's point of view to burn people out on a single program and then lose them for the next. But the PM may view that approach as the only one that can save the program.

2.7 The Mythical A-Team

Program managers want "the best people." Seems reasonable. Would you want your rocket built by "the worst people" or even by "the mediocre people?" When you undergo surgery, do you want any other than "the best doctors" and "the best medical care?" But, aside from the impossibility of me having the best surgeon taking out my appendix in Redondo Beach at the same moment you are having the best surgeon excise your tumor in Houston, how did that surgeon get to be "best?" Surely at some point some other surgeon was "best," and that prehistoric "best" surgeon spent some of his or her time training a few of "the not-yet-best" surgeons.

There is no "best engineer" team—and identifying one group as "best" will ensure a host of management problems. The "best" team will demand ever increasing ransoms in tools, time, salary, flexibility, special perks, excuses from other company obligations and so on. If you believe they are "best" and only the "best" will do, then you are their hostage. And the rest of the organization will ensure you are their hostage. Some of them will leave for other companies where they are not branded as "the not best" or the "could be better" or just the "B-team." Others will focus on more rationally managed programs, and will try to be unavailable when the inevitable call for help is uttered. The "don't go there" cliche was tailor made for the A-team mentality. Aim for a team with a diversity of talent and experience—the best engineers are engineers who consider teaching and learning intrinsic to their job. It's necessary for the company to grow into its future when the present crop of teachers retires or moves on. Also, those crusty old A-Teamers don't have the advantage of recent emergence from school or other jobs—a good source of fresh ideas. We are all quick to complain when a key engineer leaves, but we all benefit from fresh ideas. Buddhism teaches that without death there is no life, without silence no music, and without departures, no source for new arrivals.

One final flaw of the A-team. What if you are building two or many more satellites/rockets at once? Which one gets the A-Team, and who tells the client that the other satellite got the A-team?

2.8 War As a Means of Trimming Bureaucracy

Program managers do not like rules. They don't like, for instance, to obtain three competitive quotations on an assembly that has a cost less than the cost of obtaining the bids. They like to allow people to charge a hundred hours on a time card in a week and get paid for it if that's what it takes to ship on time. As the company president, I

have instituted the First Law of AeroAstro, which is simple but often in conflict with these logical programmatic behaviors. It's called, "Keep Rick out of jail." Charging a hundred hours in a single week and getting paid for it is illegal unless you've obtained previous authorization for overtime, something customers never like to give. Giving key contracts to suppliers without seeking alternative sources is also a violation of the federal programmatic kosher laws.

There is flexibility in the system—use as much of it as you can, but don't abuse it. There's always somebody who didn't make the A-team ready to call in the feds if you mess with the rules.

2.9 "Did You Really Eat the Horses Yesterday?"

—Stan Freeberg, "United States of America"

At some juncture in the life of most programs, the customer cuts off the flow of money. This is sometimes due to external circumstances, like Congress not getting around to passing the country's budget legislation. More often, it's a management move somewhat akin to blood-letting. The contractor (us) is struggling to get a mountain of work done with barely a teaspoon's worth of money. Solution? Since the teaspoon is the problem, cut the money out altogether. Like most other wars, the Dilbertian battle here is triggered over resource non-allocation.

The project manager figures that if they stop work just because of a temporary interruption in the money flow, the team will disperse and forget what they were doing. As a result, the mixture of new and old hands will make a bunch of mistakes when they try to restart after a hiatus. Ultimately, the interruption won't save money. It will cost much more money and much more time. And the PM is 100% correct.

However, even without project revenues, I have to find the money to write all those paychecks and make good on them. I have to pay the rent on the offices, labs, shops, and cleanrooms. I have to pay leases on all the fancy custom test gear the project is using. If I keep writing checks without a balance of project revenue, I violate the First Law of AeroAstro ("Keep Rick Out of Jail") and we all crash and burn. So when money stops, I tend to tell people unceremoniously to stop working and go find a paying job to work on. This usually precipitates lots of visits and phone calls from the client, the project manager, and the engineers, all of whom are concerned that I don't understand how damaging my action will be to the program. My measly defense is that it wasn't my action but just my reaction to external stimuli. When you pull the trigger on a BB gun aimed at your neighbor's plate glass window, it's illogical to then blame the BB gun for the damage.

2.10 Strategic Weapons

These are the Dilbertian wars that are fought on the ground. There are strategic war fighters involved as well. For instance, my experience in twenty years of micro-satellite projects is that they do have their interruptions, and even if they don't, the staffing needs change dynamically and sometimes unpredictably. For instance, if delivery of a critical part is delayed, several engineers may be idled while waiting for it. Also, engineers, like all of us, finish faster if we have something else for them to do next. It's just not human nature to rush like mad, work late hours and weekends, and skip going out to lunch with cronies to finish, say, a radio, believing that once that radio is finished, the next step is to be laid off.

Thus, the strategic company planner wants engineers multiplexing between several (or at least two) projects. It alleviates boredom, cross-fertilizes good ideas, prevents tribalism, and provides job security. It's also more efficient, despite protestations by clients that they want engineers 100% dedicated to their project. They may want that, but they don't want to pay the bills for it. Still, the project manager, facing deadlines and struggling to stay on a critical path, does not want to hear that a key team member is off at another customer's design review for a week.

We all know that a program run on a brisk schedule is cheaper. There are costs of keeping a program alive, regardless of whether constructive work is happening, and there are more costs if there is a launch delay. But a fast schedule means a big staff that the company won't be able to support when the project ends. That will mean laying people off. And whenever layoffs get talked about, the highly experienced people who tend to have the largest numbers of options available for finding new jobs start looking around. Pretty soon everyone is demoralized as critical talents go out the door and the remaining crew struggles to fill square job holes with round engineering talents.

2.11 The Lesson of Darwinian Survival on a Program's Battlefield

Life is a struggle for survival. People kill animals, which in turn had been in the business of eating other animals and plants, which in turn often feed on each other, or on nutrients from the soil and sunlight. Darwinian selection goes on every minute of every day, and has been for billions of years. So why do we still have such a variety of

species—or even so many different kinds of people? Smart people, short people, people with different talents. Physical strength and propensity to succumb to disease vary from one person to the next. One reason is that while one environment may favor a thinker, the next may favor a lean runner, and yet another a good walker with plenty of stored fat.

Where Dilbert misses the bus is that companies are not run by geniuses, but rather by people about as smart as the rest of the people working with them. Each of us is trying to do right by our own set of imperatives, survive in our econiche, and none of us is clueless or sinister. Even in this seemingly ideal situation of equanimity, skirmishes like the ones we've just discussed are bound to break out all around us. They are built into the boundary conditions, and we all just act them out. It's how we act them out that matters. We can play territorial games, politics, bluster and all manner of manipulation. And generate plenty of heartache and entropy in the process (and not a lot of engineering). Or we can do that least comfortable of all things—think. Project ourselves into the roles of all the other players. Understand the imperatives causing them to do things you don't like. Use thought to carefully communicate the costs of actions and to negotiate in good faith ways to minimize net cost to the program.

Within every organization, clueless, antisocial and disgruntled people may at some time randomly populate any niche in the corporate ecosystem, and you will have the chemistry for real conflict. But that is rare—I'm not sure I've ever seen a diabolical character outside of cartoons, but I admit it could happen. In everyday experience, heaping blame on deviant characters is as futile as Dilbert's model of brainless management as the root of corporate failures.

The body, when threatened, has a single, autonomic response—the so-called "Fight or Flight Syndrome" and its resulting adrenaline boost. But there we have a conflict. Should we fight, or flee? Watching squirrels surprised by an oncoming car or bicycle tells me that decision making is a problem not just for us humans, inundated as we are with information and management advice. Hire more people, or make do? Multiplex, or single-task? Break procurement rules, or obey them? Stop work at every interruption of cash flow, or press on blindly? Divert engineers to do marketing, or leave them alone and hope the next product sells itself or at least with support only from a marketing/sales group? Staff with only the "best" people, or invest in diversity?

I was nearly the last Ph.D. candidate of a stiff, and with age growing even more rigid, old-school European professor who led me to discover the answer to this dilemma. He was tough, pedantic, and demanding. No matter what I did, it seemed to always violate some rule of his for performance or behavior, either in the lab or in dealing with him or others. Over eight years of knowing and working under him, he frustrated me consistently and obstinately, and I respected him absolutely. One day, in response to his latest vicious recital of my proximate violation of his rules of proper human behavior, I gathered myself up and said to his stony face, aristocratic despite (or because of) its crooked nose "Sir, if life were just a matter of following your sets of simple rules, we wouldn't need thinking people to live it." For what seemed like a full minute, he stood there silent, proud, upper crust as always, and motionless. Then he shook his shoulders as if to straighten his permanently affixed olive suit, almost as if he would pull a hickory stick out of his pocket to address my insolence. Or was he wounded—just the thought terrified me—was it so easy to pierce this impenetrable figure? Then, for the first time in all my years with him, he broke into a broad smile.

Chapter 3

Killing the Wooly Mammoth

Blame Carl Sagan. He hypothesized that people the world around use the expression "Shhhhh!!!" to silence other people for a reason. Deep in our evolutionary past, he posited, mammals struggled for domination of the planet against the reptiles. (We won, by the way.) The universal warning of the enemy's presence among us hairy, warm blooded creatures was this onomatopoeic expression of the reptile's preeminent representative, the snake. What other firmware, wonders my cerebellum, might be wired into that mysterious brain stem of ours?

3.1 The Advantages of Small Work Groups

Count this among the many revelations we cannot experience until accepting microspacecraft into our lives. Not a nuclear submarine, not an international airline, not even the Bronx Zoo, can be run by a group small enough to eat lunch together. But the typical staff of six to sixteen people needed to build a small satellite or rocket can

communicate along all the cross-links of this diagram. The radio person can talk with the attitude control person who can directly communicate with the project leader. Ideally, there are no role descriptions, no paper reporting channels, no ICDs among subsystems "groups" (a.k.a. individuals). Rather the team organizes itself around the job to be done, divides the work according to people's abilities, skills, and daily availability.

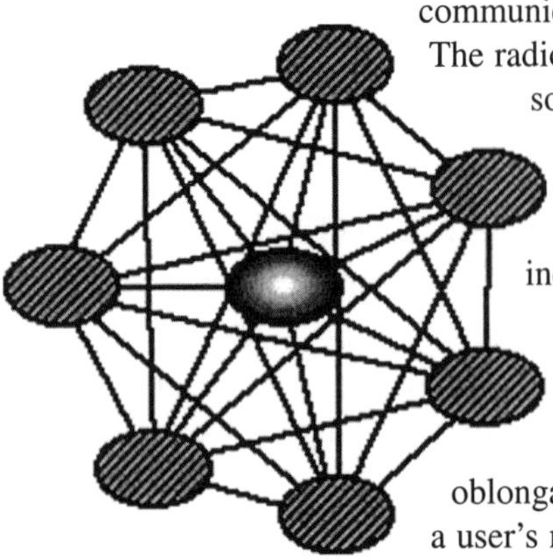

How is this possible? Enter the medulla oblongata. We don't learn how to breathe by studying a user's manual. (Otherwise we'd all have suffocated a hundred million years ago, the behavior pattern of disregarding user's manuals being deeply imbedded in the human psyche.) Digesting a Geraci's Shrimp and Pineapple Pizza is a complex interplay of muscular contractions, enzyme production, and Diet Coke ingestion, involving a complexity beyond our capacity to consciously orchestrate. But we're wired from birth for more than eating, drinking, breathing, and a few other key functions. We are a social, self-organizing species. We are born into families. We don't grow up grazing alone on the grassy plains of the Serengeti. Nobody writes an org chart, Statement of Work, and Work Breakdown Structure before we know who's going to pass the salt around the dinner table.

More to the Wooly Mammoth point, our ancestors did not get MBAs in program management at Pepperdine so they could go off and kill an animal a hundred times their mass. A group of four, five, six, maybe eight pre-cell-phone humans walked the jungles and grassy meadows of their prehistoric world, equipped with a randomly, or worse, hierarchically, distributed set of lousy tools (not that much has changed)—in their case not a Pentium III equipped with Windows 'xx and Excel, but a stone-tipped spear, some vines or hemp wrapped into a rope, and clubs. When the charging animal appeared suddenly out of nowhere, there was no mission control with pre-scripted scenarios on computer printout barking out the moves. The person in front attempted a lasso and trip, the guy on the flank of the charge line threw a spear, two if he had them, and the guy in back, assuming he wasn't already trampled, might have grabbed for the tail to distract the beast. Darwinianly speaking, groups of these mammals that interacted highly efficiently in this way survived. Not to say that loners with absolutely no social interactive skills didn't survive too—daily life is testimony to their success—but there was a definite econiche for these social beasts. Intermarriage between the socialites and the hermits is likely to have occurred at least occasionally over the eons (my parents being one particular case).

Hence the really good news: all that primordial software is alive and well in each of us. People still give birth without reading up on it first, or attending Lamaze. In fact, our society pays its top salaries to people with extraordinary instincts for team organization, like Michael Jordan and Mick Jagger. Sports teams, families, rock bands, Wooly Mammoth hunters all have lots in common. They are limited in size—certainly numbering under twenty people. They don't rely on extrinsic management. (Nobody can actually hear what those basketball coaches are screaming from the sidelines.) They also don't wear suits, but that's a topic for another day.

Self organization is what we are wired to do. We are really good at it. So good at it, that we don't even think of it as a skill. But the Org chart at the beginning of the chapter has lots of complex interconnections. Org charts like this one, WBSs, email, ICDs (Interface Control Documents) and myriad worthless management software packages are coping mechanisms we invented to extend the tools that come standard inside each of us. Look at the traditional org chart on the left. Each person, each box, copes with just a few other boxes, never more than seven or eight. Why? Because ultimately each person on a team is limited by his or her intrinsic abilities—the ability to coordinate and work smoothly with a team. Outside the boxes directly connected by lines on the chart, we have to rely on the ubiquitous memo, or whatever fancier name we give it. And as any veteran of a program bigger than fifteen people can tell you, it can be done. For a price.

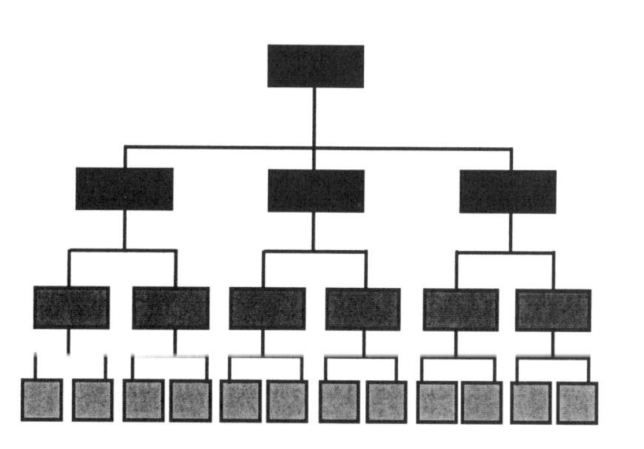

3.2 Goodbye to Efficiency

The price is first of all efficiency. It is a sobering thought to a manager or an engineer that the probability at any given moment of any professional selected at random in US society to actually be doing the job for which they were trained over many years to do, is always less, and usually much less, than one in eighteen. Why? On average, our lives divide into twenty-five years of education, forty years of work and twenty-five years of retirement. So right there, only forty of ninety total years are working years. Focusing on those forty, we only work about one-third of the hours of a day, not counting commuting, coffee making and deciding via telephone who picks up the kid, dinner, and a video. One third of the year's days are weekends, vacation, holidays, or sick time. Thus we only work 25% of the time, during those 40/90ths of our total time on earth, or about one-ninth of the time.

But even a fairly efficient organization has an overhead rate. Typically the cost of paying you $30 for an hour is about $65 to $75 dollars. The rest of that money is paying your "support staff." This staff includes not just the accounts payable clerk, the benefits administration, the company that processes your paycheck, and the marketing people who try to win fulfilling, interesting, technically challenging, and important work for you to do. It also includes people like engineers, finance people, marketing types and so on, who make your tools—including your office and desk. People writing the next release of Microsoft Word are not writing books or articles for magazines. They are tool makers. You, on the other hand, are out killing the Wooly Mammoth. So you can halve that one in nine number to account for society's back room, and there you go: one-eighteenth of the life of professionals is actually productive.

Now imagine one of us becomes 0.01% of a program team for a major spacecraft's development. Depending on your taste for long, often pointless meetings, many of them requiring days of travel, longer conference calls, even longer, always boring design reviews, memo reading and writing, and pop visits by managers asking "how you're doing," you may love or hate your job. But it will definitely reduce your time for doing productive engineering from eight hours per day down to one or two. Sure, some days you may get five or six hours work done. The rest of the time you're rebooting the PC or figuring out where the company keeps pens or replacement light bulbs. But averaged over a year, and including the one-eighteenth rule derived above, your net productivity is now about one in seventy-two. This is not really very good by anyone's metric, and certainly scores way down below the Wooly Mammoth Killers, who are basically on the job something like half their lives (though they still have a support staff bigger than their own number). Certainly our bureaucratically entangled lives beat the heck out of padding around the world of prehistory hunting for an animal possibly itself intent on hunting us. But the fact is that big programs and big organizations in general are highly inefficient not because they are irrational, but because of the growing overhead associated with management—okay, call it orchestration—of the activities of more than the number of people we evolved to manage.

3.3 Saving Money by Spending It

Things, as they often do, only get worse from here. Since it's impossible for each person on a ten-thousand person effort to know all about all the reasons for programmatic decisions, we all become rule followers. For instance, an engineer finds a brand new IC that would subsume an entire circuit board, saving beaucoup de bucks, time, mass, and volume. But he or she can't use that part on a big program. Why not? Because it isn't on the approved parts list. Why? Because, when the approved parts list was created, that part hadn't been released, and if it had, the systems engineering team didn't know about it. Or if they did, maybe they didn't have sufficient test data to know that it would work with this mission's requirements. And perhaps the part really won't withstand the radiation environment, or isn't reliable, or emits radiation of its own, or outgases, or creates magnetic fields, or requires unapproved soldering approaches not suitable for space. Who knows why it's not on the approved parts list, but it's not.

You could try to find out why not. Write a memo or meet with your manager, who will do likewise with his or her manager, and so on up the line until you reach the program office. This office is usually a bubbling cauldron of activity, with people madly preparing for the next program review, or technical review or financial review, or writing the proposal for the next phase of the program, or coping with the biggest of the brush fires that have broken out on the program, or preparing for a program visit by a high ranking politician or government person. Your parts selection problem is low on their priority list. How low is it? So low that the only way to get an answer is to play a significant ace, like getting your division chief to offer to quit if the program office won't address the question. Unfortunately, your division chief didn't get to be a division chief playing that game, so you are out of luck. And worse, logic is not on your side. It doesn't make sense to even momentarily halt progress over one person's use of one specific part.

Even if you hit them on the one calm day of the year, when just because of the sheer novelty of a straightforward technical issue compared with their harassed daily ration of crises, people find your question "interesting," the best they could offer would be to suggest to the customer that they embark on a testing and qualification program that might last years, which would wipe out your opportunity anyway. Parts reliability is a big deal on big missions. This is because (a) it is built into the medulla oblongata from prehistory when part failures actually accounted for a significant fraction of mission reliability problems, and (b) big systems have so many parts that to get an overall reliable system, each individual part has to be outstandingly reliable.

So you crawl back to your cubicle, grumble, and spend the $1M or so they originally allocated, and the ten watts, and the couple of cubic feet, for that obsolete, ten year old technology board that will accomplish what twenty-five square millimeters of silicon could now do with fifty milliwatts.

On a small program, in a contrast so stark that many engineers steeped in one size program just can't ever make it in the world of the other size program, you would be expected to know all the reliability and environmental constraints, and to make your own decision on the applicability of a part. You would have access to the systems engineer and the technical oversight team any time you wanted it. In an hour or a day, you'd have an answer, and likely you'd save $1M, ten watts, and lots of mass and volume.

But, as Churchill pointed out with respect to democracy, our organization of large programs is accomplished via a lousy system, but one that still beats all the alternatives. If you want to build a telephone company that can connect a billion people worldwide, or to provide electricity to a city of 10M people (and not a few Welsh Terriers too), or build an airplane that can move four hundred people and all their luggage from DC to Tokyo in thirteen hours, or if your goal is to eradicate some dread disease from the face of the earth, your alternatives to the "Chinese Army" org chart are the null set. After all, that obtuse and frustrating way of orchestrating large numbers of people is a wonderful triumph of our intellect, and it has done humanity great service. It is in fact a result of exactly what engineers do—take a big complex problem and "reduce" it down to a huge number of small, simple problems. This is maybe why org charts of big organizations look a lot like finite element models.

But, if you are lucky enough to have goals that can be addressed by a group small enough to self-organize according to the instincts you were born with, you'll not only have fun, but you can be amazingly productive. My own numbers show at least a factor four, but the stimulation of the fun factor makes that number more like a factor ten.

3.4 Beware Synthetically Organized Large Groups

Finally, a word of caution. While we all participate naturally in self-organizing small groups, most of the aerospace biz is carried out in synthetically organized large groups. Thus, most of your colleagues, managers, and even customers, are only familiar with the efficiency (more appropriately lack of efficiency) of large groups. So when you show them how six or twelve or fifteen of you are going to build a whole satellite, and a ground station, and all the software to make it work, and then test it, integrate it on a launch vehicle, and operate it, they will assume you are delusional, inexperienced, or just plain lying through your teeth.

Forewarned being forearmed, my learned counsel is based on an observation made by a friend from University of Michigan, who once said that whatever people are talking about in meetings, they're really talking about money. My extension toward the possibly absurd of this truism is that however many words are spoken in meetings, however many pages of WBSs and budgets, impression is everything. So be prepared to talk not about processors, block diagrams, test plans, and systems designs. Preface your technical presentation with photos and descriptions of some comparable programs that have succeeded. Nowadays there's a historical trail of missions done by amateurs, students, companies, and government institutions to point to. Bring pictures of spacecraft that were built by small self-organizing groups. You may not change many people's minds, but your viewgraphs will be prettier.

Chapter 4

What Mood Is Your Program in?

Rayleigh instabilities are everywhere. Find a stop sign at the bottom of a hill on a busy street. Just before the intersection the stop sign protects, the street gets bumpy because people try to stop their cars there. Originally there was only a small bump—maybe a pebble. Your car bumps up slightly over the pebble and lands again with the brakes on and digs into the street a little. Then it bounces just a little, slightly relaxing its force against the street. The next car goes down ever so slightly into the depression your car made, launching as you did, relaxing its force on the street and relanding, thereby reinforcing the high and low spots. Washboard roads, ski slope moguls, and the rippled floor of the ocean near the shore are all periodically varying features built up by the repeated actions and reactions of passing cars, skiers, or waves.

Developing anything that's not existed before is a road that's not without its share of pebbles. I'm not necessarily talking about a high technology satellite. Build a house—something that's been done millions of times—or just add a room or even remodel your kitchen, and you'll face lots of the same ups and downs. One of the ups

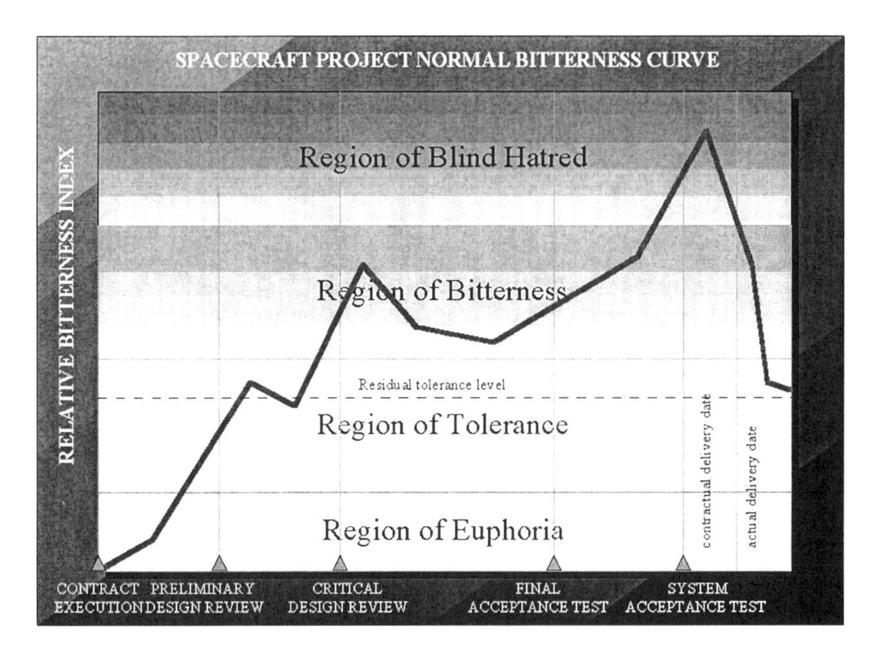

is intrinsic to all these processes—the excitement about the idea of a new thing that hasn't existed before, whether it's a center-island kitchen with enough light to see what you're cooking, or a 1 kg satellite. If it already existed, the high wouldn't be there, and you wouldn't be pouring energy into it. And there are universal downs too, usually having to do with the little glitches not expected in the original plan. You rip apart the kitchen, bring in the new gas stove, and then discover that the gas line the original builder had put in was never used and in fact is plugged somewhere, necessitating running a new line, and therefore ripping up the floor all the way out to the front door, and then digging a trench all the way out to the street. Necessitating a whole new building permit, bringing on a couple of additional contractors, more money, and lots more time. Nobody is happy. Then there's another up—partly caused by that depressing period of unexpected troubles. You weathered all those storms. If the long voyage had been without them, the cruise into the port of destination couldn't be nearly so sweet. That first barbecue is a celebration, a feat in honor of Winged Victory soaring over nature's infinite store of entropy and its angel, the ghost of Mr. Murphy.

Experiencing life to its fullest, the thrill of victory and agony of defeat, and the richness of our mood swings all combine to make great memories in our retirement, but in the business of building satellites, all of us have a goal to wipe out these excesses. We want a smooth program—no surprises, no overruns. In some of life's travails we've succeeded at least in reducing the number of mishaps. Airline and car trips are more often without mechanical troubles, more pregnancies have successful outcomes, and fewer of us live in the darkness of illiteracy. But in tens of thousands of years of building projects and in forty years of building satellites, we've thus far failed to smooth the development process. By definition, we are like dirt bikers focused on finding a road that isn't yet a road and traveling it anyway. Shock absorbers and mud in the face are part of the experience. Our twentieth-century roads have plenty of potholes.

At this writing, NASA has decided the solution is not to let its contractors work as independently as they have on a few recent programs. This end-of-the-millennium reinvention spawned those wonderful, eighth-generation Xerox copy signs you see at car mechanics. They give the price of the mechanic's time: $20/hr. $40/hr if you watch. $100/hr if you help. Why did NASA switch to tighter supervision? Partly Rayleigh at work—they were providing less supervision, programs had problems, so now we'll have more. In a few years, to reduce the product of cost times problems, we'll read in whatever passes for *Space News* that NASA is going to get out of its contractors' hair and let them perform better. But in the present more managerial phase we can ask: Why did NASA let its contractors work too independently in the first place? The answer is simple: to promote smoother, more efficient programs. And what were they doing before they tried that approach? They were being the vigilant stewards of the taxpayers' money we all pay them to be, carefully watching over their contractors to ensure smoother, more efficient programs. Rayleigh strikes again.

We are never going to cure the kitchen's mysteriously dysfunctional gas line nor Fellini's subterranean caverns stalling the construction of Rome's subway system. Murphy can be suppressed, but he won't be eclipsed—especially if we chose travel over uncharted routes as a career. Of course, we don't have to invite the ghoul over for dinner either. Winning a chance to build a satellite isn't like winning a chance to mow the lawn. When the latter opportunity arises among the family seated at the breakfast table, there's a sensible scarcity of over-promising, budget trimming, and exaggeration about the schedule we can probably cut the lawn on. If somebody wants the clippings picked up, we don't say we can do it, no extra charge, no schedule impact. The competition for lawn-mowing duty from, say, our teen-aged kids, isn't going to steal the job from us. The competition for projects—building satellites, remodeling kitchens—isn't the devil. It's a pebble—that we can amplify into a mountain.

Something like the mountain in the figure at the beginning of this chapter, e-mailed to me by parties who wish to remain anonymous. They could have been plumbers, road builders, or kitchen remodelers. But they happened to be aerospace engineers—people who spend too much time in front of computers running nifty graphics packages. They drew this rather than plumbers, kitchen remodelers, or home builders only because they happen to spend time drawing graphs. Occupational hazard. Saying that the inevitable course of a program is a gradual descent into Hell, in the best cases followed by a return to gritty normalcy is, at first glance, pointless variable to depressing and self serving. But there it is.

Streets, ski hills, and sandy beaches don't get bumpy because they are made of tar, snow, or sand. They get that way because of what's going on around them—cars bouncing up and down on their suspensions, skiers bouncing on their composite skis and knees, and water rolling into waves. Program satisfaction—what makes it so bumpy? Supervision? Lack of supervision? We've had God, Thor, the Sun, and their various families for many thousands of years. Murphy for just about a century. Our theology is nascent, if it really exists at all. But my cave wall artwork on the subject is that the energy driving the oscillations is expectations.

Lying is always bad. That's what parents teach us. We expect our academic, spiritual, business, and political leaders to always tell the truth. But the capacity to lie is genetically advantageous. The Gestapo knock on the door—the ability to say you have no idea where the residents are—that you are just there collecting the property for return to the State—is a much more likely survival strategy than to collect the wife and kids and give yourselves over. You catch more fish with a line in the water with a hook on one end and your hand on the other, than you do sticking your hand into the water.

The worlds of aerospace contracting, kitchen remodeling, and even banking are each in their own ways jungles, and survivors are hardy and few. They use whatever means they can. And in industries specialized in creating that which hasn't existed before, lying is an option. A bank really can't promise you 10% on your money if the account actually bears just 4%—but Kyle the Kitchen Konstructor can promise you a new kitchen in fourteen days, even if it might take a hundred and forty. How would you know—or him for that matter? It's a matter of opinion. How do we deal with kitchen contractors? We get lots of estimates. Kyle knows this—and knows that in order to keep his crews busy and his bills paid, he better give you the best story you'll believe. Cross too far over the credibility line and he's out. But back too far off from the edge of the credible into the dull, unattractive land of conservative do-ability, somebody will give you a better story, and he's out of the game.

The art of the believable lie is everywhere. This won't hurt a bit—we'd all like to believe that one, so we do and when your arm is sore the rest of the day and the rest of your body is sick for a week after a flu vaccination, you try to believe it's a fluke, which is what your company's nurse will also tell you. We often believe these things when we say them. After weeks of proposal writing, what engineer doesn't believe that he or she hasn't foreseen every possible contingency and has planned and costed for it? That they've succeeded in telling a compelling story and that the risks really have been all eliminated by heritage from the old program? I'm not saying contractors are pathological liars, but we all lie to ourselves and to others, if not out of intent, out of lack of information. One contractor I worked for was well established and experienced in a particular technology, and management could foresee the problems inherent in their proposed programs. They were honest, and they costed the fixes into their proposals. Very solid folks there. One of two outcomes resulted. Usually, they lost the job to less experienced firms who painted rosier, more economical pictures. When they did win, the problems they anticipated rarely occurred, only to be replaced by less well understood ones which caused even bigger cost and schedule debacles. In the first case they lied to themselves that their proposal was appealing enough to win. In the latter their chutzpah caused them to lie to themselves and the client that they really could predict the future.

The people I worked for there were outstanding both in their desire to tell the truth and in their technical depth and experience. Seeing the problems we had only makes me wonder why NASA or any other sponsoring organization can honestly even feign shock and dismay when development of new, highly complex devices isn't accomplished exactly, or even within, say, 15% of budget and schedule. Having a policy of "no overruns" is like the signs near schools that say "drug free zone." Clearly those signs are lying, and the people who put them up are well intentioned but naive. And do those signs imply that outside a school's vicinity, drugs are okay? Overruns are okay outside of NASA? Our government now has zero-tolerance for drugs—does this mean less tolerance for drugs than for murder or embezzlement? As it happens, putting a cork on a bottle does help reduce evaporation, but putting a cork on a bottle of water over a hot stove only leads to messy, dangerous explosions. In the case of program attitudes, announcing a zero-tolerance policy for overruns and delays after selecting the bidder who told the lie closest to the credibility boundary, whether out of intention, desperation, or naivete, is at best futile.

4.1 Harvard Business Review Case Study from Hell

Start with a client, maybe an agency of a government of a country that's never built a satellite before and trying to do right by starting its space experiences with a small satellite. Add some very enthusiastic technologists who have been waiting twenty years to finally fly their experiments—all motivated to help see some of their leading technologies demonstrated and tested in space—another positive facet of the program. There is a necessarily limited budget to do it all with and a world full of bidders for a shrinking number of contracts. The bidders are seldom motivated to do poorly—after all, that would not be in their own best interest nor that of the clients. But they also may not be aware of every pitfall a program may present. In keeping with the regulations common at governments trying to do the right thing, the lowest bidder who is responsive to all the proposal requirements is often virtually automatically selected. So the pressure on the bidders is to not see programmatic and technical issues that might add cost to their proposals. Taken to an extreme, the features-to-cost ratio is maximized via a very low paid, inexperienced staff ill-equipped with the costly automated tools for engineering and development that are necessary to keep the program staffing small. The proposal is seasoned with a sprinkling of a few beautiful resumes. Those elite, highly experienced people may, or more likely may not, actually devote the next few years of their lives to helping the company build something they've done before. More likely they'll be off conquering new challenges. Once selected, the inexperienced vendor, anxious to please and to demonstrate its own expertise and the potency of its technologies, will agree to virtually any client request and will try to do it all on the bid budget, figuring that the upgrades are straightforward and will help build good will.

I know only one happy outcome from this maze of circumstances. And amazingly, dialing up or down the amount of oversight is not it. This program is doomed. Not to failure, but to the evolution of moods from elation to the brink of ax-murdering depicted in our figure as deadlines are missed, budgets bloat, and people burn out. Meetings will be convened to figure out why there are so many "last minute" glitches turning up. Down the street from me there's a house that was built in twenty-four hours—the first prize in a charity raffle sponsored by local construction contractors. The contributing builders built an entire single family house in twenty-four hours—an amazing feat. The last minute glitches—cracking walls, stuck windows and doors, leaky roof and plumbing, and a driveway attempting its own journey to the center of the earth—are no more fix-able at the end of the program, or really anywhere in the program, than are the satellite's "last minute" glitches. Nothing the architects, carpenters, plumbers, electricians, roofers, pavers, or brick masons can do and virtually no amount of money thrown at the program in hour twenty-three are going to stop the house from settling, or its parts from seeping into each other before adhesives, concrete, paint, and tar have time to dry.

The happy outcome that is possible has a small catch. It is unfortunately not available to the client nor to the scientists nor to the contractor. But it is available to most of the people involved with the program—their employees, who can quit and go work for a program lacking structural flaws built in before the project ever starts. The most experienced ones will start to do this even before disaster strikes. This exodus is patchable just as are the cracks in the house's walls by offering more money and aggressively hiring replacements.

Pilots know one thing about aircraft accidents. Whatever happens, it's their fault. Value Jet's plane catches fire because someone mistakenly loads oxygen generators as cargo. It's the captain's fault, because according to the FAA's rules, only the pilot in command can have the ultimate responsibility to ensure the aircraft is safely prepared for flight. If the navigation system directs the airplane into a mountain, it's the pilot's fault for not checking the main nav system against other on board radios and radar. Same thing with program managers. Starting off with a program which, by virtue of its having been won was at best optimistically bid, things generally go downhill rather steadily. The specifically unexpectable but statistically inevitable technical glitches are only one part of the problem. Satellites are generally not purchased by people who know everything about building satellites, just as most kitchen remodeling customers are generally not plumbers and construction contractors. How can you blame the victimized customer, who's greatest sin was to ask for as much as it could get for as little money as possible?

The project management will be faulted for all the cracks in the walls and all the other foreboding signs of fundamental flaws, and when the staff exodus is underway, for over-reliance on a few star players. All caught up in the same wheel of over-expectation, selective ignorance of unpredictable and predictable problems, optimizing for efficiency, and the progressive layering of more and more oversight as problems arise, the tribal dance of blame and ill will continue until one of several outcomes occurs. Classically, that outcome is a massive overrun, but it is accompanied by emergence of a substantially functional product. The engineers and techs will be somewhat terrified, knowing all the software patches, cut traces and jumper wires holding the thing together, but thanks to their ingenuity and a prolonged debug and testing program, it works. The client pronounces the program a success, judges the contractor inept, comes up with a number of improvements in their management oversight practices for the next program, and goes on to repeat the pattern. The contractor does likewise. In order to win any new program, it announces vast upgrades in project management, quality control, and costing to ensure that adequate margins and controls are in place. All of which add to cost, so that when the next competition comes around, they are both more likely to win (with their beefed-up proposal featuring fancy management tools, plenty of grey hair, and centuries of flight heritage) and more likely to hit the cost wall earlier and harder. In the currently fashionable management fad at NASA, the only difference is that there is no massive overrun, but also no product emerges.

Assuming a program is bid at $A, and the overrun would have resulted in a program cost of $A x 2, we can say that apparently $A was "wasted" due to bad management or execution. That was then. Now in the new age of our enlightenment, NASA promises to cancel programs that have spent $A x 1.15 (a 15% overrun). At which point instead of wasting $A, as we did back in the prehistory of five years ago, or $A x0.85 now to get a working product, we waste the $A x 1.15 already invested and get nothing. This is perhaps a different outcome, but hardly rates as societal progress.

If a product does emerge and function correctly, which is usually the case, it is the result of a tight cooperation between the client, the payload, the spacecraft builders and all the other elements of the program to avoid a disaster for everyone. It is during this phase that moods actually begin to improve. Enmity still abounds, but cooperation often leads to an understanding of the difficulties built into the box the entire program is built inside of. Each element of the overall team learns that the others are no stupider, albeit maybe no smarter, than they are. And they are all working horrendous hours under difficult conditions—they are all committed to success. Although no one seems to put their finger on the true causes of their problems, they begin to realize that it's not simply evil or incompetence on the part of all the other program players. Thus the heights of rage are scaled and we begin to rappel down the back side where the rocky bluffs begin to give way to the occasional green plant.

If the satellite gets to orbit and works, there is a new focus. Just as we forget the penultimate boredom of sitting behind a wooden desk on a hard wooden chair for hours and hours and hours for years and years and years to get through elementary school and then thirty years later nostalgically sentence our kids to the same prison, intoxicated by the smell of white glue, finger paints and chalk, the team eventually forgets the pain and remembers the technical challenges they met and defeated, the interactions with one another, and the deep sense of accomplishment when the first signals came down from the satellite, and then when its payload initiated service. We do not ever return to our euphoric state—our illusions have been extinguished by the reality of experience.

In the midst of a traffic jam, we honk and shake fists at the guy in front of us who stopped prematurely when he could have gotten through the stoplight. But later, we realize—that guy didn't create the traffic jam by driving sanely. At the end of a program there is a tolerance that comes from understanding intuitively that the problems came from somewhere other than everyone-but-yourself. The problems were built in before the first block diagram got sketched, the first timecard filled out and the kickoff meeting plane tickets were purchased. And they won't be wiped out by putting more managers on the job, or requiring more documentation or adding 10% more margin onto the cost and schedule. In fact, most of those moves will make it worse. More management, more documentation, and more margins induce everyone on the team to spend more, and to take longer. The best management, the best engineering, and the best teamwork, are necessary, and insistence of greater oversight is a worthy means to create more jobs and inflate costs, but they are at best not sufficient and often totally irrelevant and wasteful.

Aerospace engineering is a tough field. Engineers are preselected hard workers. Just to escape from the University requires four to ten years of difficult course work, weekends devoted to impossible homework assignments, sleepless nights spent in labs doing experiments that never quite work right, and a life punctuated by a never ending sequence of exams. The disciplines that create satellite payloads are no better—often worse. Fields like Astrophysics, electrical engineering and telecommunications, astronomy and optics are not known as free rides to a diploma either. Good jobs in our fields and satellite programs themselves are scarce and hard won. We go in expecting tough challenges, and we find them. We promise a lot because we've always been asked to deliver a lot, and it's how we assert our superiority and our worth to ourselves and to the client. The bidding process reinforces our natural proclivity for complexity, difficulty and challenge—a push harmonically superimposed upon our natural rhythm.

We're sitting on the floor in that elementary school with the light green painted concrete block walls and felt bulletin boards, listening to a story about a scorpion and a friendly turtle. When the scorpion stings the turtle who has just saved it by hauling it off a shrinking island and across the rising river, we empathize with the poor dying turtle who asks: "Why, after I saved you from drowning, do you then kill me?" And we're shocked with the frigid logic of the scorpion's matter of fact explanation: "Because I am a scorpion."

Chapter 5

Developing a Program Plan

A recent psychological study concluded that we all believe the world was simpler and more rational when we were children—regardless of when we were children. This leads us to the conclusion that either the world is monotonically increasing in complexity and irrationality, or that a child's world is simpler than an adult's, regardless of when one is a child or an adult, or rather, despite the fact that one generally gets to be a child before becoming an adult.

I know for a fact that the world doesn't need to become increasingly complex for my life to have been simpler as a child. At that time, for example, math meant arithmetic, algebra, or maybe even linear algebra with its myriad operations on vectors and tensors. These can be complicated and hard, but at least when you get the answer, you can test it and see that it makes sense. The results are static—10 seconds, 10 kg, 10 inches, that sort of thing.

Engineering is the science of making lists. Lists of requirements, lists of components, lists of materials. Lists of pinouts, lists of commands. A spacecraft mass budget is a huge list of everything on board with an estimate or measurement of how much each one weighs. A power budget is more or less a list of all the power consuming parts of a spacecraft and what fraction of the time they tend to be turned on. And of course we have financial budgets—which are nothing other than lists of parts with their quantities and costs and labor hours multiplied by various rates—hourly labor rates, overhead rates, profit rates. Lists are the arithmetic and trigonometry of engineering—they are simple and static. Even if they aren't that simple—and a listing of all the cable pinouts for even a small satellite is long and looks complicated—they are at least static, so you can take your time to make them carefully, and if you do that, they will be reliable friends for the life of a program.

Eventually the static world of triangles and vectors gets traded in for calculus and differential equations. What's more complex about them is that they bring in time variation. The solution is not ten pounds or ten seconds or ten inches. The solution may be itself an equation that depends on time. For instance, where a mass suspended on the end of a spring but connected to a viscous damper might be after you tug at it and let go isn't a number, it's an equation like $x = e^{-kt(\sin\theta t)}$. You can't as easily just guess an answer, plug it in and try it as you could when the problem was calculating the distance from Anabru to Zurbranchburg. Life at this point apparently got more complex, and if the complexity wasn't easily understood, appeared less rational.

There is more to engineering than lists. There are schedules, which start off as lists of things you need to do, i.e., they begin as lists. But schedules are also about the time when activities need to start and end. They don't have solutions like ten pounds or ten watts. They have solutions that are typically milestones for the completion of various jobs and gates that permit the start of others.

Like differential equations, schedules add to the complexity of management. But like Dif-EQs, schedules are vital. Without schedules, there might never have been Dilbert, or at least Wally, who is usually seen defending their (lagging) development schedule. Without schedules, you can't charge for delays, you can't complain that things are late. Why expedite if you don't have a schedule? FedEx would have to survive delivering shorts from L. L. Bean, instead of expediting circuit boards from vendors. While more complicated than just a list, a schedule has the handy attribute of being two-dimensional—all the things you have to do run down the left hand side in a long column, and then the time axis spreads seemingly endlessly to the right. This provides a magnificent graphic which no program manager on earth can resist using to cover all the rub marks, holes, randomly placed pins and nails, and paint pocks on the wall left from when the previous manager ripped his or her scotch-taped schedule off the wall. This may be where the expression "Off the Wall" came from, since by then the poor manager probably was pretty Off him/herself.

Evidence of the complexity of schedules goes beyond how hard they are to stick with. Arguments about mass budgets, power budgets, and even dollar budgets are usually pretty quickly resolved. The news may not be good, but neither is it subject to much argument or reinterpretation. Schedules are a more fertile source of reinterpretation and discussion. There are overlapping tasks, some of which require the same resources, usually specific people but sometimes machines, to get done. There are dependent activities which can't happen until certain independent activities are completed. There are deadlines—the boundary conditions of our schedules. Every day you can't come in and re-negotiate the mass budget. But every day, a manager can come in and reassign people to different jobs thinking that the schedule can be accelerated by changing resource allocations to various jobs and moving completion dates around.

We have this concept of "Critical Path"—meaning the string of dependent events which are the overall determiner of the fastest the program can get done. But in most programs, just the delay of a week or two in some seemingly benign non-critical path activity can create a whole new critical path. The entire program plan may have to be recast to focus resources for getting the new critical path activities done faster and to make sure that all the non-critical path activities (the ones which yesterday were the critical path activities) don't slow down the new critical path. This is not unlike differential equation solutions—which can look like completely different functions, or suddenly not have solutions at all, just because one number, one condition changes from say a one to maybe a half or zero. In both cases the solutions are fragile. A seemingly minute change, and it's a new ball game.

We knew that life really was simpler twenty or thirty years ago, but now we know why—differential equations and program schedules. Who really knows if the solar panels will show up in four months—maybe it will be six? Or they might show up, but need some small change—is it easier to change them, or the satellite they plug into? But those changes might change all the rest of your program planning. There are ways to create non-fragile schedules, techniques for creating margins within schedules and maintaining them throughout a program, for example. Which add even more complexity.

Children live in the same world as we all do—and they confront differential equations every day—lobbing tennis balls into the back court on a windy day, playing in the waves and riding bicycles no-handed. One day, you figure out just when to jump into the water and swim with the waves to get a ride into the beach. And on one such day, a vision came to me. All the program plans for small satellites look about alike. Why? Did I stumble across irrefutable evidence of God? Or is my eyesight just deteriorating faster than I realize?

No, they are all alike just because of how we define small satellites. If you can't build it in something like two years or less, it isn't really a small satellite. And thus far, nobody's building them in two weeks or two months. To gain some intuition on the process it doesn't matter if the schedule is fourteen months, twenty-four months or thirty months. Assume there's that much squish in the schedule, depending on individual circumstances. The truth is that the resolution on these plans is no better than 25% plus or minus anyway. (I hope to sell a lot of copies of this book, and since nobody wants to believe that their own milestones aren't as fixed as the North Star and hence my program management career path is, with this bald revelation, forever blocked.) Our search for certainty is universal. You don't want to believe they sewed up your incision with whatever gauge thread happened to be in the OR that day, nor that they built your house with whatever the local lumber yard had in stock, or that your whole life was the result of botched timing on the part of your beloved parents. But they did—or could have. And while it sounds great to calculate your bill right down to the last penny for your kid's orthodonture, they could have charged 25% more or less and you would have neither screamed bloody murder nor danced in the streets with glee over the discount. That too, is life—a very coarse grained kind of a thing.

The first and most commonly forgotten step in a program is defining requirements. Berlitz has built a language learning empire on this fact—that requirements are everything. Your first day at Berlitz, you and the school leader fill out a form which asks you what you will consider a success in your language learning skill. You, knowing not one word of Hebrew, not even "yes" or "no" say something like, "I want to be able to ask directions, hail a taxi and tell the driver where I want to go, check in at a hotel, haggle over prices in the market and order in a restaurant." You figure if you could do half that stuff in this totally foreign language, with your trip just over a month away, you'll be amazed at both Berlitz and yourself. Four weeks later, your money, your ability to cram any more Hebrew into your brain and your instruction time have all been spent, and off you go to Israel—the instructor says goodbye, in Hebrew of course. "N'seah Tovah!" (have a good trip).

Two weeks later you return from Israel, show up at Berlitz, and ask for your money back. The whole way home you've convinced yourself they robbed you—you didn't learn anything—or at least nothing useful. You walk in the door, ready to be the assertive consumer.

"How did it go?" the Berlitz manager innocently asks you. Not well, she guesses from your frown.

"It was impossible," you complain. And then you describe the troubles you had understanding even simple things like the weather forecast in the newspaper. You couldn't understand any news off the radio at all. You spent a few days at a kibbutz and had no clue what people were saying around the dinner table, unless they said something expressly simple directly to you, and you struggled to try to tell people the simplest things—like your opinion on the latest scandals in Washington, which the Israelis, typically faced with more grave situations, found too easily dismissible.

At this point your local Professor Berlitz pulls out the questionnaire the two of you had completed on your first day—in effect your requirements document: "Did you order in a restaurant in Hebrew?"

"Well, yes, but that was trivial stuff," you answer.

"Did you tell taxi drivers where to take you?"

"Yes, but that's two words or maybe three. That's not really speaking Hebrew," you protest.

"OK, well, did you buy anything at the Shuk (middle eastern market), and did you haggle over prices?"

"You bet I did—it's ridiculous the prices they ask for, but with a little effort, I got some great stuff."

The instructor shows you your own sheet, your requirements document, filled out before you started the course, in your own handwriting. "I'd say you've done a fantastic amount," she says. "Six weeks ago you couldn't say yes or no in Hebrew, and—if I remember correctly, you weren't sure we could get you to the point of checking into a hotel or asking directions on the street."

"Well, those things are easy, but..."

The instructor stands up and shakes your hand "You are one of our prize students. I understand you want to do more and better in Hebrew—that's truly commendable, and typical of our best graduates. We have group classes for more advanced learners like you, people with real language ability. At this advanced point, you know, our goal is not to spoon feed you anymore, like what you had been doing when you were at the first level..."

Requirements—state them clearly, write them down, and accomplish them without distractions and temptations to pursue ever more impressive capabilities. This is your only hope of completing a program on the schedule and budget a microspace program requires. And it keeps peace in the programmatic family—no expectations inflation. Plan to have a system requirements review before the program gets underway in earnest—so that later on there are no questions in the minds of the customer, the engineers or the program managers, about what was expected. Formalize the outcome of that review into a Systems Requirements Document (SRD) that all the parties can sign off on and refer to as needed.

Conventional spacecraft programs don't allow construction, except limited prototyping, nor parts ordering, until the design process is complete. This is done for the benefit of management—nobody takes the risk that money will be spent on parts that can't be used because the design was changed. In a microspace program, we recognize that the real costs are in people's time and the budget. Everything is subordinate to that. Thus we order parts as soon as we think we'll need 'em—rather then have to wait for them and extend the schedule. We are using cheap parts anyway—why sweat it? Sometimes we get a part and it doesn't do what the manufacturer says. Other times parts are in the catalogues, but they're not really available anymore. We are shocked and dismayed, of course! Can anyone believe such a thing could ever happen? But better to find that out early, than to have a complete formal design review and subsequent freeze, only to find out the parts that compose the approved design are vaporware or museum pieces.

If engineers have the freedom to do more than design, they can actually do something productive in that long span of time between finishing a design and going to a design review. In big programs, that time does not exist—engineers design right up to the last minute. But in a big program, everything happens at the last minute, except for those things that happen after the last minute, and overruns and delays are the rule. Also, when we have nothing to do other than design, we tend to do a lot of designing. If another option is opened, particularly the fun option of actually building stuff, the drive to design to death is weakened a little bit. Often the informal little prototypes built to test concepts and parts and performance, end up as flight articles. Don't ignore miracles—you'll need them.

Though they are potentially major time and quality destroyers, go along with the PDR and CDR. Customers can't skip them, because if they ever admit to their friends that they let the contractor off without a PDR or CDR, they'll be a laughing stock. Sort of like sending your kid to school with last year's running shoes. But, in that same vein, set limits. The requirements document should spell out how many engineers—maybe one from each major discipline—will attend, and how long the CDR should last. A day is a good number. Maybe plus a half day the next morning for critiques and action items. No, you'll never go through all the overheads in one day. But the most important outcome of a CDR or PDR is that the package exists. It becomes a key reference tool. Also, you need to give engineers milestones so that they can calibrate their work. A CDR says "By this date, your design has to be done." And if there's a problem with that, you should hear about it. The reality is, hardly anything is actually reviewed or improved in a formal design review—it's a mind game played by thousands or tens of thousands of amassed IQ points. And those IQ points cost about a dollar per point per hour. You can't afford it.

Though Lindbergh managed to cross the Atlantic on his own, the airlines never have less than two people flying your airplane. This makes most people feel safer, but like all forms of redundancy, it breeds its own risks that we need to be on guard against. There is not a pilot on earth who was not at one time a student pilot, and during that time, most of them have the experience at some poignant moment of believing the instructor is flying the plane, while the instructor thinks the student is. This is most dramatic on practicing landings. Somebody has to slow the descent of the plane just before it glides down onto the runway, and when each thinks the other is going to do it, the result is quite a solid thud, followed by a bounce high into the air which broadcasts to the peanut gallery watching from lawn chairs by the flight office "WE DON'T KNOW WHAT THE HELL WE'RE DOING OUT HERE." Occasionally, planes are lost this way. The famous Eastern Airlines flight 402 that descended perfectly in control into the Florida swamp was believed by both pilots and the flight engineer to be under the management of the autopilot. The autopilot thought differently, having been disabled, inadvertently, by one of the pilots. The autopilot thought the pilot was in command, the pilot thought the autopilot was, and the first anybody realized, the jet was slogging through the Everglades.

Most couples manage a few fights in their lifetimes with their Significant Other on driving excursions that, as they whiz past the exit they should have taken, with the next one a hundred miles farther out into the wilderness, go something like "I thought you were watching the map!" "Me? I'm no good with maps—you told me that yourself."

These are the reasons you cannot afford to be without an Interface Control Document. Just as planes are not flown by one pilot and family outings are not solo performances, satellites are never built by one person, and rarely even by one organization. The opportunities for error are huge, and there is only one way to get it right. The second law of thermodynamics is lurking, waiting to increase your spacecraft's entropy. To make sure everybody knows what they are responsible for, the ICD must be detailed enough to determine whether the spacecraft team or the customer is responsible for any given connector, box, shipping expense, specification, software function or document. Like the requirements doc, the ICD can be renegotiated at any time. Events will necessitate changes. What's important is that it exists and the current version is available to everyone with responsibility on the program.

Generic Schedule and Staffing Plan

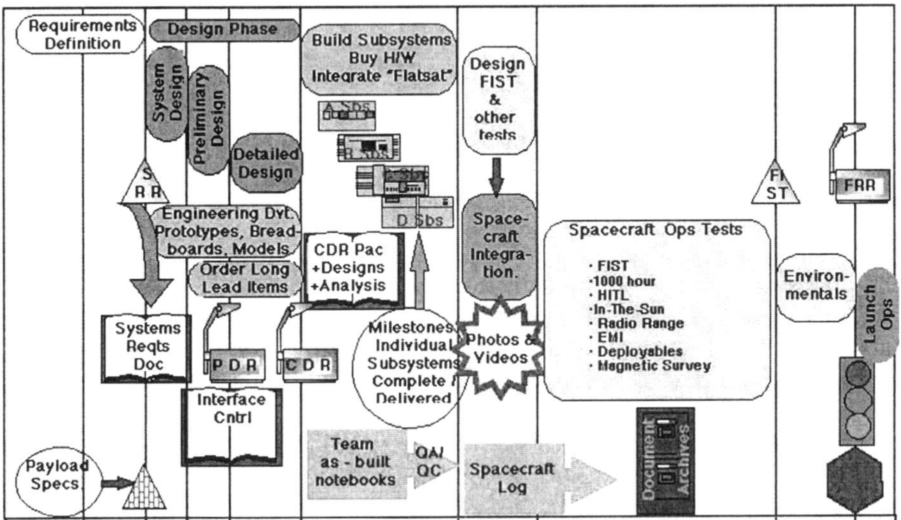

#FTE÷0.8	4	10	10	12	12	5	4	4	4
Σpers-mo	8	18	28	52	100	110	134	142	146
Months	2	1	1	2	4	2	6	2	1
ΣMonths	2	3	4	6	10	12	18	20	21

5.1 Grains of Sand in Your Program's Hourglass

A two year-ish program budgets only four months for design. If you decide that's not enough, you've decided not to have a two year-ish program. Find ways to get the design done—like stealing existing designs, using more off-the-shelf components, even renegotiating requirements so that you can get the design done. And don't forget to build an engineering team that is also committed to the schedule. A program will take about five times as long as the design interval. For every month you extend design, you extend the program by almost half a year. Multiply that by your average loading of about ten people, and you're talking half a million dollars per month of design. A two-month design overrun can double the cost of a program.

With design reviews behind you, and a lot of prototyping and some real stuff already built, you have a chance to crank out all the hardware in the next four months. Like the design interval, stretch this at your peril. In your age of innocence, meaning until your gaze fixed upon this schedule, you thought that most of satellite building was engineering (designing) and building. Now you know better—they account for about a third of the program.

At this point, while we agonize over every wasted second an engineer might spend rebooting a cranky computer and its effect on the critical path, it may be of some comfort to remember the rivers of time that have flowed before, and will go after, the development process that we are worrying about—usually only about a third of the process of getting a satellite in orbit. While we have been busy shortening the development time of satellites, the process of convincing the government to fund a program and go through their procurement dance hasn't changed much, if at all. Neither has the process of writing a business plan and getting it in front of companies that purport to invest in space programs had its clock speed cranked up much. Figure two years. Minimum. After billions of babies have been produced by humans, there is still a plus minus 10% tolerance on their development program—sometimes as much as 30%.

Since we are maybe 6 orders of magnitude less experienced building satellites, they tend to be scheduled to be done a year before the launch date. That year has only two uses: get the satellite working and then test it. The latter usually reverts the program to the former, and since it is impossible to predict how long any bug will take to squash, there will be times when even that whole long year will look puny.

Money, fuel, food, air, time. What they have in common is that they are versatile raw materials coveted by people, machines, animals, plants, and managers, respectively, more or less. Some time-hungry manager will covet thy pristine year and try to gobble some of it. Resist! You expect the satellite to work continuously on orbit for years. It should run perfectly on the ground for at least a month, what we call the thousand-hour test. You will fail this test several times and have to start over, maybe first after 10 hours, then after 50, then after 300. Without even worrying about the more glamorous thermal vacuum and vibration tests, just proving the system plays without hangups, continuously for at least 2 or 3% of its expected service life—that modest achievement can easily consume 4 months.

In addition to a year of integration and test, you'll need a few months in reserve for gestation period uncertainties. Plus, we allow a few months to ship the satellite to the launch site and fix anything that degraded during shipment. Also there has to be some time to go wild looking for things we thought got broken but really didn't—but just look different when examined in a new lab, with new test gear, new tables, new arrangements of equipment, newly mangled connectors and cables and after a few weeks or even months hiatus since the last set of tests, so that we can forget how we got it to finally work the last time. So there's a year and a half margin at the back end.

Most launches are late. Why is this? Because they can't be early, and thus given any distribution of schedule at all, the average has to be after the scheduled date. And there are plenty of reasons for the schedule to get randomized a bit—weather, equipment, part of the payload not being ready or checking out sick just before the launch. Worse yet, all of this is true of all the launches scheduled before yours. The airlines have the same problem. Planes don't leave early. By the end of the day, at least one flight will have been delayed for some reason, so many of those 5:00 and 6:00 p.m. flights are late. But of course, launch companies and airlines insert margins between their flights, right? Sure. And everyone working for them knows those margins are there and counts on them. But that's another book. Just imagine, while waiting an hour or two in your MD's waiting room for your routine checkup, all the things that could have caused his or her schedule for just one day to run 25% late. Your launch is the culmination of a multi-year, highly politicized process, and you need to plan accordingly. Two years from the time you think it's a satellite to the time it gets launched is not unusual.

5.2 Why Shepherding Is a Timeless Profession

Two years to get ready, two to build it, two to get it launched. Development is one third of the total program. And design and building are about one third of development.

So what's the rush? Have you ever shepherded four or five kids, some parents and some grandparents out of the house to go to a wedding? Each one is approximately ready—needing only a minute to grab a coat, to run back into the house for a forgotten gift, to finish putting on lipstick or to find some keys that—wait, weren't they on the counter a minute ago? Yeh, yeh, we're ready. A good rule of thumb is ten minutes per person you're loading into the minivan. Engineering is like one of ten kids that have to get done to get launched. If each is a month or so late, your launch is a year late. And besides possibly missing your launch, a year is a million dollars.

But what counts for your typical program manager, who is usually deep down inside an engineer, is to keep in mind how many other things happen besides engineering. No, a two year program doesn't mean two years of engineering. It means four months of engineering and four months of building. Maximum.

During the flurry of getting the whole to work, little things get changed. If they didn't need changing, then it would have worked right out of the box. Which it never does. This is not so terrible—unless the changes don't get documented. This is the "as-built notebooks" blob in the schedule. As changes get made they are recorded. If they aren't, bad things happen. Somebody tries to fix something, and finds out that their fix actually breaks something else. Because the something they fixed isn't the something that the documents say it is. This can be a waste of time and money—unless the fix happens to the satellite on orbit. In which case it can be fatal. If the satellite is lost in a launch mishap—almost 30% of small satellites don't survive their launch and separation from the launch vehicle—you will probably rebuild it. A few updates to the documentation to show the as-built and as-flown configuration can save not just time and money, but also a lot of embarrassment, as the team re-troubleshoots all their old errors.

It looks like a satellite—but is it a satellite? That's what two months of integration are for—to make the collection of pieces into a whole. The entity known as the FIST is not a frustrated smash of your balled up fingers through the screen of your computer, but rather the Fully Integrated System Test. Not found in any MIL or NASA spec, we invented this acronym for a test that can be done repeatedly at different critical stages in the satellite's preparation. The FIST attempts to exercise as many of the spacecraft functions as possible, given that it isn't actually in orbit but rather is sitting in the thermal vacuum chamber or on a table in a makeshift clean room at a launch site someplace. By periodically doing FISTs, the team learns to recognize anything off nominal in the behavior of the spacecraft. Like other tests, the FIST has to be designed in the cool light of logic. Don't wake up in the middle of the night with a brilliant idea, run over to the integration lab and see if your multi-million dollar satellite, about to be launched, will do X. Sounds ludicrous, but without a formally agreed to FIST, how else can you determine if the thing you're about to launch is anything like what it was when it was first finished.

Just as it's inevitable that one day you will want to know what changes got made to make the thing actually work, you're going to need to know how you got it together and all the abuse the satellite went through during integration. Compared to building a satellite from its components, making Capellini from scratch is relatively simple. If you have never done the latter, I wouldn't recommend it just by reading the documentation in an Italian cookbook. Videos are ideal for recording the nuances of the integration/reintegration process. Not that you'd want someone who's never done it to reintegrate the satellite just based on a video, but the team members do change from test to test and place to place. A video takes relatively little time, compared with writing a book on the satellite's construction, and you can even voice over a narrative highlighting critical steps and cautionary notes.

In keeping with our frank exposé on reality, another bombshell. People do things to satellites that they shouldn't. For instance, satellites don't have a big red On/Off switch. It's been suggested, but we're all too worried it will get launched switched off. But without that switch, it's pretty common to accidentally leave them on overnight or over a weekend but without leaving the battery charger connected. Monday morning, you have discharged batteries and a dead spacecraft. A day later, everything's charged back up and looks normal. But it's a different satellite—the batteries have been through one more deep discharge cycle. These things aren't fatal, but they're critical for making decisions later—like do we launch with the batteries we've got on board, or switch in the flight spares? This is the reason for the log. When questions come up, it's comforting for everyone to be able to look back on everything that's been done—

right and wrong—and make an informed decision.

Operational tests and environmentals take eight months. Program managers never believe this, and since by now most programs are months late and over budget, they convince themselves that this can all be done in four months. After all, why should it take as long to test the thing as it did to design and build it? The real answer is: It Does. Another answer is that no delay, no overrun, no additional meeting with a client to take a beating for being late is worse than having the thing not work after launch. The most powerful tool you've got, more powerful than design reviews, MIL-Spec procedures, good parts or even experienced team members, is extensive integrated systems test. The real world is cruel, filled with surprises and complexity. Modern microsatellites have so much on-board software it is impossible to test it comprehensively. What can we do about this? Test. Run the satellite for a long time, in all the possible and foreseeable states, circumstances and conditions. It is humbling how often an apparently healthy satellite freezes after one hundred or two hundred hours of normal operation. For a PC, two hundred hours of continuous operation without a reboot might be a lot. But for a satellite, that would be a reboot every week—on orbit. Every reboot can cost a day or two in data—so that would be a 20% or more loss. Plus, it's risky. If the satellite is too far gone to reboot itself, it can be lost. The only way to find those types of bugs that happen so seldom is long duration testing. That's the reason for the thousand hour test. If it doesn't fail half way through and need to be debugged and restarted, then it's two months. Testing and acceptance take time, but they'll add years to your life as you won't be standing at the launch site scared to death. The fact that you followed some MIL-Spec or bought only S-class parts may cover your rear end if it fails, but on launch day they buy you very little comfort. But if you and your satellite have been through months of exhaustive operational testing, and every bug you found was meticulously tracked down, dissected, understood, and corrected, you'll be confident your satellite is going to work. You'll go to the Flight Readiness Review (FRR) really confident that you are ready, and you will be. Those eight months are the most important of the whole program—don't let them get bargained away to save a little embarrassment or trim 1% off an overrun.

Chapter 6

The Future: A Lot of Unknowables, a Few Inevitables

One sure sign nobody has a clue where microspace is going is that I'm always getting asked that question. The Oracle at Delphi I'm not, and the Oracle of Reston, Virginia doesn't really seem to carry the same ominous resonance. If anyone actually knew the answer to this member of the set of the unknowables, then they wouldn't be asking me or anyone else.

The young Turks at AeroAstro are always imploring me to hire some wise old men and women with gray hair on their heads and experience under their belts. I try to explain to the mutinous troops that it would be a Quixotic search, like looking for a mature-looking police officer. Once you're in your forties, they all start to look pretty young for carrying that menacing looking gun. The fact is, microspace—building extremely simple, low cost and small spacecraft and rockets—is a young field that appeals to young people and is still finding its way and its niche within the aerospace ecosystem. And that ecosystem itself is radically changing. NASA is struggling to identify its role, and all government space is struggling for budget allocation. Meanwhile, commercial space is theoretically waxing. But is it real or is it Memorex? A lot of what we're calling commercial microspace looks like fins and wheel covers on the same old government contracting chassis. The answer will be obvious—if any of the microspace companies start making more than 10% profit on gross sales, we'll know the market truly is commercial.

Using the same discretion any author would have about predicting the future in a book that could be around a while, there are some inevitables that I can predict for you with confidence and on which I personally am focusing my own energies. And there are others that leave me pretty unimpressed. Let's deal with those first.

6.1 Computing Power and Memory

Have you noticed that your word processor doesn't load or run any faster with your 330 MHz Pentium III than it did with your 4 MHz V-40? This should at least temper your confidence that faster and wider processors addressing terabytes of memory are going to help you realize fantastic missions. They might be aboard your missions only because slower and smaller stuff eventually isn't even available on the market. But satellites mainly absorb data and retransmit it. The computing power necessary for repackaging, compression, and error correction is not stressing anybody's designs. The real bottleneck in data rates is not computation but downlink. If you are only sending down a megabit per second, you probably don't need a Gigaflop machine. And the same argument applies to memory. No sense in storing a whole lot more, a whole lot being a factor of ten or a hundred times more memory than you can download. Assuming you get to see a satellite at least once a week, a data accumulation rate of even 100 kbit/s is only 750 MBytes. Thus a gigabyte memory, widely available on a chip or two, is a lot of space. But how to downlink a gig? A low cost ground station antenna and even ten watts of transmit power in LEO only gives you a MBit/s. It would take almost two hours of contact time to get that data downloaded.

For star trackers and other computation-intensive sensors, present computers are overkill and all faster speed buys you is more rapid attitude determination updates. But the sensor optical integration time is already the driver, so aperture—the ability to collect stellar photons—not calculation, is the issue.

The power required by a processor is more important than its computing power. Processors tend to be on all the time, so reducing their power from 6 W to 4 W to less than 1 W makes a big difference to the typical power budget of microspacecraft in the one to one hundred watt range. You know what industry makes the most efficient processor in the world in terms of FLOPs per Joule? Swiss Watchmakers.

6.2 Batteries

We have already benefited from the laptop computer in making NiMH batteries widely available, and Lithium Ion batteries after them. This is an evolutionary improvement for microspace. Batteries used to account for 20% of mass. Perhaps in the future that number will shrink to 10% or 15%. No reason not to take advantage of that, but no reason to think it will change your life either.

6.3 Decreasing Launch Cost

Don't bank on it. We are now being managed by a generation that has seen computers shrink by eight orders of magnitude. There is an irrational exuberance, to quote one wise old sage, that rocketry will see a similar revolution. Two facts argue otherwise. Rockets have already been developed by the Americans, Russians, Japanese, Chinese, Israelis, Indians, Australians, South Africans, and Brazilians. Excepting the exploitation of very cheap labor, none of these rockets is significantly cheaper than any other. If there's a secret out there, nobody knows it. Secondly, while computers are amazing in their developmental progress, most of the things we buy have very flat price vs. time curves. Take automobiles, for instance. They haven't changed cost significantly (in real dollars) in decades. Neither have books, pianos, watermelons, silk ties from Italy, nor furniture. Rocket costs are not dominated by their electronics, but by things like tanks, engines, batteries, structures. Rockets are inherently dangerous, unlike PCs, and everything on board has to work right the first time—not a hallmark of microelectronics. There are large fluid flows and pressure to contend with, and they exist in the frame of technologies like combustion, valves and structures, that have not undergone anything like an eight order of magnitude revolution.

There are two ways rocket prices can fall. One is efficiency. By launching more rockets, and by launching them for clients more serious about cost than the Federales (who for all their goodwill, operate in a highly constrained environment fraught with lots of agendas), the price should fall. But Ariane is already doing this, as is Long March, and their costs are a bit lower than Delta or Titan, but not in a world-changing, multi order of magnitude way. The other way to lower cost is to launch smaller payloads. While physics says that the smaller a rocket is, the less efficient it will be, the engineering and management simplicities that can be achieved building smaller rockets, or smaller anythings, might buy you some savings. Launching ten kilograms for a million dollars is $100,000 per kg—the world's least "efficient" rocket. But also its cheapest by more than a factor of ten.

I'm even more pessimistic than just predicting a static world of space transportation. I see contemporary rocketry moving in exactly the wrong direction. Not satisfied to merely halve the price of a rocket (imagine buying a Ford Taurus for $19k this year, but only $9k next year), rocket developers are grasping at very thin straws to create plausible paths to impossible cost goals. The little capital society is willing to invest in rockets is being absorbed by these tenuous technologies. Things like 100% reusability (DC-X, X-33), towing rockets behind airplanes or lifting them with balloons (I can't imagine that being ten times cheaper than attaching rockets to the underside of airplanes which, by the way, costs more than launching them from the ground), and rotating wings to climb out of the atmosphere and cushion reentry, are being funded, though none of them, if they are even feasible at all, will save significantly in transportation costs.

Well, irrationally exuberant I ain't, right? Wrong. We are facing tremendous opportunities. They're knocking, but we're answering the doors we want to answer, not the doors they're knocking on. For instance, it's rather disappointing to think that the cost of putting myself into orbit will likely always be greater than the balance of my savings account. So much for leading George Jetson's life. But rather than spend my days futilely trying to lower those costs, there are other options. In the case of Me In Space, which I believe is the agenda lying just below the surface that propels most of NASA's public support and budget, a suborbital ride is probably within most people's reach. I'm sure the Wright Brothers would have loved to fly 550 miles per hour from Kitty Hawk to Paris, but they mastered the art of the doable and flew a few meters for a few seconds. I would love to feel the earth recede beneath me, see the world from space, experience weightlessness and see the stars and moon without a filtering atmosphere between me and them. On a suborbital flight costing a few percent of an orbital ride I could do all that and be home for dinner, including a stop at the 60-minute photo lab en route.

Option B is, we shrink the payload. We can now do with 100 kg what only a few years ago required 1000 kg. Here shrinking electronics can help. There are already 1 kg satellites being built, and even smaller ones being designed. So what if an orbital launch never gets under $10k per kg? If your payload is 100 g, the $1000 you spend to launch it is below the noise compared with even a day of operations or a day spent looking at the data you get. I don't care what the cost per kilogram is—what I care about is launching as close to zero kg as possible. I wish there were someone, anyone in the rocket business with a similar outlook, but I don't see any.

How much hoopla do you hear about reusable rockets, rotary rockets, airplane and balloon towed rockets, sea launched rockets, and single stage to orbit rockets? Lots. And how much do you hear about shrinking spacecraft to cut launch costs? Very little. This is what I mean about not answering the right door. The solutions to the space transportation cost reduction question are known—they just aren't sexy.

6.4 Deployables

Well, it isn't quite that simple. Microsatellites weighing a kg or less are a reality, but like palmtop computers, they don't address most users' requirements. And for one reason fundamental to both microsatellites and palmtops: aperture. Aperture is area to collect or to focus photons. If your payload needs 100 watts, it needs a solar panel of at least one square meter. Oh, with super advanced, super expensive materials, maybe you cut that down to 0.7 of a square meter. This is still a lot bigger than a floppy disk, which is where spacecraft sizes are going. And so what if a gigabyte of memory is the size of a quarter—you still need a solar array the size of a coffee table to power the transmitter to send those bits down to earth. And if the spacecraft is going beyond LEO, to geosynchronous, the moon, a planet or an asteroid, or if it's going to drift around the sun to look for spots and flares the earth can't see, it will need an antenna bigger than that projection TV (aka big aperture, small electronics) you've got shoehorned into your living room where the grand piano used to sit. Satellites may shrink, but optical telescopes, photovoltaic arrays and high gain antennas have dimensions determined by the physics of photons. Yes, you can fly 20 microsats and, with great technical difficulty, combine their apertures through phase locking. But that's a complex solution, probably more complex than a single bigger satellite. When engineers refer to physics, what they are telling you is, that's a knob they can't twist.

Enter deployable arrays. I believe there isn't a household outside of Los Angeles that doesn't already have, tucked somewhere into a coat closet, the high tech solution to the satellite aperture problem. Usually made out of woven nylon, this device is small enough to lose under the seat of your car, in a pocket of your golf bag, or in a woman's purse. It costs $10, or maybe $20 if you want a fancy one. It has no batteries or microprocessor, and even your ninety-year-old grandmother uses it with ease. Millions are deployed every day with reliability better than the best S-class parts. And it deploys, under its own power, from something the size of your hand, to about a square meter.

Why aren't there any umbrellas in space? There are in fact lots of deployable arrays in space. They are just huge, expensive, heavy, and unreliable. The latter three qualities resulting from the first. But a microsatellite doesn't need 10 kW of solar power—we have a gadget the size of a coffee mug that produces at most 100 W. Any umbrella could handle that.

Capturing solar energy is a lot easier than focusing radio waves, which is easier than building a deployable optical telescope. But there is no physical or technological reason we can't do this. It's an opportunity waiting for us to exploit it. A microsatellite weighing 1 kg and the size of a calculator could be fitted with a deployable solar array and a deployable antenna, and send millions of bits of data per hour back from the moon or even an asteroid to a relatively low cost ground station. That's what's in our future.

There are several technologies that can contribute to this revolution in our space capabilities and cost efficiency. One is flexible solar photovoltaics that are easily scrunched up like an umbrella. But many deployable arrays don't even require that flexibility—they fold up like oriental fans or flower petals. Adaptive optics are another significant technology. We can't deploy a near perfect optical telescope, but we can fit the telescope with nanotechnology actuators which can adaptively mold the structure to achieve good optical performance. Rockets can use safer and cleaner propellants to make them better adapted to carrying lots of people into space. And given plenty of electric power, we can use new devices like solid state lasers for high data rate communications and cryocoolers to make very sensitive optical focal planes.

And with increased numbers of missions being done by microsatellites, we'll see lots of scaled-down devices—maybe not all micromachines, but very small machines. Momentum wheels the size of PCMCIA disk drives (with platters about the size of a nickel), and very small magnetometers and other sensors. The way technology evolves is by leveraging innovations. Somebody starts mass producing cars, somebody else builds refineries, another highways, and pretty soon you can visit your relatives in Cleveland with your own little automobile just for the weekend. The freedom of the American road wasn't one breakthrough, but a series of innovations which, when grouped together, created a new paradigm.

6.5 Everything Else

Once you slash transportation costs by orders of magnitude by either shrinking the payload by orders of magnitude, or by carrying humans wanting time in space on suborbital instead of orbital flights, and once you solve the aperture problem with deployables, it's Miller time, big picture-wise. But while most of us hustle off to the bar or the bike path, there are a few other innovations the workaholics in the lab can try to perfect.

My favorite of the not-a-major-paradigm-shifter-but-still-worth-mentioning category is the universal attitude sensor. Every time we build a satellite, some cranky and expensive automatic control Ph.D. starts with the equivalent of butter, eggs, salt, vanilla, sugar, and special sauce handed down from professor to thesis candidate, and designs an attitude determination and control system from scratch. But for LEO satellites, two star sensors and a GPS receiver really does it. If you're in deep space, the GPS receiver can be unplugged and in its place you turn on a software module responding to ranging data from Earth. Why don't we do that now? Because there is no Hush Puppies model of star trackers, no such thing as Topsiders. You move right up to the Bruno Maglis at about $400k each, along with eight watts and several kilograms. So we resort to magnetometers, earth sensors, sun sensors, v-slit star sensors, and mix and match these to figure out complex ideas like where are "up" and the horizon. A pair of low cost, relatively low resolution star cameras armed with a GPS receiver to perceive time and position and hence ultimately the orbit elements are the core of a universal system. Add some software and a processor of modest merits to estimate attitude, position in orbit, and the location of any target. These could range from an earth station or imaging target on earth to an astronomical target or GPS satellite. And none of these capabilities is dependent upon the specific orbit, mission and configuration of the spacecraft. The whole thing reduces to a black box, which eliminates mix and match shopping for sensors and greatly reduces the effort for attitude estimation and hence also control.

Maybe the biggest impact on our future is one we don't know yet. It takes many years living with any new technology—personal computers are an example—for users to discover their most advantageous applications. PCs got started as game and learning machines (mostly learning how to work a PC) and graduated to word processing, then graphics, and now entertainment and web cruising. Maybe they will constitute the first successful attempt at video telephones. Modern microsatellites got started as a hobby and learning platform, then scientific research platforms, and now are finding a niche as the email system in orbit—the little LEOs—that provide messaging, monitoring, and bi-directional paging services.

But the future might include remote sensing using phase-locked satellites thousands of kilometers apart or in massive clusters to achieve very high resolution and instantaneous simultaneous global observation. They may monitor and repair larger satellites, something like nurse fish around sharks. Their low mass makes them ideal for space missions requiring large velocity increments, including observation of the far side of the sun, asteroid research and mining, and lunar and planetary probes. Entertainment including participatory space exploration and commemoration, including burials in deep space and even in the sun itself, will be the missions of future microsatellites. So far we have focused on scaled down applications of conventional spacecraft.

As we mature as an industry and a technology, we can hope to come to understand our niche, what we do well and how we do that, to define an identity of microspace. Today we are like kids who want one day to be pilots and the next to be marine biologists. That self-definition will result in part from identification of the missions unique to microspacecraft capabilities, including the ability to fly quickly and cheaply, to fly in large numbers, to be doable with relative simplicity, and to perform very high energy missions.

Chapter 7

Space History and a Possible Future

According to *The Hitchhiker's Guide to the Galaxy*, the history of human civilization can be condensed to 3 lines:

- How do we survive?
- What can we eat?
- Where should we go for lunch?

Had Douglas Adams predated Will and Ariel Durant, thousands of pages and two lifetimes could have been spared. One could say the same of millions of lives and hundreds of wars. Progress marches over all of us.

It's currently marching over the world's aerospace industries and institutions. Let's review five decades in five lines:

'50s: How do we get to space?
'60s: What can we do in space?
'70s: How many bigger and more expensive things can we do in space?
'80s: Who's gonna do what in space?
'90s: How can we spend less in space?

To wit:

In the mid '50s, President Eisenhower was advised that it was impossible to put a satellite in orbit, and also worthless even if it were possible.

The '60s saw remote sensing, LEO and GEO communications, unmanned space science, and manned space flight all achieve significant accomplishments.

NASA's Space Station, Space Shuttle, Spacehab, and Crewed Mission to Mars (the latter now awaiting resuscitation at the right political moment) all came to significance during the '70s.

The great international commercial space race was the significant feature of the '80s. New countries and new competition in semiprivate space monopolies commenced space activities, with Ariane, Long March, Proton, and Delta vying to launch satellites built in the US, Japan, and Europe. An ominous cloud of commercial competition appeared on the horizon of Intelsat and Inmarsat.

The '90s arrived like a New Year's Day hangover. Defense was no longer as huge an engine of demand. Nobody wants to consider a multi-trillion (with a T) dollar Mission to Mars, sticker shock and budget buster being this decade's favored Washington clichés. While many look back to our Glory Days and try to relive them with great missions—or by retreading John Glenn, the real challenge facing our community is three letters: ROI—return on investment. Trillions have been spent on space. Perhaps it is not too cruel to ask our long-haired teenager, loafing around the living room and watching MTV, to contribute more and to ask for less.

In ROI terms, small programs have a significant edge. ROI is benefit divided by cost. If cost goes down by factor 20, while benefit goes down by 2 or 4 or 10, you are a winner. It's doubtful anybody will soon make the magic crystals in space that NASA hoped would be worth $10,000 or $100,000 a gram. Even the shuttle could break even at those prices, but some very modest, sensible, and valuable programs now dot the small space landscape.

Little LEOs. Yes. In theory, three massive GEO comsats with spot beams could provide quasi-global cellular phone performance by means of small handheld and wristwatch terminals. Fact is, that's not happening (refer to big LEOs below). Small LEO satellites providing electronic mail and messaging are already in operation serving relief workers in the Third World (HealthSat and VITA); taking environmental sensor readings (Orbcomm); and linking electronic mail and geolocating people driving around somewhere on the globe's face as a satellite extension to the Internet (AMSAT). On a commercial basis, non-real-time communication from small LEO satellites is a fact of the '90s.

Big LEOs. Iridium. Globalstar. Ellipsat. These programs and their spacecraft are so large and growing that they stretch any contemporary definition of small satellites, launch vehicles, and space activity. But their roots trace to the thesis that ROI can be increased by decreased physical size, which lowers manufacturing and launch costs. Simultaneously, program risk can be decreased by spreading performance over many independent small spacecraft and building constellations through multiple launches.

I dismiss as naive Motorola's claims that Iridium has nothing to do with small satellites. Rather, it is a sign that an innovation has really matured when the innovation itself is taken for granted. Does Gap pay tribute to Levi Strauss? Today's rock and rollers admit no relationship to Moog, if they even know the word. Do Apple computer users know or care about Eniac, Turing, or even Wozniak and Xerox PARC?

Innovators build a kitchen, set the table, and stock the refrigerator, so that they can enjoy watching their children eat them out of house and home!

Space Science. ALEXIS; HETE; TERRIERS, satellites I've been involved in that all weigh under 150 kg, are now a minute dot on a huge landscape of microsatellite science missions. We take for granted now that they all have capabilities comparable to much larger, costlier, and more complex satellites of the '70s and '80s. Freja, Odin, Viking and Astrid I and II, built by Swedish Space Corporation, are proving that small platforms can do leading edge astrophysics and geophysics research. NASA's Discovery program, aimed at very small missions, was inundated with enough proposals in planetary, earth, and astrophysics research to carry out an ambitious program lasting for decades or even centuries.

Remote Sensing. OSC's Seastar is a well known small satellite remote sensing program. However, at least 5 other remote sensing programs worldwide plan to use one or more small satellites with imaging and non-imaging sensors to manage crops, detect pollution, and aid meteorology. One program not at all secret is Worldview, which is placing mini-Spot class imaging devices aboard small satellites at LEO.

Education. University of Surrey, Boston University, Northern Utah State, Technion, Stanford, University of Colorado, Weber State, Korean Institute of Technology, University of Arizona and the University of Tübingen all offer academic programs including hands-on design and/or development work for small satellites.

Ten years ago, small satellites were not even a curiosity outside of AMSAT. Nobody even thought about them. They weren't significant enough to have an opinion about.

According to my archives, there were 16 small (where small equals less than 300kg) satellite launches in 1992, 31 in 1993, and 26 in 1994. This number is staying about constant five years later, boosted by the first Little LEOs but hampered for lack of launch space comparably priced to the satellites themselves. These projects are giving our own community, and more importantly, the community of users in science, communications, education, and remote sensing a new set of options in addition to the large, highly capable, but costly conventional satellites. The march of technology is not usually linear. The hottest computers on today's market are not those with the zippiest processors, but rather the ones with modest processing that fit easily in the palm of your hand or that the 60% of American homes without computers can actually afford. Sales of Aston Martin sports cars, or even all sports cars including Porsche Boxters and Miatas are not fueling a new automotive industry. But SUVs and minivans already have. In both these cases, what technology provides is not so much minivans eclipsing BMWs, but a broader range of choices. Back in the days of the Model-T, there was about 1 choice. Now everything from electric cars to station wagons not much different from those of the '50s to minivans and SUVs rivalling small cement trucks are all consumer possibilities. You can drive a Mac that puts all but the most recent supercomputers to shame, or you can stick a Palm Pilot in your brief case and still pick up e-mail when you're at your dentist's office. Small satellites are providing choices—new ways to do missions—and the first way to do any space mission by those who have to live with four and five digit budgets.

Anybody for lunch?

Part III

A Wrinkle in Microspace

Chapter 1

AeroAstro is a small company—just about twenty-five of us altogether. Though we'd barely be a department in a bigger aerospace company, among us we span the range of satellite engineering disciplines: radio, digital hardware, machine level and applications software, guidance and control, mechanical design, finite element analysis, systems architecture, finance, and management. Keeping these factions all talking and cooperating is not easy but it is vital, so we organize occasional social events to promote interaction and a sense of community.

Virginia is cave country. Prodded by our in-house spelunking and climbing activists, myself not counted among them, a consensus developed that we all go out and bond by exploring some of the more esoteric local sites. Equipped with day packs, rope, flashlights, cameras, sandwiches, granola bars, and water bottles we jumped into a few cars one sunny September Wednesday and headed for the hills.

By 1:00 p.m. or so, most of the group were tired of crawling around on our stomachs and another consensus developed: to convince the enthusiast contingent to break for lunch and to absorb some early Autumn sun, though, unfortunately, the accumulating clouds were by now doing most of that for us. While munching on a somewhat tough, dry peach (17,000 new packaged product introductions in grocery stores last year alone, but nobody cares about producing a good piece of fruit) I watched a goat clambering around some rocks just above us. Those animals are just too tall and precarious on their skinny legs and wobbly knees. So how come they climb surer than any human rock climber? What do they understand by instinct that humans must have lost when we evolved from living in the trees to fending for ourselves on the African plateau?

I wasn't the only goat watcher. Jennifer, one of our real climbers, interrupted my mental peregrinations on the evolution of the species.

"Rick, they say the goats really know the best passages."

And I thought, "They should, they spend all their time nosing around them," but I said, "Sure, how could that animal crawl around on its stomach?"

Used to my seasoned skepticism, Jennifer shrugged me off and headed toward the goat. "Let's say 'hey' to him," I think she said, getting up and leaving her camelback pack by a tree. Not hearing the conversation, but responding to our animation, Har'el, another of our Generation X or Y-ers, grabbed a rope and paced toward the goat. Mr. Goat, scruffy and unattractive, had us in his sights. My mental model was fuzzy logic-ed about 90% that he was going to bolt and 10% that he was going to bolt at one of us.

As we approached I watched him watching us, almost motionless except he nodded his head up and down in almost a playful gesture. Goats, I thought, are not beautiful by nature. This one had whitish and brownish patches and that funny little beard. By now Jennifer and Har'el, showing a lot more courage than I have, were petting the goat. Har'el offered the rest of an apple but the goat seemed more interested in his coil of rope. Being strong, assertive, and stubborn the goat quickly achieved his goal.

The goat dropped the loops of rope on the ground but held one end tightly in his mouth and started walking easily up the hill about forty-five degrees off the fall line. Half to recover our rope and half out of curiosity, we three picked up the trailing, somewhat tangled line and followed the goat. Taking up the rear, my thoughts were first, that the goat must do this all the time with kids on a farm where he probably lives, and second that we must look pretty silly. But silly to whom? We were already out of sight of the rest of the group.

Mr. Goat had stopped to paw at some leaves, and taking advantage of the break to look around, I couldn't see the spot where our group had paused to eat. We were in a narrow saddle point between two ridges. The air had become totally calm but the transparent early autumn sky we had that morning was now a uniform slate, as if we were walking under a giant marble coffee table. Had I gotten around to saying "let's head back," it would have been moot, as our friend the goat had uncovered the mouth of a cave and Har'el was clearly determined to have a quick look.

Do goats wink? I thought this one did as he looked back at us and pulled at the rope. Realizing the inevitability of the situation I prodded Jennifer to tie the rope around the goat's neck. Surprisingly, the goat submitted willingly, and within a minute we were following a strange and possibly wild goat of uncertain intentions into a totally black cave carrying none of our gear and without notifying the rest of our group.

Luckily for the quality of this part of the narrative, my fear of dark, enclosed, cold, and dirty places had prevailed upon me, before leaving the house that morning, to stuff my wife's tiny Mag-light and a few AA batteries into the pocket of my khakis. This habit I had picked up quite naturally from years of sitting in other dark, enclosed, cold, and dirty places—specifically the cockpits of old rotary engine piston airplanes flying night freight around New England. It is truly remarkable how useless a navigation chart and a panel full of fancy avionics are while drilling through a cloud deck on a winter's night over Attleboro, when the frayed wire carrying power to the cockpit lighting is finally disintegrated by forty-plus years of engine vibration and the salt air characteristic of New England's airports.

But having forgotten, for the moment, about the mini-flashlight, I was able to focus on just how dark caves can be. The goat continued in his measured gait and we seemed to be going downhill on a twisting path. Banging my head on the odd ceiling mounted root or rock and the aching of those muscles over my kidneys from walking stooped over reminded me how small a passage we were in. The air was getting damper and cooler. Nobody said anything—we were too busy groping through the dark, unwittingly testing the dubious theory that goats really know these caves.

Proving that the cessation of stimulation is itself a stimulation, we three all gasped in unison when the goat suddenly stopped. I think Har'el's gasp may have been triggered by walking into Mr. Goat.

"Now what?" I mumbled.

And I answered -- the light. For a moment I sensed that my brain was coming back to life again after a long spell concentrating all my attention on each step and on hunching over.

As a manager, I have grown accustomed to the greeting of my brilliant proclamations by silent disdain or even some frank dissatisfaction from the troops. This prepared me well for Jennifer and Har'el's response as I said, "I've got a light," and twisted the barrel of the Mag-light. To our fully dark adjusted eyes the puny beam flooded the room in light. We, three humans and one scruffy goat, had emerged into an almost spherical room, a bubble maybe forty feet diameter deep inside the earth. Most of the bottom of the sphere was filled with cold water, the edge of which we were standing at.

"Mr. Goat has a good idea," I said. He was drinking.

"Rick," said Jennifer "I think if you take a look, it's Ms. Goat."

I shined the light on the slurping goat. "Oh. My apologies."

Jennifer looked tired. Probably we all looked tired. As we slurped and talked, the goat dozed. We covered all the normal topics one does when trapped, via one's own lack of prescience, in hopeless straits.

These included:

- what we should do
- how to get back
- why we were stupid enough to do what we had, and
- perfunctory pledges to God that if we ever were to emerge alive, we would love and honor Him in part by never doing anything like this again.

As battery #1 died, despite my sparing use of the light, we tried to discuss just what might be the goat's plans for us. The shared intellectual conviction that goats don't make plans was left, if not unthought, at least unspoken.

I flickered the Mag-light to demonstrate my success in full darkness battery change out procedure. Our conversation continued in the battery-conserving total darkness. The way we had come seemed like the only path leading to our cavern. Har'el took the light and tried to backtrack. But like so many caves, ours was a maze of passageways and all but one would dead end or loop back. No other option emerging from our subsequent musings, we ultimately resigned to the unsettling opinion that the currently sound asleep goat was our best shot.

It was almost 4:00 p.m., three hours since we left the group. They have by now searched for us and it's getting toward evening. In a couple of hours Nancy would be home from ballet class. Hearing that I'm missing somewhere in the Blue Ridge, she will definitely panic. Clearly the goat was not worried, and I was confident that he, ah she, would lead us out. But life would be unbearable—maybe untenable. Nancy has gotten a few phone calls from me out on the bike trapped in a snowstorm and from the odd emergency room with a broken arm and numerous other small catastrophes, but this time Nancy might, no, would definitely, kill me.

Was that water running? We three were treated to a serious, immediate panic of our own—we would drown in this damned cave alongside our intellectually dysfunctional goat! I twisted on the light. As the echoing sound of the rush of water overpowered our voices, Ms. Goat opened her big, chalky eyes and rocked up onto her feet.

I looked around. Best place to stand is just next to the tunnel leading in. Avoid getting hit by the high speed jet. Maybe if we floated up we would not drown—air will be trapped at the top of the dome. As I ran past the tunnel I wondered, how cold was the flood going to be? We could die of hypothermia just as easily.

As Jennifer and Har'el joined me I watched the goat and held my ears. She was just staring down at the pool of water on the floor of what might be our shared, spherical coffin. Was her peaceful gaze the Darwinian pre-death relaxation I'd read about in *Scientific American*? Unlike the goat, I was just profoundly pissed off. Of all the stupid ways to die, following a goat into a cave with a propensity to flooding. A trivial mental misjudgment, and it was the end for all three of us. It confirmed to me that individual life has as much meaning in the grand scheme as an individual water molecule has in a tureen of bouillabaisse. As is so obvious in the case of insects, our great strength is not in the individual but in the community of humanity. A community from which we three seemed about to permanently withdraw.

Mental gymnastics notwithstanding, we were frozen against the wall in stark fear. I tried to exchange a glance with Jennifer and Har'el, but I couldn't even get their attention, and the water gush had become so loud that it hurt my ears even covered with my hands. Standing there, afraid, miserable, stiff with fear, I almost wished for the end. My mind wandered over our pitiable circumstances and I thought: If a slug of water is rushing down that tunnel, air ought to be coursing through in front of it. I inched over and put my hand in front of the small jagged orifice. Nothing. I turned to cast a puzzled look to the others and in the pencil beam of the Mag-light, saw our water pool vibrate. Earthquake? And suddenly, the pool drained itself. All the water we had been hearing had been flowing out!

Chapter 2

We looked at each other blankly. The cave was silent again except for the dripping of some last few drops of water. Har'el expressed the sum of our collected wisdom, having faced death and lucked out:

"God damn."

"Rick, I thought that was the end," Jennifer was rubbing tears from her eyes.

"That thought crossed my mind."

We stared over at the goat, who was calmly ambling down to the low spot of the cavern in the middle of the area where our pool of water had been just a minute ago.

There was a sort of stone chute—I think a civil engineer might call it a sluice—at the bottom of the now-drained pond. The goat hunkered down, folding her legs underneath her and just glanced for a moment at us before sliding down and out of sight. I think that she knew we would follow. What choice did we have? Har'el, then Jennifer, then I, slid down the rough, wet and muddy chute, tentatively, feet first, in total darkness. Both hands and both feet were too busy bracing the body against the incline and protecting it from rocks and sharp edges to manage a flashlight.

Shoving and dragging a downpour of loose mud and pebbles with us, we clumsily dropped out the bottom of the chute. The scenery was not much different from that at the site of our initial enticement by Ms. Goat. We were smiling, grinning, laughing. We made it! And there was still even a bit of daylight. As we stood up and our bodies and damp clothes absorbed some nearly horizontal rays of the yellow early evening September sun, peaking out between rolling grey clouds, we became quiet. I think we were thinking one thought: We could just as easily have died in that cave. My mind traversed recollections of grisly survival stories—the stuff little boys love to read—about plane wrecks in the Andes and people floating for weeks in life rafts baking under a hot Pacific sun, and Jack London's "To Build a Fire." Conclusion: people do die, and often with a lot less reason than we had.

But we were spared all that agony and blessed with our lives. Looking around, the blue sky looked bluer, the disk of the sun was clearer, the air seemed fresher. Not seemed—this was not the elation of survival—the air actually was much clearer than when we left our group and followed the goat into the cave.

Where was that goat, anyway? None of us had noticed her since she took off down the chute. She was moving a lot faster than we were, and probably came out way before us and wandered away. No big worry, at this point we'd find a house and a phone, and call to get picked up and to let people know we were OK. We seemed to be at the edge of a farmer's field, and in this rolling country it wasn't surprising we could not see a farm house. We decided to just follow what looked like the property line, roughly defined by sections of fencing and rows of hedge. It was possible, as the sun would soon be down, we might spend the night in that field. But contrasted to the cavern, we judged that prospect at most to be a minor inconvenience.

Our improved luck was holding. A woman on horseback soon appeared over a hilltop, cantering toward us. We told her we were lost and needed to get to a telephone. I suppose we looked innocent enough. From atop her chestnut mare, she told us her house was just ten minutes walk. We all introduced ourselves just in first names—she was Hilary—the way people do when they are expressing both their informality and their belief that the relationship will be transient. Being thoroughly addicted to my own cellular phone, I didn't think anything odd about it when this tough looking country woman in heavyweight jeans and boots and a worn out, black, team jacket plucked a small telephone out of her horse's saddlebag and handed it to me.

Being far out in the country and as we were soon walking in a narrow gorge between rocky hills, I was surprised to see the signal strength showing four bars—a solid link. I dialed my home phone number and was not surprised Nancy wasn't there. Still at ballet class, I thought. But why didn't the answering machine kick in? Jennifer's line was busy—not unusual considering her three house mates share the one line. Har'el reached a wrong number, tried again and reached the same wrong number. We decided we would just try again from Hilary's house.

"Suit yourself," Hilary said, "but that's the only phone we've got."

I figured they must really live out in the sticks if they didn't even have a phone line. They were lucky to at least live near a good strong cell.

Having barely met Hilary and then so abruptly absorbed ourselves in using her phone, I felt a little friendly conversation was in order.

"How did you happen to bump into us way out there anyway?"

"No accident. I was fixing dinner when the alarm went off. All my stock are grazing the south end, so I thought it was one of Deutheimer's goats again. They're always setting off the alarm."

"Who?"

"The neighbors' goats. Nice old German couple live over there, but their goats are a nuisance. Eat everything!"

"Speaking of eating, we are all starving." (Jennifer and Har'el nodded in agreement.) "Any place around here we could buy food?"

"Not walking distance. But Gib and I were just about to eat and we've got plenty. You can fill up while you wait for your friends to come getcha."

After the day we had, and having missed lunch and skipped breakfast for the sake of an early start, I was not going to pass on that offer. After some very sincere thank-yous we all walked the last few minutes to Hilary and Gib's neat little farmhouse in a twilight of reflective silence.

Gib was a short, stocky man about thirty with a ruddy face and big round cheeks. His handshake was firm but cushy as he gave us the "hey y'all doin'" greeting I'd gotten used to, stopping for food on long hilly bicycle treks through rural Virginia. Gib's relaxed, friendly style put us all at ease, and I looked forward to a pleasant conclusion of what seemed now as just a minor misadventure. My only lingering anxiety was that our colleagues would think we were still lost out in the hills somewhere. We tried again to reach our homes while Hilary put freshly baked dark bread and a pot of stew on the kitchen table. The heavy iron stew pot was flanked by a plate piled up with fat yellow ears of steamed corn on the cob. Our judgment in goats was poor, but we had been sharp enough to maroon ourselves at a Virginia farm at the peak of the harvest. Meanwhile back to the phones, where we all failed to reach anybody—a frustration we decided to deal with over our first meal since last night.

Over dinner Jennifer told the story of our adventure with the goat. Hilary and Gib agreed this sounded typical for one of the Deutheimer's goats. Kind of ominously though, Gib remarked that a neighborhood child had disappeared almost exactly one year ago, and it was suspected she had lost her way in a cave in that same area. Gib had been a member of the search team which worked through two nights illuminated by the same full silver moon now just rising over the hills to the East. With the chill of the autumn evening, the fear of death, still very fresh, came back over all of us. What could have happened today, we all wondered, and we became completely silent.

Gib took the initiative to bring us back to life by bragging about his corn crop. "Ah think it was the weather forecast what did it—we knew jest exactly when to plant. And ah never got it too wet waterin' a day before a storm like 'seems ah always used to do."

Hilary explained they had a cold spring on the property and how many times it saved their crops in dry years. She said since some new weather satellite had gone up, their timing on planting, watering and fertilizing had gotten almost perfect.

"Yup. We didn't waste a drop and we saved thousands of bucks we used to waste fertilizin' n' sprayin' pesticide and then gitt'n' rained out." added Gib.

Licking part of the very tasty results off my fingers, I asked, "what kind of satellite was it?"

Hilary shrugged. Gib said he had no idea, but they paid $10 a month and got daily forecasts sent to their e-mail address. "They're a sight better than the TV forecasts we used to use."

Har'el, Jennifer and I thought we knew what was going on in space, working at AeroAstro, but we knew zilch about this service. I guessed that one of the commercial forecasters like Ocean Routes must have gotten into the agricultural market.

"Do you get your e-mail over the cellular phone?" I asked mainly because I was guessing the crummy cell phone service would explain why we weren't reaching anybody.

"Oh, it doesn't use the phone. It just goes straight into our computer."

Hilary pointed at a PC clone standing mute and blank on an oak table by the window.

Just then, the screen lit up and a weather map and forecast were on the display.

"Oh it does that four times a day when the new weather comes out. More if there's bad weather coming. But the weekend's going to be just fine for Terry's wedding."

"You get this on a satellite?" Har'el asked.

"I think so. See there's a little antenna on the back there."

Jennifer got up to investigate. The PC's only links to the world were its power cord and the curly-Q antenna. I had heard of wireless data services but none out here in the country. Just tests in a few cities. Maybe this was a test market area? An e-mail message came up on the screen. Hilary got up to read it.

"Gib, it's your momma again. Says since all next week's gonna be dry so she wants you and your brother to get over there and weather-seal her roof."

"Tell her we'll be over t' her place on Wensd'y. Stuff needs two days t' set, an' it ain't gonna rain b'for Satad'y next."

Gib's accent seemed to get thicker when momma was concerned. I enjoyed him a lot. Not too many like him in urban suburbia where we tend to hang out.

"Does your mother have a satellite e-mail link too?" I detected a flat irony in Jennifer's voice. Clearly country grandmas downloading wireless e-mail seemed just a bit incredible.

"Jennifa sweethaht, ain't nobody don't." Gib's answer was sweet to the point of condescension, as if Jennifer had asked if all country folk had television.

Something was missing from this picture. Were these people fugitives from *Wired* magazine? But it was getting past 9:00 p.m. and Nancy would be home and worried. We tried all our home numbers and some other AeroAstro people. While Har'el and Jennifer dialed, I asked Hilary about the perimeter alarm. Did they run wires all around their property?

"That would never work. Weather and animals and the plowing. We signed up with one of those satellite services, mainly 'cause of those darned goats. Put a sensor about every hundred yards, and if one trips, we get a warning on the computer. It calls my phone too. When you guys walked in, that's how I heard about you. Everybody around here uses it. We have the house protected too. Calls the Sheriff's office if one of those goes off."

This is some Qualcomm service I didn't know about? As I pondered that possibility, Har'el and Jennifer were still reaching busy signals, wrong numbers and no answers. I was feeling a bit paranoid with these nice but clearly bizarre people, and I wanted to get home. Asking them to drive us was a big imposition—over three hours round trip.

Then I thought, what's wrong here might just be the phone. Maybe there's some big phone system outage. Maybe all of 703 is down, or at least not accessible from where we were. I'll try to call someplace else, like my parents in Cleveland. Bingo! I reached their answering machine but I hung up before I heard the beep to leave a message. What good could they do me from there, not being even at home, and why worry them? By the time they get home, we'd have this worked out. Anyway, now we were getting somewhere. Jennifer decided to try her mom in Vermont. If she were home, maybe she could call into 703, our area code. Jennifer, standing by the sink in the kitchen, dialed the number on the cell phone.

"Mom?" She had the receiver volume up loud enough we all could hear from the kitchen table.

"Hello Jenny," It was an older man's voice. "Your mom ran out to the store. What's on your mind, Darlin'?"

Jennifer's white complexion went even whiter. She said nothing and was motionless—not even breathing.

"Jenny, you all right?" asked the voice at the other end.

Jennifer was frozen as if in terror. Her hand holding the phone began to tremble, and her eyes focused hard on the table.

The man's voice again: "Jenny dear, maybe we got a bad connection. I'll call you back. You at home?"

Why was she so afraid? Jennifer hit the END button and slammed down the phone on the kitchen counter. We looked at her but she didn't look up. Just as abruptly she picked up the phone and punched SEND, redialing her mom's home number in Rutland.

The same man's voice answered: "Jenny?"

Jennifer swallowed and gripped the phone. "Daddy?"

"How's my baby? You sound worried."

"No. No, I'm OK. We had a little problem at work. I wanted to ask mom..."

"She ran down to the store. Anything I can help you with, or is it one of those girl things?"

"How are you doing Daddy?"

"That's one thing you don't ask somebody my age," he chuckled.

"What is your age?"

"You were at the party. Seventy-two last month!"

"Daddy, when will mom be back?"

"Ten minutes. No, better make it thirty. Her friend Carol works there nights, and they'll get to talkin'."

"What Carol?"

"You know. Carol Holtz. I don't know why your mother wastes her time with that gossip."

"Daddy, you have been saying that for twenty years."

"More than twenty. Twenty's just all you can remember."

"Yeh, maybe. Tell mom I'll call her later?"

"Ok darlin'."

"Ok... Daddy? how's your leg?"

"Oh, no change. Still aches here and there, but I limp around just fine. You're not going to start worrying about the old man now, are you?"

"Love you."

"I love you too darlin'."

Jennifer looked at the slim charcoal cellular phone as if it were a moon rock. Remembering protocol she finally pressed END and quietly handed it back to Hilary, who looked down in sympathy with her and asked, "What's wrong, dear?"

All of us echoed the question in our expressions. Jennifer was silent for a minute more. Then she said simply, "That was my father. I'm sure that was my father. He called me by all his favorite names. He knew Mrs. Holtz."

Though we worked with her, Har'el and I knew nothing about Jennifer's family other than that she visited them in Rutland for Thanksgiving or Christmas every winter. I did not remember her ever mentioning her father. I guess because of the media focus on child abuse and abandonment, several possibilities occurred to me, all too ugly to bring up, leaving a murky silence while Jennifer puzzled and the rest of us held our fear and sympathy unspoken.

Did Jennifer's love of ballet sculpt her face and physique to compliment it, or was she drawn to dance because it fit her so easily? Her transparent white skin was a perfect canvas for any style of makeup, her long, straight blonde hair could be styled for any role and her thin, compact body fit any costume. And as she struggled with whatever psychological ghost her father's voice had awakened, her blank expression seemed poised to take on any of an infinity of emotions.

Like a raster scan as a computer screen paints a fresh image, her face coalesced first at the mouth, then the nose and eyes and finally in her head's erect posture atop her thin neck. She became an elementary school teacher and spoke to us with the same calm, detached authority she would wear reviewing the nines column of the multiplication table.

"My father died three years ago when I was a senior at Dartmouth." Jennifer paused to compose the rest of her explanation. "He had just retired after the fall semester at the community college. He taught agriculture there since the school opened back in the '60s. He loved to be outdoors, even in our pretty brutal Vermont winters. After us kids went back to work and school after Christmas break, the weather cleared a little and he went off to try to fish. He parked his old Willis by the road side and hiked in to one of his favorite streams, one that ran fast enough not to freeze hard.

"Mom expected him back when evening was coming and the weather started closing in again but at dinner he was still out. She called all his usual hangouts. After thirty-five years she was accustomed to his forgetting to call when he'd be out with his pals. This time nobody had seen him. Mom called the police and told them where she thought he might have gone. They sent out a car, checked all his usual spots and by 8:00 p.m. they found his parked Willis Jeep. It was dark, snowing, blowing and cold. There was no trace of Dad.

"The next day was a Friday and I drove back from school, to be with Mom. Snow was still falling but not as hard, and at least it was light out. A search party went out and combed the area on foot with dogs while a helicopter searched from the air. The snow had probably covered any tracks. Still no trace. Saturday the search team grew to almost a hundred. Mostly volunteers from town. One of the older guys stepped in a hollow spot and got carried out with a broken ankle. They worked until dusk and found nothing, but they went out again on Sunday, a bright freezing Vermont winter day. I sat home with Mom, tried to keep her calm. Most of the time I handled the telephone calls from friends and relatives. Sunday night the Sheriff and dad's old friend Walter came by to tell us what we already knew. No sign of Daddy."

Jennifer had been telling the story with detachment, but here for a moment her voice broke and her eyes began to tear. She paused to compose herself and then got back to her recollections.

"The search, which was costing the county a small fortune, was called off and the next week was miserable, the worst time I can remember. We couldn't admit Daddy was gone, but knowing he was out in those woods in that weather didn't leave many other explanations.

"We tried to keep busy with chores and shopping, but it was impossible with people calling and without Dad around. Mom was sad and would just start crying any time. The whole week and the next weekend passed in that never-never land.

"Then late Saturday, Walter called from the Sheriff's office. Some hunters had found dad's body. We all met at the pathology lab at the Rutland hospital. Dad had apparently crossed the partly frozen river and gone off looking for a new spot. Possibly he was following a bird or some other animal, or he was just trudging around to get some exercise. Maybe he got cold fishing. In any case he, like one of the rescue team members after him, somehow took a wrong step, fell and broke his ankle. Unable to walk or even crawl back to the road, he just froze to death. I made the ID of the body and during the next weeks I handled all the funeral and business arrangements for Mom. As it ended up, I took spring semester off to help out at home."

The room was silent. Should we be sorry? Or glad that her father now seemed in fact to be alive? I wondered. Hilary and Gib seemed like great people but the technological cutting edge they lived at definitely did not square with two simple country folks. Wireless e-mail to his mother? And I sure had not heard of any wireless weather forecasts. And Gib believed their accuracy a week into the future. And why no phone connection to 703 area code? A satellite-based perimeter security system? No obvious explanation occurred to me.

Hilary finally decided to say something along the lines of all this making a whole lot more sense in the morning and pointing out the rooms and furniture we could all choose to sleep on and in. Har'el and I took over the living room and conceded the guest bedroom to Jennifer. As we settled onto the two couches, Har'el, who had said almost nothing all night, smiled.

"Interesting, don't you think? It's like we came through that cave into another world."

Chapter 3

Morning brought the realities of our situation into gritty focus. We showered but put back on our same clothes, crusted with sweat and red clay. Our wallets with IDs, credit cards and cash were all stowed in our cars at the other end of the cave. We guessed a massive search would be under way for us, and Jennifer was still upset at hearing her father's voice. Hilary had been up early tending the horses and other animals. She had arranged a ride into D.C. with a semi-retired engineer/manager from one of the agencies downtown. By noon, I thought, this whole episode will be over.

Over breakfast of biscuits, local peaches, oatmeal, and coffee, I remarked that even for the country the air was unusually sweet and clear. Gib said it hadn't been so nice when they first came to the area. He and Hilary told us that a couple of years ago they had been part of a successful class-action law suit that some of the local communities organized. They shut down a strip-mine all the way over in Kentucky and forced three cement plants, two in West Virginia and one in Virginia, to implement new environmental controls on air- and water-borne emissions.

The lawyers used satellite wind maps that clearly showed that particulate plumes coming off of those four sites would, on over two hundred days per year, blow that stuff over western Virginia farming communities and ultimately suburban Washington D.C. Then they used photos taken from satellites of each of the sites every few minutes to show the federal judge just how much dust and dirt these guys were putting into the air. The expert witnesses said altogether it was several tons per hour! Part of the court settlement specified that each company must pay one of the commercial satellite companies to provide hourly photo coverage of its location to ensure there were no further violations.

Hilary said, "We noticed some improvement almost immediately. The air seemed fresher, the house stayed clean without always dusting. The next spring we had new growth around some of the hollows near where we found you guys. That summer we even had a few little carp and bass living in those ponds.

"You're eating one of the benefits of the cleanup. Every year we used to get peach blossoms but never any peaches. We assumed the trees were neutered. Maybe 'cause the sun is out more and stronger or the soil pH is back to normal, but whatever the reason now we get enough peaches to eat, can, and bring to market."

We three looked at each other. AeroAstro has for years pursued NASA as a client for wind measurements from a small satellite based sensor. NASA insists it can only be done from a satellite big enough to fill the Space Shuttle and they don't have the $500M they think they need to build it. And near real time imaging of randomly selected sites on earth requires tens or maybe hundreds of LEO satellites. Hilary's wireless phone worked fine even inside that steep walled ravine we walked through last night hiking back to the house. A terrestrial cell phone couldn't do that. I'd be surprised to find even marginal cellular service way out in these hills.

A system like Iridium or Teledesic could do that but could a small farming household afford to use Iridium as their only telephone service? And anyway, most of those services are years away. Just to check my sanity I read the date on Gib's newspaper. September 28, 1998. That checked—but there was also a time stamp—0734. What was that all about? Gib explained that there was no newspaper delivery way out here and instead it was delivered via satellite into his PC. Other than being on 8 1/2 x 17 inch pages, it looked like any other paper, except that, printed on both sides, it was only about ten pages. Hilary told me that when you get electronic delivery you e-mail in a form to select the types of articles and ads you want and everything else is filtered out. If you want more on any day you can always request it. She told us that while the wireless newspaper got started mainly for rural customers, she thought it was getting to be common in cities too. She wondered if we hadn't gotten it yet in D.C. It saves millions of pounds of paper per day, and its a lot less junk to wade through.

I thought about Har'el's conjecture from last night, "It's like we came through that cave into another world," while I looked at the publisher's insignia on the back of the page with the crossword puzzle Gib was absorbed in. It showed a small satellite coming toward you low over the horizon, and the name looked like Microcosm. Absentmindedly I mumbled a response to Hilary like, "No, we haven't got that yet," while I thought, "Jim Wertz's Microcosm?"

Gib got up to answer the door bell, and returned to the table accompanied by a wiry man of about sixty or so, with almost translucent silver hair. "Rick, Jenny, Har'el, meet Dr. Harry Lee." Between the commotion of all the handshaking and Gib's accent, I couldn't make out if Gib said Harry had recently retired from NASA, the FAA or some company with a similar acronym. Hilary told us Dr. Lee had retired to a farm out here, but still consulted back to his old job and he was going into the city for a couple of days. He had offered to give us a lift in. Dr. Lee smiled and nodded.

"Coffee, Harry?"

"No thanks, Gib. At my age bladder capacity is a major range limiter."

A real engineer, I smiled to myself. An hour or two in the car with Harry ought to get some of this mystery straightened out.

"Hey folks, sorry to be rude, but I've got a luncheon presentation to make to the boss and company."

"Ok Harry, but you promise to come by with Anita this weekend. We'll saddle up some horses."

Harry winked affirmation in his sparkling blue eyes. We all said our goodbyes and thank-yous, exchanged phone numbers and e-mail addresses and piled into Harry's silver VW Golf. He revved the little four-cylinder and we zipped off down the

dirt road. I was at the front passenger seat facing a four inch square color flat panel display showing a map of the area.

"GPS?" I asked.

"No, CD-ROM. I got it almost two years ago, before the GPS option came out."

"Mind if I scroll around?"

"Go ahead. I won't need it for a while. There aren't too many route options for the first ten or fifteen miles."

"Rick, Jennifer... Har'el, was it?" We all nodded.

"Thanks so much for the ride, Dr. Lee," Jennifer said. "We really need to get back and let people know we're OK."

"Oh, call me Harry! I save that Dr. Lee stuff for the office. It's a long drive into town and I'm glad to have company. Hope you two have enough knee room back there. Usually I'm the only soul on board this little go-cart. So, Hilary said you got lost caving?"

We recounted the whole story for Harry, the company outing, the goat, and our big scare in the cave. We told him about our inability to reach anyone on the phone, but we left out the part about Jennifer's father. A little too raw and a little too weird.

"So where are we headed?" Harry was planning to drop us off on his way into the city.

"Ah..." I started, thinking we should get to where we parked. Surely they would be looking for us around there.

"Herndon!" Har'el injected.

I darted a look at him seated behind Harry. Har'el returned my question with his "trust me" smile and a wink.

Maybe Har'el was right. We had no money and no ID and a highly incredible story. If we show up at the spot in five minutes and it's deserted then Harry might sus-

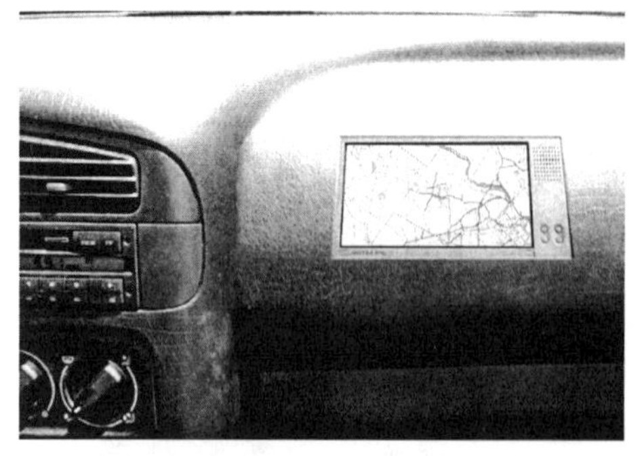

pect we're scamming him. We can learn a lot in a ninety-minute drive back home, and if Har'el's otherworld theory is wrong, and I did think it was pretty far out, then at least we're back with AeroAstro people who can get us home and reunite us with our cars and keys.

"Yeh, back to our office in Herndon would be great," I said with maybe a bit too much resolve, "and I think it's right on your way into town."

"Ok, Herndon it is. What do you kids do back there in Herndon?"

I told Harry I had founded a small company that makes little satellites and that we all worked there. Then I braced for the usual disbelief. People figure that getting a few sharp engineers together to build satellites is about as plausible as opening a family run business to supply electric power to New England. No, in my world satellites and rockets are too grand an undertaking for any but the largest companies and the federal government itself. Everybody, even people who know nothing about space, knows that.

That's the reaction I've gotten used to, from potential investors, from strangers seated next to me on airline flights, and from NASA and the big contractors to whom that myth is particularly dear.

But not from Dr. Harry Lee, retired government technocrat, who simply said, "Oh yeh? Which outfit?"

"AeroAstro," I replied, hoping he'd heard of us.

"Nope, don't know that one."

Which outfit? Don't know that one? Clearly Harry knew nothing about small satellites and was just being polite. Typical. The last venture capitalist I met with had called the small satellite business, "wonderfully arcane."

Harry got busy up- and down-shifting the Golf along the narrow and winding country road. I started manipulating the four arrow buttons for the map display, driving the cursor over the route we had taken yesterday to reach the parking lot from where we had hiked to the caves.

"How accurate is this thing?" I was not finding any of the small roads we had taken at the end of our drive the day before.

"Pretty darned good. It's a new CD and zoomed in you can even see the turn around loop in my driveway."

"So what's this?" I pointed to what looked like a wide, short piece of road overlying where I thought we had parked.

"It's a small airport. Built in about '89."

"Who flies out of there?"

"Most of the year just light planes and gliders heading into the Blue Ridge. Some crop dusters. But this time of year it's busy with fire fighting aircraft. That's what my agency, with the Interior boys, built it for. The water droppers have absolute priority on the runway. On fire alert days when they are standing by, they can be airborne in under one minute. Department of the Interior's slowly built up satellites in their constellation to where they get a fix on a fire anyplace in under five minutes.

"It can take ten or fifteen minutes flight time to reach the fires, depending where they are. But since they spot 'em when they are just starting, most times one airplane is all it takes. We haven't had a serious fire in the area—in the whole country really—in six or eight years. And that even though they've laid off most all of the smoke jumpers."

"So you worked for the FAA?"

"Do I look that old? Yeh, I worked for FAA back when they called it that. Must be almost twenty-five years now since we changed..."

The letters FAA being printed prominently on my commercial pilot's license, I was about to ask the obvious question when a loud beeper went off and a yellow arrow light pointing forward flashed on the dash. Harry slowed down and moved to the right, staring out the windshield.

"Were you speeding?" I asked.

Just then an oncoming car—a red pickup truck, really—popped into view over the hill in front of us. He was also squeezed to the right but even so it was tight passing on that tiny old road.

"What was that all about? How did that work?"

"Collision avoidance. We started it for the airliners but Transportation picked it up. It'll be mandatory next year, but most new cars already have it."

"Yeh, but how does it work?" I tried not to seem too naive.

"It's pretty simple. Each car has a low power spread spectrum transponder. A ranging system that one of the small satellite companies invented, originally to get their satellites' ephemeris, repeatedly measures range and calculates closing rates to nearby cars. If a small on-board laser radar doesn't see a target, like if it's over a hill, the alarm goes off. In heavy traffic it shuts off. But even limited to just rural and light traffic situations, it has really cut down on accidents out on the open road."

"Neat," I said, trying not to sound too surprised. "So what happened to the FAA?"

I look pretty young for forty-four and Harry probably took me for young enough to not really remember Vietnam. He patiently explained that by 1971-72, Nixon was taking plenty of heat over the cost of the war. Meantime, NASA was experiencing a crisis in identity in the letdown following Apollo. Young people were angry about the military-industrial complex. Nixon addressed all these problems in one simple stroke of the pen. He folded NASA into the FAA creating FASA—the Federal Aviation and Space Administration.

"Military people liked it since they didn't like civilian space anyway. To space enthusiasts, Nixon had two answers: that space would get supported within the new agency; and that it would grow into a huge enterprise, just like commercial aviation. To the general public what he offered was the promise of a grand space program with virtually no tax dollars. What happened in fact is the exodus from the old NASA just accelerated and by the time the merger was completed, FASA employed only a couple thousand space people."

Clearly Harry liked to talk, and we were all absorbing every word of this apparently fractured version of modern American aerospace history. "For a few years virtually nothing happened in space. It would have been a scandal except the world was preoccupied with Watergate, Vietnam, double digit inflation, $1 per gallon gasoline and wage and price controls. Space was interesting, but in those days it was labeled irrelevant, which to the average person it really was."

"Nobody launched anything?"

"Oh it wasn't that bad. The geosynchronous communications business did just fine. It seemed to thrive under FASA. Scientists and experimenters were in bad shape though. To get in space at all, researchers used funding from NSF or wherever and built their own little satellites in their own labs. They found piggyback space on commercial and some military launches.

"For a few years nothing seemed to be happening in the U.S. space program and there was a lot of controversy. Of course the Russians were scoring plenty of propaganda points with their crews on orbit. People in America, Europe, around the world really, believed America had given over the space initiative to the Russians. Some even said this proved that a capitalist society must ultimately degenerate by the forces of greed. As if to prove they did not suffer the same malaise, the Soviets, Europeans, Japanese and Chinese all accelerated their piloted space programs. Many of our space leaders from Gemini and Apollo left to support these vigorous foreign programs. But with Ford struggling to get the country under control, space still did not get much attention."

"So what was the FAA doing all this time?" I interrupted. Harry loved to talk, and I was glad to encourage him.

"FASA," he corrected me. "Not much different than ever. You gotta remember the old FAA was well over fifty thousand people, and adding a couple thousand space guys didn't get much notice. The ex-NASA people were kept busy designing a national space regulatory system. It was built to exactly mirror the air transport system.

"I mean, they established flight corridors for rocket launches, uniform construction and qualification standards for rockets and satellites and even a licensing system for satellite operators like our pilot licenses. It was all very theoretical since basically...well, ok, we actually had nobody to regulate, no "spaceports" to operate, but we did have the world's most progressive commercial style space traffic control system. Government waste at its finest, I guess.

"But the license structure turned out to be pretty useful. Rather than concede control of radio spectrum to the FCC, these guys allocated NASA's old blocks of frequencies to something called Space Services. If you passed tests in technical competency, you were basically licensed to own and operate satellites according to FASA rules. Spread spectrum was the law, and no broadcasting was allowed. That remained, as it had been, FCC's turf. Other than that, anybody could pass a series of standard tests and go off and build and launch rockets and satellites and communicate with them point to point—for private, commercial, scientific, educational or whatever uses they wanted. We figure people don't ask you on your flight plan why you are flying from A to B using the national air transport system, or whether you are driving down the freeway for business or pleasure. You pay your taxes, you pass the test, you have earned the right to use the highways, or the air routes, or in this case, the airwaves to talk with your satellites."

Harry then interrupted himself to ask if I was finished with the map display, which I was. He zoomed out to show the region from where we were all the way to D.C. then punched a button labeled Traffic. All the major routes displayed, after a few seconds' delay, in colors, and a key on the display showed red meant under ten miles per hour, orange is ten to twenty-five, blue twenty-five to forty and green indicated average traffic speeds over forty.

"I suppose you'll want to know how that works too," he chided. "Didn't you say you ran a satellite company?"

Luckily he didn't wait for my defense and just went on yakking.

"Cheap Doppler sensors along the roadsides, a few of them per mile each direction. Probably five or ten thousand of them cover the D.C. area. A little satellite comes over the horizon, reads and stores all the speeds those sensors see on the road. When I hit that 'traffic' button, the satellite downloads to me all the data relevant to the map on my display and charges my account $1. The D.C. franchise is getting about a half million hits a day. Nationwide it's a $4 billion a year operation. They are talking about Europe and Asia. I guess traffic is even worse there than here. But like all these things, fire detection, pollution monitors, weather forecasts, wireless newspapers, even simple stuff like night time illumination for police work, search and rescue and emergency crop harvest, those guys are still too bulloxed up in licensing issues. It's really a shame since all these low earth orbit systems are inherently global anyway. Right now these satellites only work over the U.S., and I guess spend the rest of their orbits charging their batteries.

"What really bugs me is these other countries' bureaucracies are costing their citizens hundreds of lives a year. Just the most basic stuff. Like these personal emergency transmitters." Harry took a pretty standard issue black Casio digital watch off his wrist and handed it to me. "The Japanese sell millions of these every year. Just hit the red button three times within one second, wait for the screen prompt then hold down the button for two full seconds and you'll be geo-located by satellite to within a few hundred feet and a rescue team will be there inside of a few minutes anywhere in North America."

Jennifer's father, I thought. That's how they rescued him. In this world, an outdoorsy guy would surely have one of those watches—probably Jennifer's mom would have given it to him. When he twisted his leg in the snow, he just pressed the button and they picked him up. Nothing remarkable. In our world, he had frozen to death.

Harry went on. "The company providing that service is getting a $10 premium per watch and over two hundred million have been sold. It's a big money maker, but it's saving over a thousand lives every year. Simple technology and Casio and Swatch are making them by the boatload. But their own people can't use 'em 'cause of outdated and bureaucratic uses of satellites and radio spectrum."

"So Harry," I was now getting adventurous, "what about the space shuttle and space station?"

"What about them? Russia built them, with a lot of help from the French. Oh the French were so intent on moving into the so-called vacuum left by the U.S. Can you believe at one point the French, Chinese, and Japanese government annual space budgets were each above $10 billion? Russia's alone was over $20 billion. We used to call them the Space Axis." Harry smiled. "Meanwhile in the U.S. we had just a skeleton regulatory staff at FASA. The Axis had grandiose plans to inhabit LEO, the moon, and ultimately the planets. And I'll tell you, a lot of Americans wanted it too. Mondale got elected in '76 partly by promising U.S. dominance of space like we had under Kennedy and Johnson. And economic times were tough when the war ended. We were in the doldrums while the Axis economies were enjoying a space fueled boom. No way would congress give Mondale the billions he wanted. How could we compete with the $50 billion per year being spent by the Axis?

"All this time we were doing a little space work at FASA, mostly giving out operator permits for scientists. But in the early '80s things began to change. The Axis was finding that their eyes and their ambitions had exceeded their means. The bigger their budgets, the more everything seemed to cost. Meanwhile, we started seeing a lot more permit applications. Some small companies started making money carrying messages and doing geo-location—tracking equipment and cargo. Pretty soon, we're processing several hundred satellites a year doing all kinds of stuff, mainly weather, traffic, news, environmental stuff like forest fire alerts and enforcing EPA rules. But there are lots of little niches. Companies put up little satellites to prospect for gas, oil, and minerals using proprietary sensors. And this company doing search and rescue from your digital watch is really growing fast now."

"And there is no government space?" I knew the answer, but I had to ask.

"Oh, there's lots of government space. Just no government space agency. You've got the Air Force doing GPS and all kinds of stuff. Interior Department is handling forest fires and resource management. As a country we are really active in space but it's always just a tool, not an end in itself. U.S. space revenues are running $20 or $30 billion per year, with almost no taxpayer expense. The Axis has spent half a trillion dollars and every year they feel they have to ante up more money. Crazy. Well, that's how those countries work, I guess. Americans would never go for that kind of socialist thing, government meddling in industry like that."

By now we were getting towards the D.C. metro area. A sign said we were twenty-nine miles from Leesburg, so we were maybe fifty minutes to our office. Looking over at Har'el and Jennifer, I think we all knew we weren't in Kansas anymore. While Harry was rambling on about the '70s and '80s, assuming we young Turks were lacking in historical perspective, which under the peculiar circumstances we very much were, I had been cruising his CD-ROM map looking at Herndon and Reston. Just as the parking area and small roads leading to our company outing site were displaced by that new airport, few of the side streets near our office or my house, which is in a relatively new part of town, were on the map. The big roads were all familiar but in place of the small ones I knew were lots of similar little streets I didn't recognize. The three of us needed to huddle.

"Harry, any place to pull over around here for a pit stop?"

"Excellent idea! There's lots of little places out this way."

Within five minutes we stopped at one. While Harry used the single toilet, we strategized. Har'el pointed out that Harry's rendition of history basically diverged from ours twenty-five years ago.

"Rick, we are in a different world. One just like ours, but something happened around 1970, and it went down a different path. The older streets and towns and landmarks are all the same, but a lot of new stuff turned out different."

"Very different. Jennifer, I bet your dad was saved with one of those emergency locator watches."

"That would be pretty incredible," Jennifer changed the subject. "I agree with Har'el. But I bet there's no AeroAstro around here. This world is filled with small sat-

ellites. There would have been no reason to start AeroAstro."

What a depressing thought! I borrowed a phone book from the clerk. Jennifer was right. Neither AeroAstro nor Rick Fleeter was in there. My brother, Tom, was in there—or at least his office was. Well, Tom wanted to be an MD since he was six or seven years old back in the '50s. That hadn't changed, and I guess he still did his residency in D.C. and stayed here. Tom's life was always a bit more directed than mine. The different course of history since the early '70s when I started college must have sent me in a completely different direction.

Jennifer suggested we head to Tom's office. Where else could we go?

Harry walked up. "You kids ready to go?" And we got back in the car and headed for Reston.

"Since we've been missing, our families are probably worried. We thought it might be better to stop at my brother's office." I told Harry the address I had memorized from the phone book. "Hey, will this thing route us to his office?"

Harry was pleased to show off the CD-ROM's capabilities. He entered Tom's address via an on screen touch sensitive keyboard. Within a few seconds, the display brought up sort of a trip ticket from where we were, which Harry had indicated by a tap on the screen, right to the passenger drop off at the medical office building. Not only was it a neat trick, but it disguised that I had no clue where Tom's office really was in the maze of unfamiliar new streets in this alternative Reston. Harry, whose enthusiasm for these gadgets was just radiative, explained that if we were lost or wanted guidance, we could toggle on the LEO satellite ranging option. but there was a $1 charge per position fix. As we got off Route 7 and navigated through Reston, we all thanked Harry for the ride and all the interesting information. Pulling into the compact suburban medical center's passenger drop-off area, Harry wished us good luck. Just in case, I asked him for his business card. Harry wrote his home phone number on the back and told me I should come visit him some weekend. We could talk satellites. I stuffed the card in my back pocket, along with Hilary and Gib's number.

And with a handshake and a wave, Harry zoomed away, leaving the three of us standing under the trees and the gray autumn clouds at the front door of the medical office I had been to a hundred times before, visiting my brother to say hi, and his partner to take care of my aging athlete's knees. This was the office that I had helped my brother move into when he opened his practice. But the office was inside a building I had never seen before.

Alone together for the first time since Hilary found us last night, we took a minute to replay the last twenty-four hours. It was tempting to dismiss Gib and Hilary and Harry Lee as just a few bit-heads who moved to the country. Information age survivalists holed up with an arsenal of exotic gadgets. Ok, the satellite phone, e-mail, and weather could easily be bogus. But what about Jennifer's talk with her dad? Harry's story sounded genuine. How could he have found the location of Tom's office from just the address if the CD-ROM were a fake?

But the overwhelming fact was that the major routes like Leesburg Pike were pretty much the way we knew them and the old Reston landmarks like Lake Anne were still here, but all the newer streets and buildings were completely different. And how to explain that neither AeroAstro nor I am in the Northern Virginia phone book? Maybe, opposite of Jennifer's father, I am alive in my world, but dead in this one. My brother is in for a shock in that case.

We agreed that whatever the explanation given we have no place else to go, no ID and no money, my brother was our best, and possibly only, shot. We walked in, found his suite via the directory and as we walked in, I told the receptionist I was a cousin of Tom's and was in the area and stopped by to say hello. If I really were dead, I didn't want him to freak and maybe call the police before I got him to see me. She buzzed him and led us back to his office.

"Dr. Fleeter is in with a patient. He'll be with you in just a minute."

Chapter 4

The first floor room was somewhat larger than the third floor office that I remembered, but its contents were familiar. Tom's diplomas and medical board certifications hung on the wall, and our grandfather's delicate antique examining instruments were neatly arranged in a glass case.

Tom rushed in, dressed in his white lab coat over a button down blue shirt and club tie.

"Hey, Rick. I thought BJ said you were Sandy. She said my cousin was here. Launch get done early? Who are your friends?"

Employing a strategy which helped me achieve near pinball wizard status on all the college boards, I tackled the easy one immediately. His question also provided me a handy test.

"You remember Jennifer and Har'el from the company?" In my world Tom came to most of our parties and met all the AeroAstro-nauts.

"Sorry. But how often do I show up in Oregon?"

I thought to myself, "I work in Oregon?"

"Anyway, nice to meet you, Jennifer, and... what was your name?"

"Har'el."

Tom nodded acknowledgment.

"Tom, we have something we need to talk over..."

A voice on his office intercom announced Mrs. Vanoker was ready to see him in room D.

"You're busy—you breaking for lunch soon?"

"Half an hour. Why don't you cross over to the hospital and I'll meet you in the cafeteria."

"Ok—see you over there."

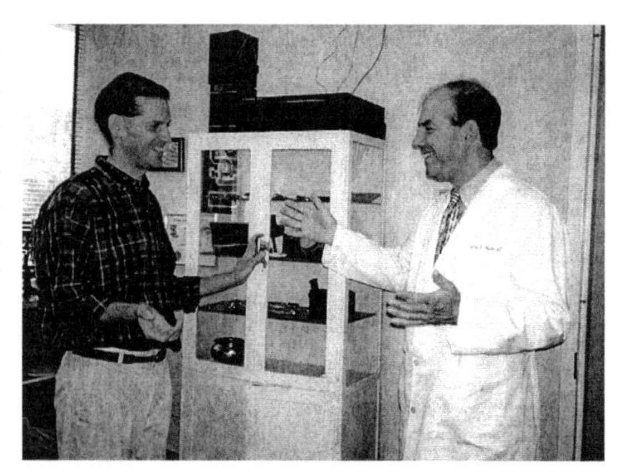

Tom had left us with a few minutes for exploring, so we walked over to the main street and then to the Town Center. Not too much seemed different. Even the cars were familiar—mostly. But among the Fords, Chevys and Toyotas were some very strange vehicles. Boxy structures, lousy aerodynamics and no obvious place for an engine. Har'el said they were probably electric—silent when stopped at a light and no exhaust pipe. The lack of any exhaust smell, even in a mix of conventional and electric cars in the heavy Reston Parkway traffic was very noticeable—a lot like when smoking was banned on the airlines. Jennifer wondered if this world had a better battery technology.

"Maybe they do, but I have a different theory. This society is serviced with really abundant environmental feedback from space. Remember Hilary's story about their lawsuit. My guess is these people recognized the need to clean up the environment. The satellites made it clear just how big a mess people were making of their precious air and water. Once the need is clear, the innovators can take over. Look, already in our world we have electric cars that easily go a hundred miles on one charge. Nobody commutes farther than that. And we all park at offices with electricity, but nobody is set up to let you recharge during the day. We don't have a technology problem even in our world. We are just limited in our desire to improve. This world is much more aware of what effects it is having on itself; it has feedback. And that fuels motivation and ultimately innovation. That rapid feedback loop is the difference." This being just a minor variation on one of my standard refrains, the other two politely ignored me as we wandered back to the hospital.

My brother was (as usual) late, so we strategized. Neither Jennifer nor Har'el had any friends or relatives in the area who would necessarily or even probably be here in this alternative world. Jennifer's family was in Rutland. She lived here with college friends and their friends. And Har'el's family was scattered everywhere between the Middle East and Hawaii. It sounded like a major undertaking to track any of them. Outside of Tom, I doubted I would know anyone in this area either. We realized we were total strangers right in our own neighborhood.

When Tom finally arrived he apologized for being a bit late. "I'm the orthopedic specialist on call on the 'net this week and just as I'm leaving an e-mail pops up from a clinic out in Indonesia someplace. Not where you want to be with a compound tibial fracture. It's like 2:00 a.m. and this guy drags his leg in, under his own power. Motor-scooter accident. Must have been drunk. Do they drink there? The night nurse sent me over some crummy x-rays via the satellite, but they, along with her description, were good enough to stabilize him for tonight. I sent orders for the views I'll need to finish up. He'll get 'em in town on Monday and e-mail 'em to me."

"So they have an Internet link way out there?" Har'el asked.

"Not quite," Tom was a little sarcastic. "If you have a phone you are overqualified for this service that I volunteer for. It's some satellite thing. Remote clinics have a little satellite ground station run off a laptop. They have a scanner for x-rays, and a voice thing to narrate the pictures. Doctors in the rest of the world volunteer in various specialties at various times. The network routes the data to the right specialist who is on duty. Pretty clever."

It figures, I thought. I spend my life trying to convince people to use small satellites and one Reimann sheet away, they are everywhere. I could have followed a goat through that cave ten years ago and saved myself a lot of trouble.

"Should we get something to eat?"

"Great. But just so you know, we don't have money with us. It's your treat. OK?"

Tom grunted. He's used to having a flake for a younger brother. When we got back to the table with our trays, I broached the subject. "Listen, Tommy, what we need to talk with you about is pretty serious and pretty bizarre. You will think we are crazy, but please, give me a chance to tell you the whole story."

Tom just sat and soaked up our story. Regular interruption being more his style, I was worried. Probably he thought I needed a psychiatric referral. In fact, reviewing the last twenty-four hours out loud, I was tending toward about the same diagnosis. We leave a company outing to follow a strange goat into a cave, wherein we almost die. OK, that could happen. We are met, on emerging, by a woman on horseback with a satellite telephone.

"Big deal," says Tom, pulling a similar phone from his shirt pocket.

"And they had wireless e-mail and weather forecasts. Even their newspaper came by e-mail."

"Yours doesn't?" Tom asked.

I replied with my own question, "Do you have a gadget in your car to show the traffic speed on your route?"

"If I didn't and I ended up late to an emergency call, do you think I might get sued?" was his rhetorical comeback.

Until this moment, until hearing the combination of impatience and worry in his voice, I thought that we three were normal and this new world was alien or illusory. Now I realized...This world is normal and we are aliens.

"Ricky, you don't remember that you are in Virginia for a test launch for your company? You launch rockets out of Oregon some place."

Fourteen years growing up with Tom have provided me certain highly sharpened instincts—one of the less useful being arguing with Tom, and I blurted out "No! I build satellites right here in Herndon." He glared at me.

"Look, if I'm crazy, OK. But what about these two?" I gestured toward Jennifer and Har'el. But Tom didn't know either of them, and they looked as dirty and disheveled as I must have. I was losing ground fast.

"Tom, my thinking is that up till about '70 or '71 things were the same. You remember the pink linoleum table with the fluted steel edge in the kitchen when we lived on Throckley? We had a tree house in the back. You played clarinet in the Wiley Junior High School marching band. We called in that great swim meet spoof to the Cleveland *Plain Dealer* in '69." None of these details of our shared childhood impressed Tom. He knew I was his brother. If I were just an impostor with my facts skewed, he wouldn't be so worried.

Tom was silent and sullen. I think he believed I had been reprogrammed by my younger companions. or I had snapped. In any case, I was in need of serious psychological help, and Tom, who practices a very tangible type of medicine, knows that not everybody with that type of illness gets better.

Suddenly, I had an idea. "Hey, Tom. The Rick Fleeter you know—where is he today? Didn't you say he was in the East on business?"

"Yeh, launching a test rocket somewhere on the Virginia coast."

"Great. Call him up and see if he'll come by your house for dinner. Put us together face to face and I bet we can make some sense out of this mess."

Tom said nothing. He pulled his phone out of his pocket and punched a few numbers. "Hello Rick? Tom. How's it going with the new rocket?...Good... Yeh, great flying weather, I guess...Yeh...Hey, any chance you could get up here for dinner? I know, you're trying to get back today...well, look, it's pretty important, worth putting off the trip back. If you can make it up here...yeh, you can stay overnight at our place and fly home tomorrow from Dulles. I know it's painful—but trust me—it'll be worth the extra day...Great...Everybody'll be glad to see you. You want me to call Kathy? Ok. See you around six."

"Who's Kathy?" I asked innocently.

"Your wife."

"Do I have kids, too?"

"Two little girls."

"Great. Hope I meet them some day."

Tom lightened up, knowing his real brother was apparently OK, whomever I might turn out to be.

"You guys are kind of a mess. I've got surgery this afternoon. Do you want to get cleaned up at my house and I'll meet you there tonight?"

"That would be great," we chorused.

Tom led us to his car and drove us to his house. He introduced us to the housekeeper, Rose, who greeted me with a warm, "Hi Ricky." Not surprisingly, I was an excellent likeness of myself. Rose gave us sweats to wear while she laundered our clothes. We showered and Har'el and I shaved. By the time Tom's wife and kids showed up, we looked and felt virtually human.

"JoAnne?" I recognized her immediately. History diverges for seven or eight years but Tom ends up marrying the same woman. The power of self determination. Tom lived a much better planned existence, I repeated to myself.

"Hi Ricky."

"Hi Jo. Did Tom tell you we would be here for dinner?"

"Yeh. Three of you?"

"One more is arriving around six." We helped Joanne and Rose in the kitchen and played with the four kids, Josh, Joel, Jane and Jennifer (the girls were twins), until Tom and Rick walked in together around 6:30. They had pulled into the driveway at the same time, and apparently Tom hadn't said anything to Rick. I greeted them and, as weird as this was, the surprise was a welcome relief to Tom and the three of us. Clearly I was neither demented nor reprogrammed. Cloned, maybe, but not crazy.

"Rick Fleeter?" I asked my double, staring at me in the doorway. Rick was in jeans, a broad-striped cardigan and what looked like a cross between running shoes and hiking boots. He wore a pager like Tom's on his belt.

"Yeh. And who are you?"

Half kidding I said, "Back where I come from, Rick Fleeter."

"And where do you think you come from?" this version of me was coming off a bit belligerent.

"Cleveland, originally, but the last eight years, Reston."

Tom interrupted "Sounds pretty wild but he seems genuine—knows all the details of the family, way back to the '50s."

Rick's expression showed mixed anger, frustration and suspicion. I wished Tom had filled him in.

Har'el spoke up. "Look, the last twenty-four hours have not been exactly a picnic for us either. Could you two please try to figure this out instead of just glaring at each other? Tell you what...I'll pick a year and one of you tell the other what you remember, OK?"

Starting around 1960 Rick and I recalled all kinds of trivia. School teachers, piano and cello recitals, swim meets, even the first time he, or should I say we, kissed a girl.

In a few minutes it was clear we shared one life. Rick's suspicion gave way to curiosity, and everyone joined in with Har'el's game over dinner. Finally we came across a major schism. Early December, 1971 we were a high school senior and I remembered flying to Providence to spend a long weekend on the Brown campus hosted by the swim team, which was recruiting me. At first, Rick didn't remember Brown at all. Then an old recollection came to him.

"Yeh. I did apply there. But '71-'72 was that super cold winter. Providence was shut down and I couldn't fly in. Same thing with New Haven. Remember we wanted to go to Stanford but Mom and Dad said it was too far away? I just pointed out that if I went there, they could come visit me and at least escape the cold. They relented enough to at least go look. We all went and seeing how beautiful and warm it was, and it's a great school..."

"I know," I interrupted, "I did my Masters there in '77. But I came from Brown and later went back to Brown for my Ph.D."

"No, I went to Stanford and just stayed on for the Masters. Did it the same year, '77."

"So what'd you do after that?" we asked each other in almost perfect unison. We laughed. Everybody laughed.

Har'el's game had worked. My credibility as a, if not the, Rick Fleeter was restored and we all began enjoying the novelty of the bizarre situation.

Dessert was a truly excellent selection of fruit. I started munching on some melon while Rick told us about his career. Like me, he had wanted to work in aerospace but unlike me, his luck in finding space work, even with the Stanford degrees, was poor. Through a Stanford Prof's recommendation he went to Boeing where there was a propulsion group with some interest in rockets. It would be a holding pattern, since in post-Nixon, post- Vietnam America there was basically no space program. But it was a start.

As it turned out not a very auspicious one. Six months to the day after Rick and his wife Kathy had come to Seattle, a VP met with all the members of the group and told them the company was closing down its rocket operations. "Our whole group was pink-slipped right there." Rick still seemed wounded by what happened. "A couple of us headed down to the cafeteria. Munching on our cherry popsicles we talked about what could have been—proposals we had made to Boeing management to make the rocket business a commercial success like the airplane business."

"I guess that would be like advising IBM to get into personal computers back in the days before Apple." I was both reinforcing Rick's feelings but also checking history. The PC movement was apparently either under way or inevitable by the time our histories diverged.

"Worse than that," Rick clearly enjoyed recalling his company's shortsightedness. "You think of Boeing as this paradigm of American commercial success. That is the company's PR image—carefully groomed to maximize favors from Congress like Rand grants, government financing for their customers and legislation protecting them from foreign competition. And that is over in commercial aircraft. The military planes and rocket guys honestly believe the CBD is a commercial marketplace. They didn't see the government buying any rockets in CBD and concluded there was no market for them." [*Commerce Business Daily* (CBD) publishes government procurement needs for anything from Mil-Spec peanut butter to satellites. *-ed.*]

While Rick nabbed a few hunks of watermelon I remarked that in fifteen years working in our aerospace world, that was not just business as usual, but really the only way to stay in business.

"My world too," he agreed. "Back then we were radical. Or really just naive. Whatever we were, three of us moved down south of Portland where we could live cheap, borrowed about $100k for expense money and opened not a rocket company, but a high speed freight transportation company. We got hold of some surplus missile stages, and mated one of them to a liquid upper stage to make an ultra cheap two stage rocket. Really an ICBM—too heavy to get to orbit, but we could toss a few hundred pounds half way around the earth. Our marketing plan was based on rapid freight delivery to Australia, a straight shot from Portland, and I had some friends over there. We built the rocket and made arrangements to parachute the upper stage and payload onto the huge Woomera range. But we didn't have a single customer."

"Wait a minute. Three of you built a couple of two stage rockets for $100k? Just a guidance system could cost that much."

"We borrowed that first $100k, we all invested our savings, probably another $100k or so, and we all worked on it for about two years without pay. That's probably worth over a half million right there."

"Yeh, but what about parts? You needed an engine and tanks, valves..."

"We did it all on surplus stuff and contributions. The engine was a NASA '60s era design. We scavenged the last two from a storage shed inside the old NASA Lewis campus at Cleveland Hopkins Airport. The government even paid us a few cents per pound just to haul all that old scrap outta there. I drove my pickup round trip Portland to Cleveland for that stuff."

I looked at this Rick Fleeter sitting across from me. Were we really two incarnations of the same person? Our faces, hair and eyes were identical. We were the same height. But Rick must have weighed ten or twenty pounds more than me, his voice and clothes and skin were rough and weathered. Even his manners were coarser. And he was more aggressive. Or his aggression was at least not so well disguised. His pleasure in frugality, though, was very familiar.

"You have two daughters?" I was continuing my train of thought, not his.

"Rebecca and Emily. Six and four. And a boy due next month. Maybe we'll call him...Peter."

Everyone smiled at this little inside joke based on a family myth that my father had named me Peter, Peter Fleeter, but Mom made him change it to Richard.

"And what about Kathy?"

"She supported us all those years working as a visiting nurse in rural Oregon. Now it's my turn to work and she's mostly home with the kids. She goes in to the community hospital a couple of shifts per week now just to keep up with the field and get a break from home."

I thought about my first wife who had supported me for a while, also working as a visiting nurse, but in rural Rhode Island. But I wasn't driving to Cleveland in a pick up truck in pursuit of a dream to build a rocket and to make good on an apparently ill-chosen career path. No, I was pursuing a dream to be a real scientist doing a Ph.D. in Thermodynamics. And running away from a mainstream aerospace career, which in my world was reasonably attainable but totally undesirable. A life of meetings, presentations, program reviews, writing program requirements documents and using only ancient Mil-Spec components.

Our two lives were definitely woven on the same loom. But different thread ran into the machine and the product had a totally different color and texture. I was mesmerized following the road not only not taken, but not even on my map. "So you were saying you guys actually built a couple long range suborbital rockets to move freight to Australia?"

Rick paused, and got one of those "it's a long story" looks on his face, as he was undoubtedly pondering where he should start. A moment later, he began slowly, to relate the tale.

"I traveled all over the U.S. promoting the service. People thought it was a good idea, but nobody bought a ride. Probably it was the price: $5 million for five hundred pounds. At $10,000 per pound it might not have been so bad, but without a customer base we needed one client to take the whole flight. For about a year we traded off. One guy would market while the others worked on the vehicle and launch site. We aimed our marketing at the car companies, medical stuff like organ donation and international banking. Our research showed if we could save one day delivering a critical part needed to restart an assembly line, the $5 million was a bargain.

"We had taken on lots of shareholders since everything we couldn't scrounge or make ourselves we purchased with equity. Slowly they, and our families, began to conclude there was just no market for our service. The other guys got jobs in town and I took on what consulting work I could find. At least that kept me circulating around the country so I could do some low level marketing.

"Then about a year later I got a call from Lockheed. They were finishing a $6 Billion proposal teamed with Bechtel to build a huge geosynch launch site at Cape York in northeast Australia. Huge project with competing bids from all over the world. Well, somebody got the date wrong and thought it was due the next Monday. It was already Thursday here and about one a.m. Friday in Australia. And now it turned out they had only until close of business Friday in Canberra—about sixteen hours—to

submit their proposal on time. Lockheed had a company jet—several, in fact, but they were still awaiting its arrival at Burbank airport in L.A. and then it needed twenty hours to reach Canberra.

"The Lockheed Proposal Team had phoned the Australian Space Office in the early hours of Thursday morning U.S. West Coast time when one of the proposal volume managers realized their mistake. The proposal was due on Friday in Canberra, not the next Monday, and asked, pleaded, for an extension of the deadline. ASO's legal office faxed a reply back at 4:00 a.m. our time Thursday—I guess about 7:00 p.m. or so in Australia so they must have deliberated a bit internally there—explaining that granting an eleventh hour extension would expose them to protests, possible lawsuits and diplomatic embarrassment threatening the whole future of the Cape York program. By phone the ASO guys urged Lockheed to do what they could. Well, I guess Lockheed's proposal team leader got on the phone and woke up not just Lockheed's flight dispatcher but the Bechtel proposal lead in San Francisco who flew their input down on a Bechtel prop jet, finishing a lot of it by hand on the plane and in the limo from Burbank into Lockheed HQ.

"That was around 7:00—10:00 p.m. in Australia with nineteen hours to go 'til closing of the ASO and the deadline. I know 'cause the LA location Bechtel proposal rep, a guy named Roger David, was a Boeing guy who got laid off about when I did. Roger knew about our rocket transport business and called me from the limo.

"Naturally the Lockheed and Bechtel Brass nixed going on a rocket. We'd probably have better luck selling a rocket ride to an ad agency than to an aerospace company. These guys don't even believe in their own technology. How do they expect other people to use it if they don't?

"Lockheed's brilliant strategy was to jet the proposal into Canberra to arrive on Saturday and work the problem politically. Maybe they could argue that they got their submission in before business opened on Monday which was close enough. Anyway, Lockheed's Chief Counsel was already looking for loopholes in the proposal wording to argue that even if they were a day late, they were actually on time.

"Roger and I did a little deal. He signed a purchase order to us for the rocket flight. No way did he have $5 million authority so we did it for $500,000, which I thought I might be able to make stick in court. It was probably less than Lockheed would spend on their lawyers to keep their bid alive. I promised Roger if Bechtel canned him over this we would hire him–assuming the success of this flight got us into business. Roger had made a few well placed friends at Lockheed during the year or two he had been working this program. With their help he pirated one copy of the proposal—over four hundred pounds of paper—and took it with him as checked luggage on a commercial airline flight to Portland."

The story was fascinating to me, but what surprised me was that Tom and Joanne didn't seem to know it either. But I recalled during my own ten years living in the West family connections grew pretty weak. Being near them in Reston we shared a lot of experiences—bike rides, family issues, Tom rescuing me at the Reston Hospital ER and me doing things with his kids. Apparently those kids weren't so completely enthralled. As Rick was about to continue, the twins interrupted.

Chapter 5

"Daddy—Come on! It's time!"

Joanne prodded them, "Can you invite our guests too?"

We climbed the steps to the girls' neat bedroom gaudily decorated in primary colors. Next to their bunk beds on the desk was a PC. On its color monitor was sort of a kid-pix version of orbit tracking software. A red panel flashed an analog clock counting down the seconds—'til what? Joel, who was the oldest and apparently knew the system, sat down at the keyboard. While we watched the seconds tick down, I noticed a cable from the computer to the window sill where it was connected to a black box with a Curly-Q antenna like the one on Hilary and Gib's computer. As the clock chimed zero, it evaporated off the screen which faded into, what else, the icon of a satellite approaching us over the horizon that had been printed on Gib's newspaper this morning.

"Send my message to Sanji!"

"Send mine to Ruan!"

The twins were intent in their demands. Joel opened two files that each looked like a picture postcard. Each had a picture of one of the girls and a short message, one addressed to a long name beginning with Sanji and the other beginning with Ruan.

Little Jennifer pleaded, "I wanna do it!" and she climbed onto the chair with Joel. She clicked on the card's address block and then on a stamp icon with the word SEND. Resembling the feather from Forest Gump the postcard wafted up to the shimmering satellite where it grew a Gumby-esque arm with a fist that knocked on the satellite's door. The animation, accompanied by tapping sounds made by the computer, continued for a while.

Joshua commented, "The satellite is busy." He looked bored but the twins were enjoying the show. Then the door opened and a figure made of IC packages appeared from inside the satellite, accepted the card and latched the door. The girls shrieked with satisfaction.

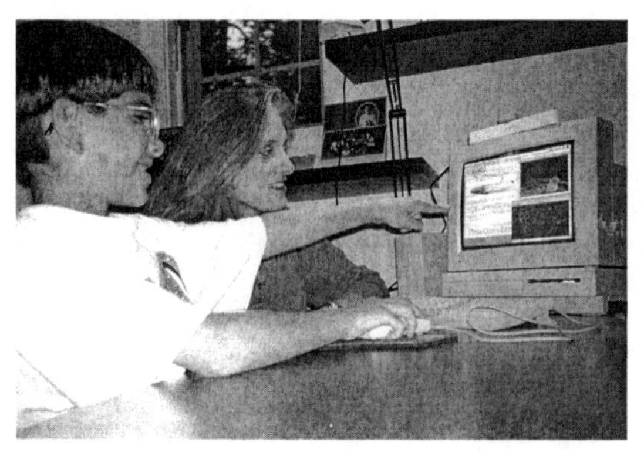

Jane repeated the process with her postcard, generating an equal level of delight. When the little IC person latched the door, and the computer uttered a reassuring CLUNK, which sounded to me like a digitized sampling of a Lexus door shutting, the girls smiled and clapped their hands. Jane let me in on a secret. She whispered to me "Sanji is going to Reston Day Camp with Jennifer and me next summer."

"Maybe, Janie. We'll see." Joanne added, but her eyes had that what-they-want-they-usually-get look of parental exhaustion.

Joel and Joshua were sending some more complex-looking files. Our Jennifer guessed it looked like somebody's homework.

"Yeh," answered Joel, "it's a story problem I worked out for our friend Carlo," As he spoke, an hourglass characterization of the sun appeared on the screen slowly setting behind a crescent blue earth.

"What's that?" Tom prodded the kids.

"Satellite is going over the horizon," Joel quipped very matter-of-factly. He was getting too grown up for parental prodding.

"Short pass off to the east," Joshua added, mainly, I think, to show off his virtuosity in satellite ops. "The next pass will be better but it's after our bed time."

We old guys thanked Jane, Jennifer, Joel, and Joshua for their demo, and headed back down stairs. Joanne followed in a few minutes. For a while the group just chatted. We needed a break from absorbing a day full of facts about the peculiar world we had unceremoniously slid into by following, of all things, a goat. I wandered around the house comparing it with Tom's house in my world. The lack of difference was striking. But poking around the garage I noticed the riding lawn mower and their minivan were both electric. I was almost flat on my stomach on the floor of the minivan looking for the hood release when I heard my own voice whisper, "It's in the glove box," and it was. But hey. Who was that? Rick of course.

"Tom says you say you live in Reston."

I explained about moving east after working in the classical aerospace world at TRW and Jet Propulsion Lab. I told him that I started AeroAstro because of my frustration with the huge resources and long time it took to do even simple things in space—and out of my belief that space could be so valuable if we could just learn to use it efficiently. Rick understood the feeling. He'd had the same experience at Boeing. "No electric cars back there?" Rick noticed my curiosity about the minivan. "A few. They're a curiosity and the car companies claim they're impractical and unnecessary."

"No news there. GM and their crowd are still saying the facts don't support a need for electrics. Most of the automotive press says they're unexciting. But a lot of people, the ones that can afford them, are buying."

"Because of air pollution?"

"That's what started things. The network news started running specials with color enhanced maps of hydrocarbon and NOx concentrations in major cities. Pretty scary stuff. Most of it was blamed on cars. People started to see that they were breathing all this stuff, and killing themselves and their environment. The few people who bought

electrics to make a statement started word of mouth about how nice they were. They always start. They're quiet. Very safe. But mainly, you never stop for gas and never need an oil change. Here were the car and oil companies saying how inconvenient electrics would because of the hundred-mile range but the fact is you just plug in while parked at work or at home. You never have to stop and deal with gas stations."

"You live in a pretty interesting world" I said. My mind was busy digesting the last twenty-four hours and trying to understand my double. "I suppose a bunch of schools got together and built a store-and-forward messaging satellite?"

Rick's look was questioning. He didn't know what I was talking about.

"You know, that little demo the kids just gave us."

"Why would they build a satellite?" Rick still had no clue what I was getting at.

"You know. To carry messages to those kids in Indonesia."

His look changed from vaguely quizzical to vaguely impatient. "I guess building your own satellite is one way to do it. You could string a wire from here to Cleveland to call mom up too." Now it was my turn to look quizzical. "Hey, there's thousands of businesses using that technology. Tracking equipment monitoring pumps and electric lines, measuring river height, sending data around." I remembered Tom getting an X-ray to diagnose before lunch today. "I think some company donated some unused time on a couple of their satellites. A parent wrote that cutesy interface over the company's mail application. A lot of schools have copied the approach and reused that shell so now almost every kid gets a little satellite time. The schools loan out the RF boxes sort of on rotation."

"And what is the connection with Indonesia?"

"You really are from another planet. What do you say we see if there's any watermelon left." The garage entrance to the house led into the kitchen where we shared slices of a late season melon.

"Has emerging into an alternative universe stimulated my appetite or is the fruit really better here?"

Rick shrugged, having no way to know what ours was like but Har'el, who was rinsing dishes, agreed. "It's the weather forecasting—that was Gib's read. That plus less pollution."

"Weather forecasting has definitely improved. We have a lot more data profiling the atmosphere than we did ten years ago. But I was reading that soil moisture has turned out to be pretty important. Companies have been selling soil moisture sensors to farmers for planning planting and irrigation. They stick these sensors in the ground and read 'em out with LEO satellites. Simple. And big farms really tailor watering to local moisture trends. They get optimum growing, better fruit, and save water.

"Somebody got the idea to gather all the soil moisture data into a single uniform database and provide it to the people running the big weather forecast models. Apparently it really helped a lot. A lot of atmospheric water comes from the ground's moisture. Since that discovery, companies have gone out and put ground moisture sensors in nonfarm areas like cities, forests and national reserves. They sell the data into the weather forecast system and the profit motive has pretty much taken care of the rest. The whole country is well covered with those instruments. Now they're talking about upgrading them with other sensors like air temperature, humidity, wind speed and direction."

By now Tom, Jennifer and Joanne had joined us in the kitchen. Tom said he had seen in the *Post* that some company was launching a fleet of satellites to prevent frost damage mainly to citrus and coffee crops. "They use infrared light," he remembered.

Rick, being in the space business, had read about the project. "They plan to have about two hundred fifty satellites randomly dispersed in three-hundred-mile, thirty-five-degree inclination orbits. Each carries a ten-kilowatt source tuned to the atmosphere's four-micron IR window. A rechargeable lithium-ion battery runs the light for just a few minutes. At least five of these satellites are able to illuminate an area like central Florida at any given time. And since the light is invisible, harmless, low level IR, it's safe and there's no light pollution."

"It sure is low level. One automobile—the regular gasoline kind—can put out that much heat. How's that going to save a big piece of Florida or Brazil from a frost?" It was my turn to play skeptic. "And what about clouds? That wavelength light is strongly absorbed by water."

"Those are the reasons nobody has done it already. Not enough power, was the war cry of the skeptics. But it turns out there are some special things about frost that make it work. First of all, frosts in warm climate areas almost always happen on clear nights. You know, earth radiative cooling. Clouds just are not a factor. Second, it doesn't take much heat if you put it right where you need it. It's the moisture right on the fruit surface that's getting that little bit of heat. And with that many satellites, at least five are lighting up the area continuously all night."

"Yeh, I've done this calculation. You're still talking milliwatts per tree, max. I don't think it will do anything."

"I'm not the expert. I just hear that the slight heating inhibits condensation. But the more important effect is in slight warming of a large area near or at the ground. The atmosphere on cold nights is very stable with dense cold air hugging the ground and warmer air just above it. Heating the ground over a broad area creates convective motion which brings warm air down to the ground. That's why cars or kerosene smoke pots or even helicopters don't cut it. They're fighting the atmosphere instead of working with it. And they are too localized. You can't invert hundreds of square miles of air with a couple of helicopters. Even if they do some local warming, more cold air just slides in and displaces the warmer air and the net effect is zip."

"Ask a simple question...but what I really want to know is what happened with your deal to fly the Lockheed Cape York proposal over to Australia?" Har'el's curiosity was shared by all of us.

"How late is it getting to be, anyway?" Rick and I share a propensity for going to bed early.

"Not even nine. The kids aren't even tired yet." Tom is not into sleep.

"OK but after this, I'm going to bed. I want to get in an early run before I head for the airport."

"Gee, an early run sounds good to me," I interjected. "Tom, you got any spare running stuff?"

Tom is noted in the family and among close friends for his monosyllabic responses. Commonly referred to, somewhat unfairly, as grunts, my hope is one day to exploit their potent information compression capacity for doing interplanetary missions with little satellites and cheap ground stations sporting limited gain antennas. From this particular coded message, approximately alliterated "ygmnp" expressed with the Parisian nasal accent, my forty-plus years of interpretive practice allowed discerning three messages:

#1 Yes, albeit grudgingly since putting on shoes that other people, even your own brother from another, nearly parallel, universe, have sweated into is unappetizing.

#2 Restatement of the too-often implicit assertion that while Tom's earned his living through steady application of hard work, his brother(s) always have time for a run or swim or bike ride while he has to be scrubbed and ready to do arthroscopic surgery at seven tomorrow a.m., and despite this lack of diligence we seem to get by all right.

#3 An older brotherly satisfaction in enduring stuff like point #1 in order to perpetuate the state described in point #2.

My thoughts turned back to Rick's story. It seemed to not be such a pleasant subject for him.

"It was an ironic start for us. We were radically anti-paper and here our first flight wasn't a heart or kidney being rushed to a transplant recipient, or even a critical component of a stalled assembly line. It was hundreds of pounds of paper. Our competition wasn't another rocket company, or even another mode of transport. It was lawyers ready to argue the rules of the game. And even though the customer was huge, and had plenty of cash and billions at stake, we settled for a price one tenth of our break even point.

"Well, Roger, my friend who was working the proposal for Bechtel, showed up at the airport late morning, and I met him and all his boxes of papers. It was around 1:00 p.m. when we got to the launch site, already 4:00 a.m. on Friday. We had just thirteen more hours to erect and launch the rocket, make an hour's flight to Australia, find the payload in the reentry zone, transport it to a chartered jet at a nearby airport, fly into Canberra Airport and rush the package to the ASO. We estimated from launch to arrival in the ASO would be a minimum of three hours, so we pretty much had to launch by late that night. At any rate, we were only permitted to launch in daylight hours, so bottom line—we had maybe six or seven hours to achieve ignition on the pad.

"It seemed like everybody who had ever helped us building the rockets showed up to help launch it. Kathy must have called a bunch of people, they called friends and you would have thought I had the President of the United States with me. Big reception. Nobody had ever launched a rocket from Oregon before.

"The press was there in force. They wanted pictures of fire and smoke. Time was passing and the crew, which had a few years to forget how it all went together, was struggling. We cannibalized lots of parts from the second rocket rather than trying to fix anything. Most of the rocket groupies had gotten bored watching us struggling and they packed up and trickled away. It was nearly sunset, around 7:00 p.m. or so, 10:00 a.m. in Canberra, when we triggered the countdown timer at t-30 minutes for the last time. Kathy had refiled our flight plan with FASA every two hours and they had advised that this was our last window of the day. We couldn't work after dark out there anyway.

"About twenty of us were left by the time the ignition command was finally sent from the computer in the blockhouse out to the rocket, which was venting oxygen into a nearly completed sunset. To all of our amazement, its black and white image on the monitor showed a rocket ignite and lift, slowly, and then faster and faster, away from the pad. Once it was clear of the area we all rushed out of the block house and watched the rocket climb out over the ocean. We had never launched a rocket before, we all realized. It was beautiful, but I was scared and worried. I remembered that we had no insurance. We hadn't paid a premium in over a year. What if it went off course?

"We had no tracking past line of sight, so there was nothing we could do except call the 'crew' at Woomera—James—and tell him the ETA. It was almost 11:00 a.m.

Friday at the Australian Space Office in Canberra. By noon Australian Defense Forces radar had seen our rocket streak across Cape York and disappear from the screen over Woomera. A helicopter we managed to rent in a big hurry that day just for this mission started searching for the bright orange and blue parachute the rocket was supposed to deploy. No luck. No beeps out of our locator beacon either. Time was not on our side and Woomera is as big as Oregon.

"Without spotting the parachute in the air or hearing the beacon, we were in deep yogurt. I was in touch with the Australian crew chief, and the only guy we could get hold of on zero notice, James, who was on a radio phone in the helicopter. We knew what must have happened. The parachute didn't open, so the chaff didn't deploy and that's why Melbourne Air Traffic Control never saw it. The resulting hard crunch broke the locator transmitter. Nice theory, but what to do next?

"James and I were stuck. We kicked around crazy ideas like dispatching their Air Force for a massive search and rescue effort but nothing we could get done in time occurred to us—we had just four hours to find the payload and get it to the ASO.

"The radio link with James started to get noisy and then dropped out completely. Kathy and I sat there in silence. Gary and the other guys walked outside–maybe to mull over their thoughts or maybe to give her and me a few minutes of privacy at the comms console of our makeshift control room. I was frustrated as hell. We had made a shipment from the U.S. to Australia, the result of so many years of hard, unpaid work. Everything worked perfectly—tanks, engines, pressurization and guidance systems, all that on our first flight. And the customer's payload was lying there a one hour plane ride from its destination in Canberra. That flight—delivering that payload to the ASO—would have proved our concept and our business were viable. The failure, along with having expended rocket #1 and gutted #2 for parts meant we couldn't fly again without at least the revenue from this flight. Or in simpler terms, this was the failure of a dream which we had nurtured for so many years.

"The most frustrating part was at that moment it wasn't yet dead. It was alive somewhere in Woomera range. But with less than four hours to quitting time at the ASO our dream, our potential livelihood, our investors' money and most of all our revolution in the commercial use of space were all being destroyed with each passing minute.

"I would have given anything to somehow change things. If, if, if. If we hadn't flown at all, at least we would still have hope. If we hadn't waited so many years, the parachute would have been freshly packed. and would have worked. If we had repacked it. We knew it was due, but we couldn't get somebody who knew how to do it out to the launch site fast enough. I had made the decision to launch without a repack. Heck, there were twenty more items on the checklist we couldn't get done and still launch in a few hours' notice. A lot of the guys wanted to scrub. But I had pushed to go. I rationalized that if it had taken so many years to get a chance wouldn't it take years more to get another one? And we would be even less ready then, when more of the crew had moved away, forgotten that much more or just lost interest.

"Gary, who had been as much a part of this as I had, and had at least as much invested, and who was a more patient guy, had voted to postpone. 'We can launch next week. Who cares about Lockheed's four hundred pounds of paper? The whole Cape York thing just perpetuates big expensive, bureaucratic pork barrel space anyway. And who knows if they'll pay us? If we launch next week we'll have time to do it right. You don't need a payload from Lockheed to prove the rocket works. And if it screws up, Lockheed will only be too happy to tell the world that cheap commercial rockets are science fiction.'"

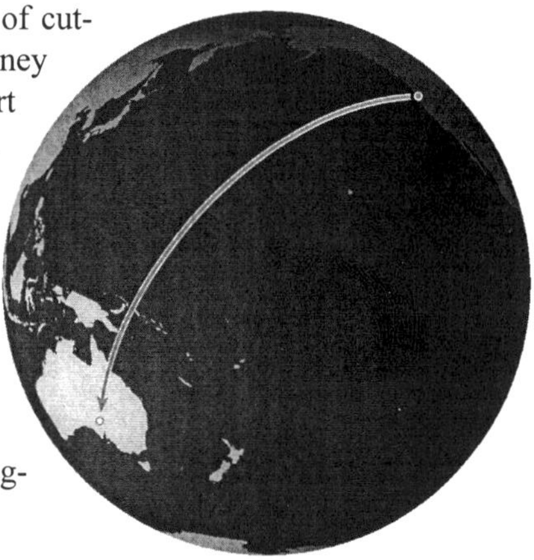

"I asked angrily if after all these years of cutting corners 'cause of lack of people and money all of a sudden today we were going to start doing things right? Gary gave me his 'you really are a hopeless case' look. I knew what he meant. I had always been the one pushing for corner-cutting. Not him. I said that lots of people knew how to send rockets half way around the world. ICBMs have been doing it since the '50s. People will refute our cost numbers saying its only cheap 'cause of all our free labor and scrounging for spare parts. We'll launch next week empty and the world will not even bother to yawn.

"I still don't know why Gary went along with those idiotic arguments. I guess he tired years ago of trying to change my mind—a waste of his time. Plus, in reality, we were both sick of this project, which had drained our savings and sapped our energy over so many years. How many days had we talked, over a lunch of melon (mine) and a deli sandwich (his), about why we stuck to this aerospace business? Most of our Boeing buddies were in the Bay Area doing computers or working for Hewlett Packard in Colorado. We were the only ones still working space stuff. Maybe the failure of the rocket would bring this highly unrewarding chapter of our lives to a decisive conclusion. Shit. What difference did it make now anyway. We screwed up, and failure—anathema to us bright hyperachievers—was something we would have to learn to live with.

"Gary and a couple of other guys straggled back in. They said nothing and their gazes were at the floor only. Eye contact was too painful. 'I guess we oughta secure this mess for the night,' I said. 'I'll be back in the morning to start tying things up.'"

"Yeh, I can take off work tomorrow," Gary volunteered.

"I'll take the car home and pick up something to eat on the way," Kathy looked exhausted.

"It was the end of maybe our last thirteen-hour day at the site. We went onto the pad and hosed down the launch site to clear out the residual kerosene and debris. It was dark and still as we worked by the light of our car headlights glaring in our eyes while not really illuminating the pad at all. The launch tower looked to me like the

claw of a buried dinosaur reaching from its concrete tomb, reaching for the stars in defiance of its ancient extinction."

Rick smiled as he remarked, taking a stress relieving break from telling this incredible history, "You might say I was somewhat depressed."

"The monuments to our failure would be around to haunt us for a long time. The state had built a road to the site, and run electric and water lines. They had named the street Rocket Fleet Road as sort of a pun on my last name. At the end of the road were the giant pad we had poured six feet deep out of concrete, the vehicle assembly and payload integration building, and a huge steel scaffold to support the rocket on the pad. From my visits to ancient sites like Woomera I knew all that stuff would still be there forty years from now, rusting and decrepit. People would come by and look it all over, with valves and gauges and pipes running everyplace and wonder what somebody had tried to do there. The answer would be that a bunch of ex-Boeing guys moved down from Seattle and tried to start a rocket launching company. What a crazy idea.

"While we had washed down the pad, said goodbyes and sat cooling off in the shed, our little enterprise, which had never become a business, was evaporating. Three hours 'till closing at the Australian Space Office, and our payload was still lost in the Woomera desert."

Rick looked very melancholy and kind of vulnerable as he thought about the day. "The physical work of securing the pad improved our mood like running cold tap water on a bad burn. Momentary relief but you know you'll be feeling the sting for a long time. After handshakes and pats on the back, the rest of the crew headed home. Gary and I shuffled back to the control shed guided in the pitch black by our memories and by the feeble light of my trusty key chain flashlight. 'If I had known we were going into the night time launch business, I would have at least gotten a new battery,' I said. Gary responded only with a drag on his cigarette.

"Back by the console I fished a Diet 7-Up out of the mini fridge and handed Gary a beer left over from the day's festivities. He lit another cigarette and we sank into the two operator's swivel chairs: government surplus gray metal with '50s green vinyl. The phone rang. 'Hello?' I answered not using the company name. In our minds the company was already dissolved. 'For you. Jana.' When he got off the phone, Gary said, 'Kid's sick again. Gotta stop by the drug store on the way home.'"

"Worse than shouldering this defeat together was going to be the drive home and facing our wives, our families, and ultimately ourselves. But Gary had medicine to deliver and Kathy could use some help with the baby. We should get going. We stood up with all the vigor of a pair of ninety-year old men, switched off the lights and locked the door behind us. As we stepped down the three metal stairs to the gravelly asphalt, I heard the phone ringing inside. Should I bother climbing up, unlocking and lunging for the phone? It's probably Kath, and I'll be home in twenty minutes any-

way. Well, maybe she needs something from the store too. Gary waited on the top stair holding the door open while I snapped on one fluorescent lamp and grabbed the phone. Before I even said hello, I heard the bleep of an intercontinental connection and a lot of static. I said to Gary, 'James from the helicopter again.' All I could hear was, 'xzxz...package...xzxz'."

"'James?' I yelled into the receiver so loudly that I could hear my voice echo in the dark woods outside.

"This time I heard him clearly above the noise of engine blades, wind, and radio interference: 'We have the package. Now proceeding from Woomera direct to Canberra.' James' voice was in the clipped style of a military pilot as he struggled to be heard.

"My mouth didn't even have time to blurt an instinctive WHAT? before a crescendo of ear splitting static and that bleep again, and the line went dead silent. I looked at the damned phone in my hand then at Gary. My stomach felt like I was at the state champion swim meet in Columbus about to step up on the blocks for the finals of the hundred-yard breast stroke.

"The body and the brain are complex little chemical systems and there's a limit to how fast you can totally change mental state. I looked up at Gary in the midst of all that chemistry and diffusion through my brain's cell membranes and said in a totally flat tone, 'That was James.'"

"Yeh? What does he want?"

"Says he has the package and they're heading direct to Canberra."

"His head still cast downward, Gary looked up at me through his eyebrows with a glance that said, 'Are you shittin' me?'"

"I replaced the receiver with a little flourish and broke into a wide grin. Gary smiled, which for Gary indicates the satisfaction meter is pinned. We were both stunned, and there was nothing to say anyway. I wanted to call Kath but I didn't want to tie up our only line. Gary sat down to think and savor the moment. He lit up another Winston. Midway through his puff, the phone rang again.

"I picked it up. Bleep. Crash of static. 'James again,' I said to Gary. He picked up the extension at the second operator's console."

"Rich?" (James' appellation for me.)

"What's going on there?"

"I'd be happy to tell you if you would quit hanging up on me in fucking mid-sentence."

"Australian humor," I said to Gary.

"'I'll ignore that,' crackled James' voice on the noisy radio phone. We could clearly hear the helicopter blades. It sounded like James would be reporting the rush hour traffic conditions any minute now. I said 'It sounds like we have a decent connection.'"

"This phone call was potentially going to change all of our lives and possibly alter the course of space history and I'm sitting here exchanging signal reports, I thought. 'It is a good connection,' said James. 'We're at seventy-five hundred feet now cruising toward Canberra. High enough to get a good radio link through Melbourne.'"

"James told us his story. Our last conversation about somewhat desperate and far fetched means to find the rocket nose cone holding Lockheed's proposal in its payload bay had gotten cut off because the radio link was unreliable while he was flying so low over the Woomera range looking for the crashed rocket. But hovering out there he got an idea. A spent rocket stage hitting the desert floor without a parachute to slow its descent would create quite a bang. He climbed to altitude and called a friend on the faculty at the University of New South Wales in Adelaide. The University maintains a network of seismic sensors for detecting earthquakes all over Australia. Australia happens to be the world's most seismically inactive continent, but the sensors get used to triangulate quakes in the Pacific and Antarctica. James' friend put him in touch with the head of the seismic detection center, Professor Tooey, who confirmed an apparent event, a small shock, at about the right time. A rough triangulation suggested the epicenter was inside the Woomera boundaries.

"As a favor to James' contact, Professor Tooey and a post-doc named Shankland processed the data. While they worked, James refueled at Woomera's airport, which consists of one asphalt landing strip, two refueling trucks and a quonset hut. While a '50s vintage Texaco tank truck pumped Jet-A into the helicopter James called Shankland on the pay phone, got the coordinates of the impact point and plotted them on his aeronautical chart.

"James tried to phone us right then but we must have been outside washing down the pad. Back in the air he flew West out the 255-degree radial of the Woomera VOR to a DME of seventy nautical miles. Descending through about four thousand feet in the empty desert sky, he spotted the fuel and Ox tanks of the crashed upper stage of our rocket. They had made a shallow crater in the sand and apparently ignited a small brush fire which had gone out—not much brush to burn out there. A few hundred feet to the southwest were the debris of what had been the nose cone. Among the shards of graphite epoxy composite and mangled aluminum load bearing members was the wooden crate we had nailed together from two by fours containing Lockheed's twenty-six-volume-plus-attachments Cape York proposal wrapped in plastic and blankets and wedged into the crate with packing peanuts and sliced up bits of foam rubber from a mattress I used to sleep on at the launch site. I think what in fact had held it all together was about a hundred feet of good climbing rope we had tied over and over all around the crate.

"There was kind of an eerie story about that rope. A bunch of the guys working on the rocket were serious rock climbers and gave us the rope, which was ready for retirement, to use one day when we took off to do some spelunking. We were out in the hill country to the East. One of our guys, Angelo, was crushed when the mouth of the cave he was scrambling out of collapsed. That rope was still around his waist when the rescue team dug his body out the next morning.

"We kept the coiled rope at the site in his memory. When we needed a strong but flexible support for that makeshift wood crate we figured Angelo would have been happy to see that rope launched on our first flight."

Chapter 6

While Rick took a second to finally take off his hat and flannel shirt (wearing a T-shirt from his launch company underneath), I reflected on this semiparallel between our worlds. We—Har'el, Jennifer and I, had just reached this destination tied to a climbing rope after a day of spelunking. Coincidence or not, it strengthened a feeling. I took off my flannel shirt revealing my Brown T-shirt.

As a child I loved to visit my German grandparents and great grandmother in Ohio. I enjoyed their accents, their way of keeping house, the breakfasts of toast, butter, honey and a soft boiled egg in the shell and the smell of percolating coffee. As a young adult I worked in Germany. My first visit there was for several months at age twenty-two. Now, I had this same feeling, that I had never been there before, but that I was completely at home. The linens on my bed, the way topics were discussed at work, the look of people. Half a dozen times I thought I saw my mother, who had not been in Germany for thirty years, from a distance. A stranger's figure and stature and hair and skin would be a perfect match. I heard my grandfather's argumentative Ja ja... from the tables of so many diners at the restaurants. Learning German was easy even though my grandparents had never spoken a word of it to me, except maybe Guten Morgen Richard! But their English carried the syntax, the pronunciation and the Germanic roots of its vocabulary with it like pollen in the September wind. I breathed it and it was in my system.

So it was with this Rick and his stories. I had never launched a rocket, been fired from Boeing or fathered two daughters. Superficially our lives were orthogonal. At least since about 1971 or 1972. But I knew exactly all his feelings. The tremendous guilt in having rushed and then having the parachute fail, even if it was the right decision at the time and the tendency to champion the less technically rigorous side of many judgments. The inability to ignore the compelling logic of an enterprise despite its impracticality or even impossibility. And a tendency to be a David to a procession of life's Goliaths. I felt I could alight upon this life without any particular acclimatization or accommodation. It fit fine right off the rack. Rick interrupted my thinking as he continued his tale of their first flight.

"Our original plan was for two large helicopters each with two crew members. James was the crew chief. But with essentially zero notice, we were lucky just to rouse James himself and one small two-seater. The crate was too big to fit inside the little helicopter and too heavy for one person to lift. They don't fit winches on those little helicopters either. So there was James basically stuck, out of radio contact, alone with a tiny flying machine and this huge box. James' second brilliant thought of the day was to untie some of dearly departed Angelo's rope and tie it to the helicopter's skids. Not exactly an approved flight configuration but he was inside the craft's gross weight and CG limits, and so off he went dangling four hundred pounds of paper, packaging material, wood and rope beneath him."

"Off to where was the next question. There hadn't been time to schedule a charter jet from Woomera to Canberra. If he went to Adelaide or Melbourne there was no guarantee he'd find a suitable charter and also no guarantee he wouldn't get a stiff fine and his license suspended for flying over people's heads with his jury-rigged freight sling. Anyway, given the sorry condition of the now untied box James worried it might break up on landing and he'd have no way to repack and get back under way in time."

"Having just filled both the main and aux tanks at Woomera, James opted to slog on to Canberra nonstop. He was cruising en route in clear skies with a light tail wind up at seventy-five hundred feet when he finally reached us on the radio phone. His ETA at Canberra was 4:30 Friday afternoon—thirty minutes' margin but with landing, untying, possible hassles with airport authorities and a long drive at rush hour into central Canberra, no way could he make a 5:00 p.m. deadline for delivering the proposal to the ASO. We had overcome much bigger obstacles that day and Gary gave James the solution, simple but elegant, over the radio phone.

"At 4:05 p.m. James requested and received a routine VFR (visual flight rules) clearance to enter the Canberra Terminal Control airspace traversing the area West to East. He complied with a routine request to set his radar transponder to 5406. He proceeded overhead Canberra International Airport and headed directly toward the ASO building in midtown Canberra."

"At 4:20 p.m. James declared a smoke-in-the-cockpit emergency to Canberra and, just to make sure everyone at the ATC knew this was serious he propped both doors open, creating a huge noise, lending his coughing and gagging an enhanced urgency. In case that wasn't dramatic enough he set the transponder to 7700 (the international distress code) thus setting off alarms in air route traffic control centers from Queensland to Tasmania.

"You'll never guess where, in the dense urban landscape, James found an emergency landing spot. He set the helicopter and the crate, still hanging by some pretty frayed climbing rope, down on the top level of the concrete parking structure of the Australian Space Office. He let the crate settle first, then he backed off a few feet so the front of his skids cleared the crate by a few inches. Before getting out to untie the box James stuck his Halon fire extinguisher into his coat pocket, crumbled an old sectional chart under the passenger seat and ignited the map with a match from the survival kit. By the time he sliced off the rope and tossed it in a dumpster, smoke from the smoldering seat was blackening the windows. James discharged the Halon into the cabin, smothering the fire with its foam and then tossed the empty bottle onto the helicopter's floor.

"With his fire now extinguished James headed into the ASO and phoned Canberra tower to report, with total honesty, that he had landed safely with no harm to anyone or any property and the fire was out. He got off the phone fast saying he had to speak with the property owner and would call back in a few minutes with the address where he had landed so that an aviation safety team could meet him and complete a formal report.

"He then went to the room number the proposal was addressed to and reported he was submitting a Cape York proposal. The clerk checked the wall clock—4:45—and said he better hurry up. They were closing at five sharp. Clerks who accept aerospace proposals have certain props always on hand. An accurate clock with a time stamping gadget attached to it, always made of thick steel and painted gray, a large, windowless room whose door is fitted with a few mean looking locks, a key chain with even more keys than the door has locks, stacks of legal looking forms ready to be time stamped and, most importantly for James at this moment, a good solid dolly.

"James just cast a glance at the dolly and the clerk nodded. As James pulled open the door, she said, 'Better move, I'm off in thirteen minutes.' 'Great,' James said with a wink, 'I know a lovely spot we can go for a quiet drink. You like tequila?' He disappeared out the door down the corridor, up the elevator and out the fire door to the roof of the parking structure. In front of the soot-blackened bubble of the poor little rented helicopter was the splintered four-hundred pound box with Lockheed's multibillion dollar Cape York proposal sealed inside. He had to wedge the crate between the polycarbonate cockpit and the dolly to wrestle it onto the wheeled forks. As he tilted it back James heard a pronounced scrape as an exposed nail galled a deep trough into the helicopter's smooth bubble.

"No time to worry that now, James struggled the hulk back through the fire doors, down the elevator and the long corridors and shoved it through the door of the proposal clerk's office, breaking the door's glass window on that same exposed nail. He picked a few of the largest pieces up off the floor and gingerly dropped them into the cylindrical gray metal receptacle which is also standard issue for this type of office. They shattered as they fell to the bottom of the trash can with a not at all subtle clink-crash which echoed off the linoleum floor, metal furniture and bruised sheet rock walls to ensure that anyone who might have been asleep during the initial shattering was nonetheless made completely aware of James' arrival. 'You'll want to watch that bit,' James offered politely, referring with his eyes to the offending nail.

"'I'll need it unpacked and deposited in the closet,'" the clerk quipped, further reference to James' arrival not being particularly necessary. James, in a full sweat from prying the proposal documents loose from the box and their packaging, returned to her window. He'd made it, he thought, with minutes to spare.

"'Very well then,' she didn't even look at him. 'Just fill these out.' She handed him a stack of forms to complete.

"James looked at the clock. 4:56. 'See here,' he pleaded, 'the instructions were to deliver to this room. The RFP [Request For Proposals—*ed.*] said nothing about unpacking and depositing in your closet.' James had never in fact seen the RFP but the argument seemed sound enough. Already today he had dashed out of work, rented a $180 per hour helicopter on his Visa card, logged three hot noisy hours flying to and all around the Woomera desert, motivated UNSW's Geophysics department to find the payload via seismic triangulation, illegally attached and flew the crate four more hours to a faked in-flight fire and forced landing, set the heli on fire, put the fire out and, combined with puncturing the bubble on a nail rendering it unairworthy to say the least, made himself liable for several thousands of dollars in repairs plus another several thousands of dollars to disassemble the aircraft, truck it to a certified repair facility and reassemble it there. Finally, he had manhandled the proposals' four-hundred-pound bulk onto the dolly and into the Cape York proposal office where he unpacked it. And all with four minutes to spare. He was yet to face the aviation safety investigators' debriefing later that afternoon, and for some reason he was just not in a proper mood for missing the 5:00 p.m. deadline owing to the necessity of filling out the proper paperwork.

"James grabbed the forms brusquely and was about to make a statement to the clerk he was sure to regret almost immediately when he noticed they were all time-stamped and properly initialed at 4:50. As he stared at the forms in a mixture of relief and disbelief she explained 'No mistaking that proposal came through my door at 4:50. Every person in the building knows it.'

"'I could kiss you.' She looked at James, face sweaty, crusted with desert sand and dust, and unshaven, hands sooted and smelling of jet fuel. Clothes torn, partly soaked in perspiration and charred from the fire. 'I'll take a rain check on that one.' She

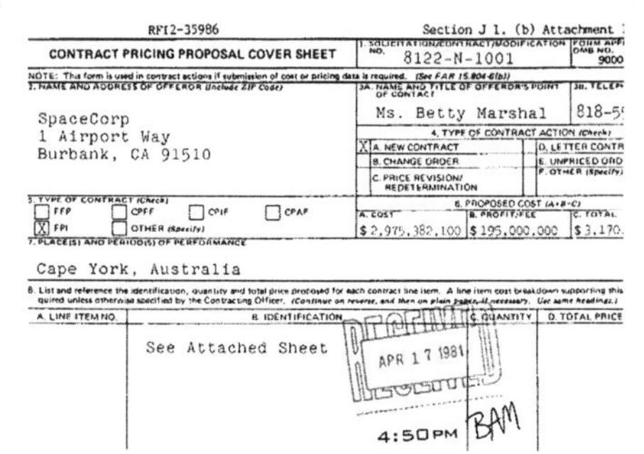
RFI2-35986 Section J 1. (b) Attachment

CONTRACT PRICING PROPOSAL COVER SHEET

1. SOLICITATION/CONTRACT/MODIFICATION NO. 8122-N-1001

FORM APPROVED OMB NO. 9000

NOTE: This form is used in contract actions if submission of cost or pricing data is required. (See FAR 15.804-6(b))

2. NAME AND ADDRESS OF OFFEROR (Include ZIP Code)
SpaceCorp
1 Airport Way
Burbank, CA 91510

3A. NAME AND TITLE OF OFFEROR'S POINT OF CONTACT: Ms. Betty Marshal

3B. TELEPHONE: 818-5

4. TYPE OF CONTRACT ACTION (Check)
[X] A. NEW CONTRACT
B. CHANGE ORDER
C. PRICE REVISION/REDETERMINATION
D. LETTER CONTR
E. UNPRICED ORD
F. OTHER (Specify)

5. TYPE OF CONTRACT (Check)
[ ] FFP [ ] CPFF [ ] CPIF [ ] CPAF
[X] FPI [ ] OTHER (Specify)

6. PROPOSED COST (A+B=C)

A. COST	B. PROFIT/FEE	C. TOTAL
$ 2,975,382,100	$ 195,000,000	$ 3,170.

7. PLACE(S) AND PERIOD(S) OF PERFORMANCE
Cape York, Australia

8. List and reference the identification, quantity and total price proposed for each contract line item. A line item cost breakdown supporting this ... quired unless otherwise specified by the Contracting Officer. (Continue on reverse, and then on plain paper, if necessary. Use same headings.)

A. LINE ITEM NO.	B. IDENTIFICATION	C. QUANTITY	D. TOTAL PRICE
	See Attached Sheet		

APR 17 1981

4:50PM BAM

looked at the huge crate. 'Looks like a good bit of effort went into this one.' 'I imagine,' said Jame,s a bit sardonically.

"'Rumor I hear is the Senators won't fund it anyway. Need all those billions for limos and junkets and girlfriends.'

"'A bit cynical are we?' James made it more a statement than a question.

"'Just saying what I hear in the course of a day's work around here. Who's proposal is this anyway?'

"'Lockheed.'"

"'Yeh? I heard they wouldn't be making the deadline.'

"'Well, we made it. Guess you can't always believe the rumor mill.'

"'More than a rumor. They sent a fax about four this morning that they would be airborne out of Burbank within the hour. You can't get here from there in twelve hours. Can you?'

"'We really lit a rocket under their courier service.' James said seriously but with a twinkle in his blue eyes.

"'Yeh? Well you look like you were selected to hold the match.' Having already exceeded the level of her interest in the whole Cape York launch site business, the clerk returned to her desk and began locking up for the weekend. She was slipping on her jacket when James handed her the completed forms.'

"'A security guy is coming up to help you clean up and lock the proposal in the closet.'" She stepped carefully over the remaining smaller shards of broken glass littering her floor. 'Careful of that nail, dear.' And she was gone.

"While awaiting the guard James tried to call the launch site back in Oregon but the switch board at the ASO was already closed. It wouldn't have mattered anyway since it was past 3:00 a.m. on the West coast by then. We had sat around till after midnight, hopeful but anxious, and finally left for home. By the time the Australian Aviation Safety team was finished gathering data for its "Incident Report" including photographs of the wounded helicopter and a detailed and tape-recorded debriefing of James, it was too late for him to catch a flight home. It was lucky the crash investigators hadn't shown up till almost seven, since by then there wasn't a soul left at the ASO to notice the coincidence of James' emergency landing exactly where he had made his rushed last minute delivery. He figured to stay over Saturday and get the wounded helicopter packed up and moved before it attracted any unwanted notice. He tried to phone from his hotel at midnight—9:00 a.m. Friday for us—but the line was busy. Gary and I had been awakened early by calls from the press looking for confirmation of the story, details and photos. We had driven back to the launch site to meet reporters and even one TV crew. Kathy and Jana plus kids and babies were handling the phone. In fact we had very little to show or tell. James was not at home and both the ASO and air safety offices were closed. The dismembered remains of rocket #2 were not impressively photogenic, but we did have a nicely charred launch pad and a bunch of brochures with a good photo of the rocket.

"A reporter asked if it was our helicopter which had crash landed at the Australian Space Office. Crashing had not been in the game plan. Maybe that's why no word from James? To confirm that would ensure James got in real hot water so we just stuck with what we knew: The payload was flown successfully to Australia and as far as we knew was delivered on time to its destination. That last part was admittedly wishful thinking but we were happy to play along with the media's desire to make a heroic story out of what they were calling an historic event.

"By noon word had gotten around and the crew which had left so sadly the night before was back with wives, husbands and kids, beer, sandwiches, frisbees and big broad smiles. A reporter climbed part way up the launch tower and photographed about fifty of us squinting up at her into the sun and grinning ear to ear like drunks. which about half of them were at that moment.

"Canberra is a town with a few stately and expensive inns, and inside one of the most historic and opulent, James awoke to the sound of singing birds and to a tray of fresh rolls, coffee, and orange juice. A few hundred bucks for the enormous suite was lost in the noise compared with his 007-style escapades of the previous day. Plus, the renovated old lodge, the first in Canberra and the venue for numerous historic visits by foreign heads of state had not only a vacancy when he called at midnight but also a limo—what turned out to be a Rolls Royce Silver Shadow managed by a quite proper British chauffeur who sniffed at James, 'Will there be no luggage tonight sir?'

"Despite having needed the limo for lack of enough cash for a taxi, James did have his Visa card and seven years' education in two of England's finest public schools to sustain him, and he replied to the old gentleman simply, 'I should think my man brought them 'round earlier, thank you.' The driver's chin was properly raised and eyes shut as he held the right rear door of the Roller open for James, whose sojourn to the lodge was further punctuated only by the sound of Vivaldi on cassette and the taste of Champagne from the wet bar.

"After slipping into the complimentary fine terry robe and pouring himself a generous morning cup of steaming strong coffee, James made two phone calls. First to the Concierge for bathing trunks, slippers and the picking up of his laundry, then to Oregon where this time it rang through to Jana. After relaying the gist of his story, confirming what in the course of our celebration had become a foregone conclusion, James headed out to the inn's spa for a massage and a swim."

Chapter 7

It is getting late at Tom's. Rick's eyes, like mine probably, look dark and tired. But it is such a great story, a major element of that encyclopedia of special events every family collects about its members over their lives. We all just sit in that particular silence of a busy home after the dishes are done and the kids are asleep and we listen. This seems to be the first time Rick has recounted the whole story to his brother and he himself is reliving its excitement in the retelling. Now he smiles with what I take as nostalgia remembering this dramatic realization of his dream.

"That weekend and the next week were just amazing," he continues. "The governor of Oregon congratulated us on TV and we even received a telegram from the White House. All these people from space agencies around the world and from our own FASA called about sending delegations to discuss cooperation for future, grander missions. Meanwhile James spent Saturday at the ASO getting the helicopter torn down and hauled off for repair. He was there when the Lockheed proposal shipped by company jet arrived in the company of two guards, a notary and a video camera to document and legally certify the fact of their presence. They were almost a full day behind us and it was pretty clear from James' experience that the ASO would not have accepted a bid even five minutes late, let alone twenty-plus hours late. James has a way with people and a few excellent connections in Australian politics and business, and he spent the next week making us all famous and popular in the press there.

"That honeymoon lasted, unfortunately, only about a week. Gary and I had exhausted all of our vacation long ago taking days off to promote the rocket and to work on it. That Monday morning we both walked in and quit our jobs figuring now, finally, we were in business. A week went by before we realized we had no income and some pretty substantial bills to pay. $20,000 just in liquid oxygen and kerosene. FASA charged $10,000 for flight plan and radar services, and we had a another $10,000 bill for insurance. We had paid for a lot of rocket components on consignment, mostly valves and the command destruct radios required by FASA in case the rocket went off course. With the rocket now used, not to mention destroyed, all those little loans came due to the tune of another seventy or eighty thousand bucks.

"The good news on James' end was we didn't have a jet charter to pay for. On the other hand, saying the airworthiness of the rented helicopter was compromised, and the Australian Air Safety Bureau billed us $20,000 Australian. Including the certified mechanics' time to disassemble and transport the aircraft to a qualified repair facility outside Sydney, replacement of the canopy, a complete electrical rewiring and X-ray of all major structural elements and all rotor blades, plus the basic rental including daily charges during time out of service we needed to come up with almost $100,000 Australian—about $75,000 U.S. James' $1000 hotel bill plus air fare home was definitely down in the noise.

"All told we figured we needed $200,000 to $250,000 to clear all the payables. But we had $500,000 on the receivables side from Lockheed. The surplus $250,000 would pay Gary and me for a long while to rebuild the other rocket and market our service. Riding on our high profile success, finding new buyers should be easy. Things still looked so promising, our main concern was competitors entering our emerging market.

"A few storm clouds did start threatening to rain on our little picnic. But we had misread how other aerospace companies, steeped in decades of government contracting, would react. No way would or could they compete. Their cost structures, complete with all kinds of fancy tools and buildings, loaded with layers of management and sporting whole departments to create and track lists of qualified components, to enforce approved procedures even for the lettering of labels on parts bins, and to maintain libraries of military specifications, all of which were required and paid for by their government customers, made them ten times more expensive than the few of us techies working out of a corrugated aluminum quonset hut. Anyway, these companies didn't do billions of dollars in sales $1 million at a time. They sold $100 million or even $500 million rockets and satellites to their governments.

"Competing with us head to head on cost and service was impossible. We knew that and we knew we could build a business now that we had shown the world what we could do. Which is to say we were hopelessly naive. Problems started popping up like pesky weeds in a garden. First some fairly obvious questions got asked about James' emergency landing. Sure, in retrospect it didn't take a rocket scientist to see that landing was no accident nor to wonder how James carried that big crate with his tiny 'copter. But there was no reason for suspicion when the investigators came and the report had already been filed before the whole story came out in the press there. But somehow the case got reopened. I guess we were lucky too, that due to our popularity in Australia we got James out of trouble, but only after incurring about $20k in legal fees.

"Back at home a lot of questions started getting asked. The parachute failed on that one and only flight. How could we be sure the guidance system might not fail next time and cause a crash into Portland or some little seaside town? Maybe Honolulu. Was it worth putting thousands of lives at risk so two garage shop amateurs could play around with rockets? Maybe we did have some command destruct radios, even a fully qualified redundant set as we claimed (but, they insinuated, who knew since the design wasn't reviewed, presumably by them), but they could fail, right? And their range certainly didn't reach Hawaii. Never mind that by the time we were out of radio range the rocket was too high and going too fast to hit Hawaii. The press was focused on these sexy, doomsday questions, not on logic and engineering. Didn't we have to admit that we could have been just a little off course and might have hit Sydney or Alice Springs? Think of the international repercussions if a U.S. entity (suddenly we were an "entity" like General Motors or Tonga) hit a foreign city with an ICBM.

"I remember that exact discussion going on the network news one night. And the anchor person asked if, like with airplanes, weren't a few risks well worth the enormous benefit to the world of having the ability to get vital supplies transported almost instantly. On two big monitors behind her were two gray haired men, a retired Air Force four-star in his blue uniform covered with brass and impressive decorations, and of all people, the VP from Boeing who had laid off our whole division. He had since retired—I heard he was a victim of his own downsizing. But he had left with a pretty sweet package and now was being paid more than ever as a PR consultant for the company. These two "impartial" panelists agreed vociferously that rocket transport was a fantastic advance. In fact, both Boeing and the Air Force have been studying and planning such a capability for at least ten years, they reported factually, referring, I surmised, to the work we had done before they fired us. Then old Dr. Goldman pulls out a sketch I had made years ago. Last time I had seen my cartoon showing the launch site, the rocket trajectory, and the recovery and delivery operations at the receiving end was in a meeting where he held it up as an example of wasting corporate resources to set the stage for the termination of our group.

"'We just feel," the two old idiots went on, "that given these significant risks and the vital nature of this type of service, that government and industry must cooperate to ensure its safety and reliability. Many lives were lost needlessly in the early days of aviation before companies like Boeing, supported by government R&D, combined with military and civilian agencies to create the safe, efficient air transport system enjoyed by all of us today." Pretty depressing stuff.

"Luckily we weren't the only ones not buying that BS. But it definitely was not helping. Potential investors always asked how I planned to compete against the almost infinite resources of Big Industry and Big Government. The fact of what we had accomplished while they just talked and made viewgraphs was not convincing, and the assertion they could never touch our low cost was countered with examples of low cost cars and consumer electronics made by big companies. I would insist that aerospace companies had consistently failed to compete in commercial markets. No one could accept that Boeing was anything less than a paradigm of what's great about U.S. competitiveness.

"But no sense dwelling on negatives. None of our detractors had ever built a rocket and they weren't likely to start now. Gary and I had gotten this far with almost zero investor support, and in fact we kind of liked it this way—nobody to answer to.

"The continuous stream of visitors at first buoyed our spirits. Japanese, French, Swedes, South Africans, even a few Americans! But there were problems there, too. All our illustrious callers said they wanted to cooperate with us, but we came to realize they were mainly using us as free tutors to teach them about flying cheap rockets that they planned to build without us. How could the government of Brazil admit it needed to team with two guys in a garage in the Oregon woods to get it right? Luckily, we could hide behind U.S. export control laws, and we really could not give these spies for our competition anything substantive. Unluckily, our detractors, whoever they were, called for an investigation of our possible violation of arms control regulations. While we knew we were innocent and in fact very happy to shield ourselves with the rules, we were forced to start paying out $5k or $10k a month for legal representation.

"After six months, $50,000 and two trips to D.C. to give testimony to big committees, all they came up with was that we had illegally exported certain parts of that first rocket to Australia. We finally were exonerated by showing that U.S. airlines don't get export licenses for the airplanes they take to foreign airports. Plus there isn't much export control on scrap metal which is all our rocket was after crashing into the desert floor at five hundred miles per hour.

"All these things were really just little annoyances. Our real problem was that Lockheed refused payment of the $500,000. They had a pretty long laundry list of reasons not to. They claimed Roger did not have that level of purchasing authority, which he probably didn't. Plus he had let a sole source contract to us without any evidence of seeking competitive sources or auditing our claimed costs. We protested that there were no competitors, then or now, and there was hardly time for an audit. As it was, we made the deadline with less than four minutes to spare. We questioned the need for an audit, given we offered a fixed price service at 10% of cost.

"Unfazed by our logic, they contended we had failed to comply with their myriad subcontractor requirements including presubmission of an environmental impact statement, although we had done one and the state and federal governments had approved us—we just hadn't presubmitted it to Lockheed per their internal procurement procedural requirements. And there was 'compelling evidence' that we had violated the law in delivering the service. Of course we also had no written contract, though the fact that Roger showed up at our place with the proposal ready to fly weighed in our favor.

"Legal battles, our lawyer Keith had advised, are like war. And two guys with no money are no match for the fire power of a tens of billions of dollars multinational. In our settlement, which still cost $10,000 to negotiate, we agreed they owed us nothing and they agreed to cease claims that we had violated any U.S. or Australian laws or international treaties. One final bit of irony was that on the day we signed that settlement, one year to the day after the launch, Australia announced the cancellation of the Cape York procurement. The clerk at the ASO had been right. The money was just not there."

"Well, that was Lockheed's problem. But we were all disappointed, since they were the leading bidder and in fact had just submitted their BAFO [best and final

offer—*ed.*] Payment or no payment, even Lockheed would have had to admit we had delivered. Now the whole exercise was moot.

"At that point Gary and I had gone a full year without a pay check and the company debt, with interest, was over $500,000, all of it on our credit cards, on notes we had cosigned for the Company, and in contingent payment agreements with lawyers and suppliers. We had hosted hundreds of visitors and spoken all over the country and the world about what we had done, but we did not have one single order and our only customer hadn't paid a cent and never would.

"At that time the president of Eastern Airlines was a retired Apollo astronaut—I don't even remember which one. He read about us in one of the popular space groupie magazines like *Omni* or *Space Frontier*. To make a long story short, Eastern bought us out for $1, an agreement with our creditors where they got about 30¢ on the dollar spread out over five years and we agreed to continue pursuing this rocket business for them at a flight engineer's salary, about $2k per month each. Gary and I were elated. 'Till Eastern came along, we thought we would have to find conventional engineering work and spend the rest of our careers paying off a half million dollar debt. Not to mention having to shut down the company. And on its side, Eastern had bought a fledgling but proven new line of business for peanuts—$30k a year in debt repayment service and $48k a year in our salaries.

"Literally the day after we signed, Eastern filed suit against Lockheed on behalf of the old company claiming Lockheed had acted in bad faith to suppress competition and destroy a small business. Eastern is no Lockheed, size-wise, but Lockheed realized Eastern could afford lawyers, which Gary and I sure couldn't. Eastern ended up somewhere short of the full $500k, but given the reduced debt they had negotiated and our small salaries, the business was immediately in the black.

"The acquisition and settlement got minor coverage in the aerospace press, but the big story was brewing out of a phone call I got during those two weeks of negotiations. Overnight package delivery was reshaping the air freight industry right about then and the older, more conservative companies like UPS were struggling. UPS' survival was hardly in doubt but some of the lesser known firms were desperate—desperate enough in at least one such company to hire a young, brash MBA and give him license to take really radical measures to save the company.

"Tim West was that new Stanford MBA graduate. He had wanted to work in aerospace. He got an engineering BSc at MIT and MSc at Stanford, then worked for Lockheed long enough to realize how stagnant things really were and convinced them to send him back to biz school. When Tim got out, Lockheed really didn't have a slot for him and he ended up at a freight forwarding company called Embassy Air Freight. A lot of big-name freight companies would use Embassy to handle pesky small loads to obscure foreign destinations, mostly in Asia. Embassy consolidated multiple small loads into one or two big ones and then contracted with a freight airline like Tiger to carry the loads trans-Pacific.

"These forwarding companies don't sell to consumers, just to freight companies, so most people have never heard of them. But in its better days Embassy was a $500 million company with offices all over the west coast. Now they were getting shut out by the new hub and spoke services which were cutting delivery time to twenty-four hours by eliminating use of consolidators like Embassy. When Tim came on to help a select team perform radical surgery on the company, they were down to maybe $150 million and had consolidated to just their home office in Portland.

"Still an aerospace freak at heart, Tim had been following our brief rags to riches to rags romp through the annals of aerospace history in the local paper. He even showed up at the big post-launch party. Australia had been one of Embassy's biggest markets and Tim got the idea to initiate overnight rocket service to Sydney, Melbourne, Canberra, and Perth. On the phone he told me that the best service now required thirty-six hours for a Sydney delivery: three hours for pick up and queuing in the western U.S., fourteen jet aircraft flight hours, three more hours for delivery there and a sixteen-hour time difference. A package sent on a Wednesday afternoon wouldn't arrive 'till Friday morning. Thursday and Friday packages were delivered the next Monday. This was exactly Lockheed's problem with their Cape York proposal. By Thursday the commercial carriers wouldn't deliver before Monday and even their own jet only got there by late Saturday.

"Tim's idea meant three hours pick up and queue time here, one hour transit time via rocket, three hours delivery time in Australia, and sixteen hours lost to time zones—twenty-four hour service to Australia. That would catapult Embassy to prominence in the air freight business. Pick up by 3:00 p.m. for guaranteed next-business-day delivery to every major Australian city. Japan, Taiwan, and Hong Kong would be next. A huge market and a chance for some revenge on the mainline companies that had abandoned them.

"Top management hated the idea. Over a hundred years in business, Embassy had never competed with the retailers like UPS which were their customers. They should kill their own business to gamble on rockets? And what did they know about the retail side of the business? Nothing. It was suicide. But in a highly contentious board meeting, the elite reorganization squad of which Tim was a member got the project approved. They pointed out that the company's hundred-year niche was, in fact already had, evaporated. The day before the board meeting, Embassy's sole remaining major account had faxed over a notice of intent to terminate their contract for

Embassy services effective in thirty days. Embassy was already near the limits of its bank lines of credit and owed large sums to most of its major suppliers, particularly Tiger. Without decisive action the company faced insolvency by years' end. A dramatic entry into an exciting and new line of business, yet one so closely related to Embassy's expertise, could at least impress their banks and shareholders to extend a little more credit for long enough to come up with a viable long term survival strategy. Tim's team leader, the retired CEO of CAFE [Consolidated American Freight Enterprises, which twenty-six years ago he had turned from near bankruptcy to become the U.S.' largest trucking conglomerate—*ed.*] admitted it was a desperate, maybe even crazy move, but the only one which might save Embassy.

"The re-org team had been given 1 vote on what had been a twenty-nine member board. Fourteen board members went against the proposal and one, recovering in the hospital from a bypass surgery, abstained. The proposal passed fifteen to fourteen. Within a week Eastern negotiated the first regularly scheduled rocket transport service in history. Sunday through Thursday, five evenings per week for a minimum of one year, two hundred and fifty-five launches out of our launch facility at 6.30 p.m. Total contract value was $1 billion. There were options for additional years, to add return flights and for other destinations which could bring the total over the next three years to $8 billion. It was a bold move and short term it worked. Eastern and Embassy share prices went way up, and the deal was front page news in the *Wall Street Journal*. The day after that coverage it got picked up by the *Times* and the TV networks.

"The other airlines and freight companies took a wait and see attitude, saying it looked like a money loser and wondering in the press if it was safe and if it would even work reliably. You want your package dumped into the Pacific Ocean? Rockets aren't reliable, they said. The aerospace giants exercised their usual zero sum gamesmanship with their complaints that such a service done by an airline inexperienced with rockets was unsafe. And they said it was totally unfair since Eastern was meeting only commercial standards while their costs had to be higher to meet military specifications.

"Eastern just went about their business like this was simply another route acquisition from a rival airline, one that happened to use a slightly different model of aircraft. They brought in a sort of gate team to handle propellants, payload and other materiel logistics, route managers, as well as a small finance team—everything you would need to run an air freight line from one place to another. It was either brilliant or stunningly naive. Gary and my role was mainly on the technical side to advise on all the storage and ground handling equipment and procedures. And we taught a team of Eastern engineers and aircraft mechanics how to build and fly the rocket. We worked with the company's buyers to get the best prices on propellants, tanks, all the parts of the rocket. That was a lot of fun. Suppliers who wouldn't even talk with us before were vying for the chance to take us to lunch. Two hundred and fifty flights a year backed by Eastern and Embassy. The press likened it to the airline boom of the '40s and '50s and to the success of geosynchronous Comsats in the '60s and '70s. The next big commercial wave in aerospace. And everybody wanted to ride that wave.

"It was a little frustrating to see our dream become absorbed into Eastern's bureaucracy, of which Gary and I felt like two relatively unimportant members. And to see other airlines latch on to the concept and compete with us and contract with people like Lockheed for their rockets. But for all their faults, Eastern did good. They ultimately improved the vehicle, started recovering both stages and opened new routes, the first being Australia to the U.S. West Coast.

"Embassy and Eastern made real money and the pioneer spirit of starting something new reshaped them and the whole industry. Gary and I never earned a dime, outside our regular salaries, for creating the concept or making the first flight. And we didn't get much recognition either. Eastern had us agree to decline any PR whatsoever. They didn't want to associate their company with two wild inventor-types who weren't above a few quasi-legal hijinks. They wanted a squeaky-clean genesis legend. Today's consumer, they told us, has no appetite for the Wright Brothers or Mark Twain except as historical fiction. Banks won't finance it and customers aren't comfortable buying it. We had to admit from our own experience that they were right."

"What about James?" Har'el asked.

"We wanted Eastern to have him run the air delivery network in Australia. But they already had relationships with local carriers. They did get James a right seat [copilot—*ed.*] slot on a Citation running from Woomera to Perth and he did that 'till something more interesting came up. When work brings him out to the U.S. he drops by for a beer, but that's about it."

"So what are you doing out here in the East?"

"South Africa service trial shots. We have had service throughout Europe since the year after Australia got going—ten, twelve years now. But it's gotten to be a competitive route and not too profitable anymore. South Africa has good potential for us. We got off a successful shot again today. Mainly it's the local logistics we need to exercise and they're getting better now. We will be in there five days a week starting next month."

I didn't say much. It was just a fantastic story, and I was soaking it up. I wanted to find out if all these hundreds of launches hadn't really cut orbital launch costs too, and who was doing that. But it's late, we're tired and we have an early run the next morning. "Unbelievable. We have nothing like this at home. I've got a million questions for you. But I'm dead. What time are we running?"

"6:30. Sound good?"

"I'll be there."

Chapter 8

Later I laid in bed in the guest room next to Tom and Joanne's bedroom trying to assimilate the events of the last thirty-six hours. Is that all it had been? Just one and a half days since Har'el, Jennifer, and I walked into a cave and literally out of our world?

As bizarre as it was to hear these stories from Harry Lee and from a nearly perfect double of myself, to see a Reston built to the same concept but with an unrecognizable layout of streets, the really eerie sensation was the similarity. Having dinner at Tom's as if he were my brother, but knowing my real brother Tom is in another Reston probably searching for me and convinced I am either dead or dying. From the next room I heard the 11 o'clock news on the TV. Doesn't he ever sleep? This Tom is so similar to my own Tom who is up and on rounds every morning at 6:30, sees patients all day, plays with the kids in the evening, and still has energy for e-mailing, TV, and a shrimp pizza at midnight. I fell asleep quickly, not surprising given the time and the intensity of the day. But I woke up agitated and sweating after less than an hour.

I sat up. What was happening? The room was dark and quiet and a red LED alarm clock read 1:17, the colon winking once per second. I had done nothing wrong. But I was filled with guilt and dread. It was a nightmare and I began to reassemble its fragments. I dreamed I was in bed with Nancy, my wife, except I knew she was not Nancy but instead another woman who just looked like Nancy. No one else could discern this woman I was sleeping with was not my wife, so wasn't it just innocent voyeurism on my part to play along as if I too hadn't noticed? For a while I enjoyed this new partner whom I could will to disguise as my wife to assuage guilt, but at the same time I could exploit for the excitement of an illicit affair. The elation was not sexual, but more like being eight years old and, home before mom or dad, sneaking one of their cigarettes from the drawer of the night table in their bedroom, lighting up and blowing the evidential smoke out through the window screen. But knowing and not knowing cannot coexist and neither can you toggle between them like the two states of the red LED colon of the alarm clock. The ceremonial glass I shattered at our wedding signifies the irreversibility of knowing, the responsibility which attends our passage from innocence. The glass cannot be unbroken, and I cannot unknow. In bed in my dream in my mind's eye, I saw the polished white linen napkin with the crystal wine goblet wrapped inside. I felt it crush more than shatter under the hard leather heel of clumsy dress shoes I had agreed to wear instead of my trusty Adidas—a peacemaking concession to Nancy and Anita. That was the sound of breaking glass that had woken me up.

Too shaken to attempt sleep I slipped on the running shorts Tom had loaned me and headed for the kitchen in search of a Diet 7-Up and ice. In a way I wasn't surprised to see Har'el and Jennifer were already there, I guess for pretty much similar reasons. They were talking softly but intensely over two glasses of orange juice.

"You guys couldn't sleep either?"

"Jennifer's worried about her dad."

"If he's really alive I have to see him."

"I guess we could..." I let my voice trail off to let her speak.

"That Christmas before he died didn't go well at all. Well, it had until the weekend before I went back. My boy friend at the time drove to our house from staying at his parents' in Maine. He was in Air Force ROTC and very hard core pro-Reagan, Star Wars, and all that cold warrior stuff. At the time I thought my dad, since I thought he had been in Korea, would enjoy talking with Randy. Turned out later my mom told me he was way too young for the Korean war and strongly antiwar during Vietnam. But he had been to Korea later—that was my confusion about him being in Korea. He was there helping farmers to reclaim land and return it to agricultural use. I knew he was back in Southeast Asia working for the army when I was born and I had always assumed he was a soldier. He never never talked about it. Mom says what he saw this second trip really affected him. But I was just a baby when he came back. I didn't know.

"Sunday morning after breakfast I was upstairs packing and I heard them. A loud argument. I came down with my purse over one shoulder and my suitcase in the other hand. I'm a slow packer and I was rushing to get my stuff in the car. Randy had been waiting for me for half an hour. He said to me 'Got everything?' I nodded, and he commanded me, 'Get in the car. We're late!' I did feel sorry that I held him up so long and he kind of scared me. I looked up at dad standing in the doorway. He was glaring at us, at Randy really, and Randy started the engine. I don't know why, I just got in and we took off. That was the last time I saw my father. No kiss goodbye, no smile even. I have to go to Rutland tomorrow and see him—if he's really alive."

Har'el spotted a phone with a pile of phone books next to it. Some of the airline names listed are familiar to us, and I'm surprised to see Eastern is in there. They had disappeared maybe fifteen or twenty years ago in our world. Saved by the rocket business? I wondered. Har'el dialed their number and yes, they fly Dulles to Rutland connecting through Boston Logan.

"I'd like to make a reservation for that flight...one person" I nudged Har'el and held up two fingers."Sorry, make that two people...return?" Jennifer and I shrugged. "Could we leave that open for now?" Har'el hung up the phone. "OK, you're on at 9:35 tomorrow arriving Rutland at 11:50."

"Wow, that was easy" Jennifer's expression was more relieved than excited. We were quiet for a while and I sipped my drink. Here we sat in a world where people we knew were dead were in fact alive, Eastern airlines still existed and people shipped transcontinental freight almost instantly on daily rocket flights, but Diet 7-Up tasted exactly the same.

There was plenty to talk about, like how we planned to pay for these tickets, what Jennifer would say to her parents to get us picked up at the airport and when we were coming back. Not to mention whether we should take this whole story to the police or to a scientific organization. Topping my list was how to get back. Here we were at best a curiosity and at worst a nuisance. At home we were causing a lot of grief. I worried about Nancy especially. She was always warning I'd kill myself on one of my crazy adventures. She would be sure the inevitable had happened. But we were tired and it was so much easier to talk about logistics. What time we would have to get going and whether we should buy a round trip ticket with an open return. Chitchat was relaxing or at least boring and we pretty soon all headed back to bed.

I found Rick outside stretching at exactly 6:30. The sun was filtering through the trees and the air was fresh and still cool. I felt better, less worried and more curious about our whole adventure. Yes, people at home were going to be very upset at our disappearance. But we would come home. My strategy was that if we were back within three or four days they would still be searching for us and we could be "found" out in the woods. But it wasn't so obvious how to get back.

As we started in a slow jog, I asked Rick "Summer of 1971, were you fixing bicycles?"

"Yeh, in the garage in the back with Neil and Rob."

"Were you captain of the swim team that year?"

"Sort of."

"Why only 'sort of?'"

"Oh, I was in an accident on my bicycle."

"The white Mercier?"

"That's the one. Got squeezed by a car passing too close on my left and the front wheel dropped into a drainage sewer grating."

"I remember in Cleveland they had those iron grids with slots aligned with traffic."

"Yeh. Nice and smooth for cars but just wide enough to swallow a tubular tire and rim. I flipped over the handlebars and hit the pavement pretty hard. Hit my head but luckily took most of the force on my arm, which I broke. I sat out half the season."

"When was that?"

"I remember the date. September 9, 1971. Two days after my birthday. A Thursday. It was the 1st day of school."

"You sure do remember."

"Well, it was partly the legal stuff that I remember it from. We sued the State of Ohio and Cuyahoga County for these unsafe gratings. I must have been asked the date of my accident a thousand times."

"Win anything?"

"The legal fees absorbed most of the money. But I am the reason Ohio doesn't have those gratings anymore. It was a moral victory."

"Hey. Thursday isn't the first day of school. I remember Tuesday was always the first day, the day after labor day."

"Usually, yeh. But remember? That was the year of the meteor hit."

Chapter 9

"A meteor hit somebody?" I had no clue what he was talking about.

"Not somebody. Sometime overnight that Monday night. Remember, mom made a big picnic for our birthday one day early 'cause of school starting, and then you went out to Chagrin Falls with Sara?"

"Not really."

"You, I mean I or we or whatever, really liked going out there. The falls were cool on a hot night, and we liked the popcorn balls they sold in the country store by the bridge."

"Yeah. I do remember. We took my '65 Ford. It was a nice drive and we were both depressed about going back to school—a last little escape before a tough year. It was a warm sticky night. Like summer still."

"Right. So do you remember that late that night there was a civil defense alarm and the whole country went on a nuclear war alert?"

"No! And I think I would have remembered that."

"You sure ought to."

By now we were warmed up and Rick was starting to run a bit faster. He took a break to breathe and I looked around at a Reston I'd never seen before. Lake Newport surrounded by newer homes I didn't recognize at all, even though I lived in one of them. My street didn't even exist. Instead there was a new townhouse development with inlets and docks and small sail and pontoon boats. But as we come around Lake Anne, the first of the town's lakes to get built up, things looked almost identical to the Reston I knew.

"You've got to remember this," Rick continues. "You dropped Sara off about 12:30."

"Sounds about right."

"And you're driving home listening to WMMS on that FM converter gadget in the Ford. Then the radio civil defense beep sound comes on. You figure it's just a test and tune around the dial. All the stations are running that same tone. As you pass the high school the civil defense horn there is blaring too."

"I definitely have no idea what you are talking about. Is there a civil defense horn at Heights?" [Cleveland Heights High School—ed.]

Rick is looking a bit impatient.

"Look, I remember the night out with Sara on the last night before school. But if there had been an air raid horn blasting on my way home, I think I would remember and I don't."

"Well do you remember school was canceled on Tuesday?"

"Nope."

"Huh. Well, it was. When I got home, Mom, Dad, and Carol were up watching TV. They said an enormous nuclear blast, the largest ever, had occurred in the very early morning in the Pacific. Kwajalean Island was the apparent target since it is a highly sensitive U.S. base for testing ICBMs. There was almost no real information beyond that."

Rick was breathing hard. Pretty difficult to run fast and tell a long story. "Let's walk" he sort of coughed, out of breath. "Lack of information never stopped commentators from exploiting a juicy story. Everyone agreed it was a nuclear attack on a U.S. possession by China or the U.S.S.R. They said the enemy went for Kwajalean because of its military significance and its isolation. There was a limited loss of lives and no immediate fallout danger to any major population centers. Seismologists put the impact point a couple hundred miles north of Kwajalean, but the seismic shock was estimated to be a thousand to ten thousand times larger than the U.S.' most powerful hydrogen bombs.

"With the destructive demonstration of such a super weapon, the TV news people anticipated China or Russia to claim responsibility and demand the U.S.' immediate surrender. They called it 'nuclear Pearl Harbor.' Then Nixon got on the tube and said the U.S. would take no rash action, but that our military was on full red alert awaiting further intelligence. He ominously reminded the world of our undiminished ability and resolve to respond at whatever force level might be necessary right up to full scale nuclear war.

"That triggered a national panic unprecedented in history. We were facing annihilation. Grocery stores were looted and emptied. Riots broke out. The phone system ceased functioning from massive overload. Most normal law-abiding people spent that terrifying night boarding up windows, storing water in bottles and bathtubs and praying. Many people took advice from the TV and piled dirt up around their houses to block radiation. And a lot of people jumped into their cars and tried to head into the country figuring cities would be the primary targets of the incipient attack. The interstates became gigantic parking lots, jammed with cars and campers.

"We were up all night digging, putting up water, cooking and freezing food. Then about 6:00, 7:00 a.m. on Tuesday, the tenor of the news started to change. Almost twelve hours had passed since the explosion and there was no political ultimatum. In fact, Russia and China, at an emergency overnight session of the UN security council, said they had no idea who had launched the attack. The DoD [U.S. Department of Defense—*ed.*] admitted they could not detect any abnormal radiation in the cloud which had lofted over the explosion area. They also confirmed no missile launch had been observed. Still, the weapon could just have been carried on a small ship, and there was always plenty of shipping traffic in that part of the ocean.

"Later that morning Carl Sagan got on TV and I think was the first to publicly advocate an alternative, that the earth had been struck by a meteor. A small asteroid, less than fifty feet diameter, could cause a shock that big if it were to collide with the earth. The Nixon hawks scoffed, but Sagan made some good arguments and the science community started to find reasoning to support him. The dawn hours of autumn are a time of maximum meteor shower activity. The lack of radiation. A Professor Shoemaker at Arizona showed on TV about ten other sites on earth where over the past few million years similar impacts probably occurred. He said this was just a natural event like a big volcanic eruption or a major hurricane.

"By noon, astronomers from the Navy, the Naval Research Lab, and Naval Observatory, where they keep the world's most accurate clocks and star maps, came up with the proof of the asteroid theory. The earth's rate of rotation about its axis had slowed slightly. Less than 1% of one part per billion but still a perceptible change, coincident with the time of the event. A change in rotation rate has to come from an externally applied force. It can't result from an explosion because you can't create or destroy angular momentum. The barely measurable change in the earth's spin rate matched the impulse from a rock about sixty feet in diameter hitting earth at about ten kilometers (6.2 miles) per second. This size and relative velocity is typical for hundreds of asteroids with orbits crossing earth's. Shoemaker estimated that collisions of this magnitude were likely to happen about once every twenty thousand years. In geological terms, not a big deal at all."

As we walked across the wooden bridge and past the fountain in Lake Anne Rick smiled, reflecting on the whole episode. "Our government at work" he said, somewhat cynically. "Nixon seemed to almost welcome the crisis. He'd be reelected for sure, I guess he figured. By that Tuesday afternoon, the major panic was over. But the problems it caused sure weren't. Literally millions of people were stranded out on the roads without gasoline to get home, food or water. Thousands died out there. There were several hundred suicides, including a few families who decided to die together rather than face Armageddon. Hundreds more died in robberies as people fought for supplies they thought they would need to survive after the coming attack. Property damage was in the tens of billions of dollars. And the psychological damage, especially to little kids, was just impossible to estimate.

"All that stuff we did to ourselves. The meteor impact itself did even more damage. Several islands within a few hundred miles of the contact point, Kwajalean included, simply disappeared. The blast wave killed millions of fish which washed up for weeks on beaches all around the Pacific Rim. But the big killer was the tidal wave, which reached Australia, Indonesia Southeast Asia, Japan, Alaska, the U.S. west coast, and western South America during that Tuesday.

"Apparently, a tsunami is a hard thing to forecast. It depends on how the wave is focused as it propagates into shallow water toward shore. California got almost nothing more than a foot or two higher tide than normal. Western Canada and Alaska had much higher waves, and Anchorage was virtually ruined. Most of the population had been evacuated but still thousands were lost.

"The worst U.S. damage was in Hawaii. There wasn't time for an effective evacuation and I think two hundred thousand people died there. And about twice that many died on the west coast of South America, mostly around Santiago, Chile. But the biggest loss of life by far was in the ASEAN countries, especially Indonesia. They have almost two hundred million people living mainly in seaside villages on the Indonesian archipelago. There was no electronic communication system connecting most of these people to the outside world and the tsunami hit was a complete surprise. That actually may have been a blessing. Wave heights were estimated in open harbors over a hundred and fifty feet. Many of the eastern islands were completely submerged. Some smaller cays were never seen again. No one really knows how many were lost, but sixty million was the estimate. And that was just Indonesia. Though they were the hardest hit, there was damage and loss of life all over that part of the world. It was just unimaginable. But for the first day or two, the U.S. was fixated on our own problems, which were, by comparison to the world wide loss of over a hundred and fifty million people, trivial. By Wednesday afternoon there wasn't much left to do other than call relatives and see if they were all right, and swap stories with friends.

"I guess the politicians figured that resuming normal life would help take people's minds off the whole debacle and they sent us back to school on Thursday. It was on the way home that afternoon that I fell off the bike. Sort of the least important event of the previous few days!"

"I have to tell you," I interrupted, "that none of this happened to me or my world. I vaguely remember the picnic and a date with Sara the night before school started. No doubt I probably drove home at 12:30 that night listening to WMMS. But there was no civil defense horn, no mass panic. No huge losses of lives or natural disaster. And I did not break my arm, either."

By now we were standing in Tom's back yard. "Hey, Jennifer and I have a reservation to fly to Vermont at 9:30. I better hustle. What time are you heading off to Oregon?"

"My flight's at ten something. I'll take you up there."

By 8:15 or so we were all showered and dressed and standing around the kitchen drinking orange juice and cutting off pieces of bagels.

Jennifer asked, "Har'el, you don't want to join us in Rutland?"

"Nah, I think I'll relax and explore the world down here in D.C. a little bit. I'll ride with you out to the airport and maybe catch a bus into town from there."

"Rick," I said, "you know we have no money or ID."

"Yeah, Tom and I talked about that before the run this morning. Here's a credit card, $50, and my driver's license. That's my address," he glanced at the license. "Mail it to me when you—" He pauses. When I what? We both wonder. This isn't your basic cross country trip on the airlines visiting relatives. We had been somehow displaced into an alternative earth that branched off from our own, probably when that big meteor hit.

"Thanks. You'll get them back." I said in a low voice.

"Don't worry about it."

My mind wandered. I was pretty sure this had never happened before. People invading parallel worlds. The scientific importance of what we did could be enormous. Har'el was right. Should we go talk to somebody? Who would believe us? We could just be very enterprising twins with a few accomplices. "Should we get going?" I asked. "We're liable to miss our flight."

Chapter 10

"I'm ready to roll." Rick pulled rental car keys out of the front pocket of his jeans. We said goodbye to Joanne and the kids and headed out to the car.

"Rick, last night the girls were sending messages to friends in Indonesia. Was that...?"

"Related to the tidal wave? Indirectly. An international relief effort was focused on helping the countries hardest hit by the wave."

"What wave?" Har'el and Jennifer hadn't heard the meteor story yet.

"I'll tell you later. Let me find out what I can from Rick while I've got him."

"It was all very political. Politicians grubbing for votes tripping over each other to be helpful to the victims. Not just here. All over the so-called First World. They let big reconstruction contracts to large engineering firms at home to go to a target country and build roads or bridges. Neither the governments nor these companies knew or cared what the locals wanted or needed. Just show your constituency how much you care, denominated in tax dollars, then buy big companies' support in the next elections with fat, loosely managed contracts."

"The good things never change," I said trying to match Rick's cynicism.

"Oh, and all the world's aerospace companies were ready to 'help' with their own well-padded solutions. What the surviving half of Indonesia's populace needed was direct broadcast television to carry lessons into the class room, of course. And the fact that our networks and advertisers could capture another hundred million viewers. That's just good old American capitalism at work. So, Uncle Sam handed out $1.5 billion for two geosynch satellites, ground terminals, and launches so that the 0.1% of Indonesia that can afford a TV now has something to watch. Meanwhile 99% of their kids still don't have even one book, one pencil, or a piece of paper to write on. But they have a billion dollars in space hardware forty thousand kilometers overhead.

"The U.S. amateur radio community lead by the ARRL [American Radio Relay League, the U.S.' oldest and largest amateur radio organization—*ed.*] had a lot of experience in emergency message service and in satellites. They proposed to set aside a small portion of their spectrum allocation expressly for communication in affected regions. Anyone in those areas with or without a ham license can use those set aside frequencies. With donated time and materials they built, via amateur groups in the U.S., Argentina, Germany, Mexico, and Japan, five small messaging satellites. The French launched them into LEO on donated space and the International Red Cross started a fund to purchase and distribute tiny ground terminals. Back then these were old teletype terminals and Heath Kit style radios.

"Even though the terminals needed electricity and you had to be able to read and write in Roman letters to use them, they were really popular. Over a hundred thousand terminals got set up for the service. First they were mainly in institutions like hospitals, schools and village centers. Most of the messages were about needed supplies and requests for instruction on making things like primitive stoves and water pumps work. When little computers like the T-80 came out, battery/solar powered satellite terminals got popular and individuals and families could get them. The user population soon got into the millions and an expanded spectrum was allocated. The hams built more satellites and the system grew up. Nobody knows how many users are on the system. It's estimated from the noise floor measured in the spectrum space that world wide it's at about fifty million and growing. You've never heard of the Astronet?"

"Me? No." Har'el and Jennifer nodded in agreement.

"It's a big thing now. Institutional users have big setups with bandwidth up to megabits. For hospital work mostly, sending X-rays and what they call Remote Diagnosis Data."

I remembered Tom's patient with the broken leg—in Indonesia.

"But for most of the fifty million, Astronet is like the old short wave radio broadcasting services, only digital and it's two-way. Villages get forecasts and severe weather warnings. And not general consumption broadcasts. These are specifically directed messages. Broadcasting is banned just like the rest of the satellite services except for the radio and TV DBSs."

"So that was how the kids sent up their e-mail last night?"

"Yeah. A lot of peripheral uses of the Net have gradually gotten approval as the memory of the '71 tragedy has faded. Twenty-seven years already! International education was a natural. The original mission of the early satellites which eventually became the Net included bringing basic educational stuff to people without access to traditional information services like radio, to telephones, and to geosynch satellites. Most of this was kind of adult stuff at first. How to irrigate a particular area or how to care for a family member or even, if the one cow a family depended on for milk showed certain symptoms, what to do until help could be sent. But pretty soon a form of classroom-like education got going. Volunteer teachers gave instruction to groups of kids in villages and the kids would come back across the Net with questions. Kind of a non-real-time dialogue would get going. Many of these teacher-student dialogues continued for years."

"Teletyping with kids who couldn't read?"

"For the younger ones and those who couldn't read the digitalker was the answer. Just digital voice compressed to six or eight hundred baud. No big deal. Anyway, the system got so famous that kids with access to normal classroom learning wanted to try it. Older students in the earlier client countries became teachers to American students and eventually to kids all around the developed world. One thing every kid learns now is just a little bit about satellites. The net has designated one older satellite just for pen pal stuff. That was the one they logged onto last night."

"What airline are you going to? Rick looked at his watch. "It's gonna be close. I'll drop you off, then get rid of my rental car."

"Eastern," said Jennifer.

"I'm not going anywhere," added Har'el. "I'll drop you all off then take back the car."

The morning was still and warm, and the heavy smell of jet fuel seeped into the car as Rick pulled to the curb at United's door.

"Thanks Har'el. Why not use the car today and drop it off tonight? It's already paid for anyway. Compliments of GOES."

"Who?" I wondered if this world, so sophisticated in using satellites, had as lame a system as our dowdy GOES satellites. And what did they have to do with Rick? Probably another long story.

"Good Old Eastern Spacelines."

"That would be very cool. Thanks a lot, Rick, I'll try to keep it in one piece for the rest of the day."

"Good idea! You're not insured, you don't have a license or any ID and I'm not going to be able to help you from Oregon."

"Exactly my kind of challenge," grinned Har'el.

My discomfort was a bit more than vague at that remark. Har'el had gotten us into all this armed with just his two feet and a length of rope. Who knew what havoc he could create let loose on an alternative earth with a car and a full tank of gas?

I fished the $50 out of my pocket from the cash Rick and Tom had loaned me. Against my better judgment, I handed Har'el a twenty. "You'll need this for gas, or tolls, or food, or something." With his patented wink, Har'el's sparkling blue eyes and broad smile said thanks. A minute later as we stood on the curb he zoomed off with only a hint of rubber deposited in his wake.

"You guys better run," advised Rick. We were more than the same age. We were the same person. Even so, his beefier build and the fact that this was his home earth, not ours, caused him to treat me as a kid brother.

Jennifer extended her hand to him. "Rick, I really appreciate all your help, the airline ticket, money. I'll find some way to pay you back."

"Totally unnecessary. Good luck with your Mom and Dad."

"See you, Rick." I shook his hand and looked myself right in the eye.

He smiles. "Yeah. Every morning in the mirror."

Chapter 11

Sunday afternoon the twins and Joanne were booked at their cousins' and brother's house in D.C. The rest of us were in Tom's electric minivan heading West. It was my first ride in an electric and it was remarkable only in what wasn't on board. No gear shift, not even PRNDSL. Just an On/Off switch and a Fwd/Reverse knob. Full time 4wd was standard since each wheel had its own motor. You didn't start it, and at a light I couldn't shake the feeling that the engine had stalled. And that feeling didn't really dissipate when pulling away either. No sound except a faint whir mistakable for an air conditioning fan. No shifting. Finally about sixty mph some wind noise came on which reduced the odd feeling of non-motion.

In the nearly silent minivan, I recalled the events of yesterday. Contrary to our expectations, both our adventures were almost non-events. Jennifer and I had spent the afternoon and night with her parents and the local Jennifer. So little had changed in the few years since her dad's accident, her mom seemed to forget we were aliens. No one else was forgetting, not being used to having two versions of one person over for dinner.

Jennifer came to Rutland filled with emotion to see her poor father who had died cruelly and senselessly in that early January snow storm and to clear up the unpleasant aftertaste of their last abrupt goodbye when she was whisked away in her hawkish boyfriend's Camaro. But to her dad and mom, no big event had ever happened. The accident when he broke his ankle was a distant and forgotten episode having left no

trace except for a slight unevenness in his walk. No more memorable than a child's strep infection cured by penicillin. His rescue was itself a non-event.

The lack of tragedy in her parents' lives was a void filled by their Jennifer. Unlike herself, this world's Jennifer lived with her two children and her husband right down the street. Both parents telecommuted, he to a company in Boston and she to San Diego. The kids ended each day with two hours at their grandparents' and I got the feeling that what Jennifer's mom and dad wanted most was a vacation from their new career as baby sitters. Like mine, Jennifer's local equivalent mistrusted and resented her. Maybe out of fear that she would advance a claim on her children or her home.

I shared with Jennifer a vague feeling of guilt, like after a long period of neglect I had come to call on once dear friends. But in the many years' lapse of contact it turned out those old friends had found new ones and they themselves had evolved into new people with whom I shared no other intimacy than that of a faded and now irrelevant childhood memory.

Early Saturday, Jennifer and I sat on blue and orange vinyl chairs bolted into a row of twenty identical uncomfortable copies under the fluorescent drop ceiling of Rutland's air terminal, watching the first snow of the season out the grey-coated windows.

"They're both alive and well. That's the main thing," I suggested.

"But they're not my parents."

"Not since 1971. Biology is one thing, but a quarter century apart makes a bigger difference. It's as if you were raised by adopted parents and then as an adult found your biological ones. Physical resemblance and a few personality traits in common don't comprise a family."

"But you related to Tom."

"Yeah, but Rick and Tom never see each other. Tom and Joanne never even knew that whole story of the rocket business Rick and Gary started. And I wasn't invading Rick's home turf either. Given their little bit of contact I was almost as close to being that Tom's brother as the other Rick is."

Jennifer and I had succeeded in locating people planted just in the same place as our own families and with the same genetic composition. But we had played no role in training those neural nets and we were just oddly similar aliens.

"Turn here." Har'el was using the dashboard color LCD, directing Tom down the same country roads Harry Lee had taken two days ago. He had spent Friday at old man Deutheimer's farm. Geographical splintering was a primary characteristic of his family, one consequence of which was that Har'el had attended Gymnasium in Göttingen, Germany for two years. He lived there with his father's brother, who was an aerodynamicist working in the laboratory started by Prandtl and then Schlichting and now part of the German air and space research institute. Deutheimer was from the Harz mountains nearby Göttingen and mentally had never made the transition to America. The two passed the afternoon over Weissbier (a Berliner spezialität), Brotchen (bread rolls) and Ziegekäse (goat cheese) talking in German about the Heimatland und die guten alten Tagen—homeland and the good old days.

Deutheimer's goats were his pride, and after a sufficiency of Weissbier had been consumed he personally had introduced Har'el to each of them. One smallish and brownish goat bumped Har'el hard enough to push him over, then gave him a familiar wink and bucked her head up and down.

"Ruhe, Fraulein!" Deutheimer quieted the unruly little animal. "Ach, Katrina. She is always mit the devil in her. Just like my daughter. They always so happily together were playing."

"You have a daughter?"

"Had, I am zo zorry zu zay. Little Katrina. Ve named zeh goat after her. After she disappeared." The farmer's mildly inebriated eyes sparkled with a trace of tears.

Har'el remembered Gib and Hilary telling the story of a girl's disappearance. "So that was your little girl? I'm very sorry." Deutheimer nodded. "Do they think she got lost in a cave?"

"Maybe. Ve just don't know."

Frau Deutheimer called from the house. "Ernst! Komm bitte! Lunch is ready."

Back in the farm house Har'el shared a dinner of potato, kraut, and cucumber salads, Bockwurst, heavy dark bread, and light amber beer with the middle aged German couple. He arranged for Tom's family to come for a visit Saturday afternoon, saying the kids, meaning Tom's kids, would love to play with the goats and see a real farm.

Bouncing down a few miles of unpaved road, Har'el turned to me in the back seat and opened his backpack. The water bottle, fruit and flashlight were from Tom, but he also had some tape cassettes, a magazine and a floppy disc. The cassettes were thicker and more square than our standard and the discs were almost square and flexible.

"Josh and Joel gave me these." The boys smiled proudly. Souvenirs from another planet. "Nobody will believe us when we get back, but they'll have to believe this stuff."

"Maybe." I was reminded of people with supposedly alien materials from UFOs into which they had been abducted. Nobody believed them either.

"Look, I have pictures, too. Of Tom's family, of Town Center and a few street shots showing some of their electric cars. The tapes and discs I figure are written with a totally different format. How could we fake all that stuff?"

"You're right, but we'll still be labeled nut-cases," piped in Jennifer.

"OK. The driveway is a dirt path just over this hill on your left." Tom eased the van down the narrow rocky path and pulled up next to the farmer's old VW pickup truck. The dry weather Gib had predicted was definitely holding. The van was as coated with fine dust as were we.

Deutheimer came running out of the house to greet us. He squinted in the bright sun as he pulled a big Bauer style hat on to shade his face and unusually large ears.

"I vas just hearing z' o-pe-ra. La la la La la la La la la LA… dipe... Dah! Hochzeit des Figaros. Mozart. Nicht zu schlim für ein Östereicher, you know?"

Tom smiled and put his hand out to Deutheimer. "Tom Fleeter."

"A pleasure. I'm Ernst. People around here call me Ernie."

Tom introduced the boys to Mr. Deutheimer and Har'el introduced Jennifer and me.

Tom asked "Mind if I plug in? It's a long drive and it's getting low."

"Sure, sure! Right here."

Tom unreeled a cord a few feet from the nose of the van and plugged into a weatherproof outlet mounted on a wooden post.

"Komm! You must all after zo long a ride nur ein bischen hungrisch sein, Ja! Frau Deutheimer has made her spezialität—Apfelkuchen."

After we sampled the apple cake made from their farm's own apples and butter and made a little polite conversation with Grettle Deutheimer, we headed outside to tour the farm and meet the goats.

"Und vas ist mit der Rucksack Herr Har'el?"

Har'el explains that we're going to do a little day hike out in the hills.

"Wish I could join you, aber mein eins-und-zwanzig cows auch vas zu essen wollen."

His cows need tending, Har'el translated.

The stunted little goat named Katrina found Har'el and began to gnaw on his rope almost as soon as we got out of the yard. She was already pulling us along as we said goodbye to Tom, Josh, and Joel.

"If we don't come back," I started to say.

Tom finishes my sentence "I'll know you made it home."

"Right," I smile.

Tom had to go watch the kids who are wondering off with the other goats and Katrina pulled Har'el toward the hills.

"At least you have beautiful weather. You won't have any trouble." Tom said.

Remembering the fright we'd all gotten the last time, I only wished that I shared his confidence.

Tom looked at me and said, with his patented matter-of-factness "So? Get goin'!"

And with a quick exchange of smiles I turned to follow Har'el and Jennifer. It was a warm and sunny afternoon and our path—the goat's path—was across open unshaded hillsides. Katrina led us across dry creek beds, over small rock outcroppings and over some pretty steep grades. After almost two hours of steady exertion we came to a cool pond surrounded by apple trees. The goat dropped the rope that she had held in her teeth even though Har'el had tied it in a neat collar around her neck, and she started tugging at the lower branches of a tree.

"I think she wants an apple," intuited Har'el.

While he climbed up a few feet to reach some of the fruit, green with just a trace of red beginning to surface on the skin exposed to the sun, I decided a little cross species bonding would be a good idea. After all, without this goat we were marooned in this alternative society. And attractive as it was, cleaner, safer, and less political than our own, we were quite literally redundant here. Plus, even if Katrina didn't lead us home, it would be nice if she didn't choose to abandon us out here in a hilly, uninhabited and desolate corner of rural western Virginia.

With completely selfish motivations in mind, I petted the goat's rough neck and risk my fingers to feed her half a granola bar. I noticed I could barely squeeze my fingers under the rope collar Har'el had very solidly knotted around her neck. It was my turn to intuit.

Behind that knowing smile and playful personality was a loyal young man solidly intent on our return home. Obvious? Har'el had been the adventurer who started us on this voyage. He thoroughly enjoyed all its twists and trials. And he had no apparent double here. Only twenty-two years old, possibly, no, undoubtedly, he had not even been born in this version of earth's unfolding history. His parents may or may not have even conceived a child when he was conceived, but even if they did, the millions of tumblers in the genetic lock could not possibly have all coalesced to the same combination that created him. Har'el's motivation, given he could live here as well, maybe better, than home? I watched him deftly maneuvering among the tree branches stuffing apples into his opened shirt.

The motivation was us, Jennifer and me. He had led us here, and it was probably pretty clear after our midnight conversation back in Tom's kitchen that we were feeling alienated and worried about our families back home. He had taken the responsibility to see us safely return, and his skill, strength and insight were focused since then on carrying us back.

Accompanied by a flurry of leaves and small branches Har'el descended out of the tree and poured about two dozen apples onto the ground. Jennifer brought back three water bottles filled from the spring that fed the pond, and we sat feeding ourselves and the goat and relaxing in the shade of our tree. Har'el reached into his pack which he had left at the base of the tree and pulled out about an inch thick stack of brochures.

"Before I went to see Deutheimer yesterday I stopped by Harry Lee's and BS'd with him about rockets. He knew all about Eastern Spacelines but he had never heard of Rick Fleeter or Gary. In his garage he had all these files stacked in cardboard boxes. Stuff from his office he cleaned out when he retired. Amazing quantity of paper. License applications for satellite and launch systems, all kinds of federal regs and drafts of international agreements. Real concentrated boredom. But in one file he had some pretty sharp marketing jazz he collected from all the commercial launch operators."

"Look at this stuff!" I was already flipping the pages while Har'el narrated.

* Orbit-1. $1 million puts 1 ton in LEO.

* SpaceExpress: Daily departures to all major inclinations.

"What, these guys launch every day?"

"Not quite. They are kind of freight forwarders—agents. They book space on all the available carriers and retail it."

"These operations are for real?"

"Look inside the Orbit-1 brochure." I unfolded it to find a yellowed 1993 calendar. Almost every weekday showed one or sometimes two launches. Most were for two hundred pounds, others were fifty and some went up to twenty-five hundred pounds payload. Each month had a photo of a rocket with a performance curve—altitude and inclination vs. mass to orbit. Or a shot of one lifting off a pad near the ocean, somewhere tropical. Florida maybe? The back had all the pricing, insurance and reliability numbers. $250k for two hundred pounds to orbit. Insured. $750k for an insured thousand-pound ride. And $1M for a ton. With a thousand launches over the past seven years they had compiled a pretty consistent record—above 99% reliability.

I wish I could have asked Rick about these guys, I thought.

The rest of the brochures were for a few smaller launch services and space service providers. Mostly the usual communications and remote sensing stuff, plus some vaguely described agricultural services. But one brochure was just agricultural services including freeze protection and overnight illumination for twenty-four-hour harvest operations. They also offered a law enforcement and disaster management package, selective lighting, laser tagging and tracking.

"Pretty far-out stuff."

"Yeah, but it looks pretty real."

"You think it's worth $179 an hour to light up a hundred acres?" Jennifer was studying the price list.

"Probably beats losing the crop to a storm or deep freeze. Whatever, these brochures ought to catch some attention back home."

"Yeah, people will either think we're funny or crazy, but very good with a Macintosh."

"We'll see" said Har'el. "I think we can open some eyes."

Our guide, Fraulein Katrina the she-goat, having downed a few apples including stems, cores and a few leaves, was slurping water from the pond. The scene, with these fruited trees, the clear brook feeding a tranquil pond, the little goat, all set under the blue hemisphere of the sky and illuminated by a yellow midafternoon sun, reminded me of a biblical promised land. Not an Eden of naive perfection but a reality humans can achieve with their efforts if and when they choose to. The world we were now on course to reenter so far had made other choices. Pursuit of grandiose and costly human exploration of space. A fixation on combustion of fossil fuels. Daily destruction of hundreds of tons of trees for newspapers which become the next morning's rubbish. Stubborn unwillingness to perceive other countries as fellow passengers aboard space ship earth. Closing our eyes to megatons of soil, rock and chemicals lofted into our atmosphere. Disregard for millions of kids growing up without books or teachers. We have all the technologies possessed by the inhabitants of this promised land. We just haven't yet made the choices.

Typical of me, I spent my few days on this alien earth homesick for my own and now, standing at the threshold of our return, already I was homesick for this place.

Looking up from my empty gaze into the pond, I saw Har'el and Jennifer lined up behind the goat, rope in hand. Joining the parade I saw where we were heading. At the opposite end of the pond from its source the water drained down a three-foot wide bed of smooth pebbles which gradually slips underground. Without even a look back at us the goat slid down the rocky chute and disappeared.

Har'el held his end of the rope taut, while Jennifer and I used it to brake our slide down. We called to Har'el from the pitch black hole we were then standing in, soaking wet, and I felt the rope go slack. He abandoned us! With the pack with food, lights and fresh batteries. Just then his body, feet first, slammed into mine.

"Sorry about that. No brakes."

I imagined, now sprawled on the wet ground, Har'el's reassuring wink and smile. He somehow fished the flashlight out of the sack and we got a look around. Here was another narrow, low-ceilinged, wet cave with a maze of passages. Har'el petted Katrina and checked the collar, and we started on our way, the goat first, then Jennifer, Har'el, and I were at the rear with the backpack. As we walked Har'el shined the lamp forward, illuminating our way while casting our distorted shadows on the walls and floor ahead of us.

This time we expected a long voyage, but after only about an hour of steep up and down hill hiking, then maybe thirty minutes of very steep climbing in a narrow and twisting tunnel, we noticed some natural light seeping into the cave. It was not bright sun, but a grey foggy glow. We had been walking in an expectant silence, but now I heard something. Not an ominous rumble or a rush of water, more like wind in trees. Rain. Or more like a steady drizzle. I noticed water was running down the walls and floor of the cave and rushing over our shoes in a miniature torrent as we walked single file behind Katrina, steeply uphill toward the dim grey light.

The air filtering towards us from the illuminated slot was even wetter than what we'd gotten used to inside the damp cave. It carried tiny water droplets which collected little by little on our skin. As we climbed they combined with our sweat and soon we were well soaked, hair matted down and clothes stuck to our bodies. Maybe the others understood but I didn't, and it took until we reached the slot for me to realize. It was our exit point from the cave. Standing near it we heard the steady fall of drizzle and rain and felt the cool damp air of the early evening shower. Weak remaining daylight filtered by clouds, fog and rain drops left the light from the slot even feebler than our little flash lamp. The natural light's grey monochrome made our battery light seem yellow and our skin tones sallow.

Katrina stopped a few steps short of the slot and just stared blankly at it. She looked every bit like a brainless farm animal and I wonder if she really led us to a new world and back? Or was she just a sort of a living Ouija board allowing us to go where intuition leads? Did I read intelligence and recognition into her huge brown eyes?

My thoughts were interrupted by a brilliant flash and an immediate loud thunder clap which echoed deep into the cave and shook its walls. I peered out the slot and saw sheets of rain falling through a heavy fog. Cool water pouring in through the cave's narrow orifice soaked my shoes and socks and the bottom few inches of my khakis.

"Maybe we should wait it out in here?"

Jennifer: "Definitely."

Har'el: Nods

Katrina: Paws nervously at the ground.

Har'el: "Animals don't take well to thunderstorms."

Jennifer: "Me either."

Her gaze up at the ceiling made us all look up. Chunks of mud and small rocks were being loosened by water percolating through the passage's roof just inches above our stooped heads.

"I'd feel safer..." BOOM! I was silenced by another thunder crash amplified by the long narrow cavity, one end of which we stood in, and which led semi-infinitely down into the blackness we had spent the last two hours climbing out of.

The cave rattled and heaved, resonating with the storm's acoustic energy. We were treated to a painful shower of pebbles and wet mud.

"Like I said…" I had to yell to be heard over the white noise of the rain echoing in our rocky tubular cavern. "I'd feel safer outside in the rain. This leaky cave with water pouring in from every place is making me claustrophobic."

"Maybe we should retreat back into the cave," Har'el suggested, looking at Katrina who had already stepped several feet farther from the grey lit slot.

The roar of the storm and the wet, dirty cold reminded me of the cockpit of the '30s vintage biprops we used to fly freight in around New England. They were loud, the heaters never worked, and at a hundred and sixty-five knots you could get thoroughly soaked traversing a decent storm. I remembered my instructor, Al, no nerves and always flew by the book. Came in to Islip one February night with at least an inch of ice covering the airframe. Came down the glideslope like he was on rails. Reached the inner marker, just two-hundred feet AGL, [Above Ground Level—ed.] but didn't see the runway through the swirl of big wet snowflakes. Oh Al! It had to be right there in front of you. Just go for it. But the book says no visual contact at the marker means go around. Al rammed the throttles of the old Beech 18 to the firewall, put the wheels up and started his go-around. The nine-cylinder Pratt and Whitney radials roared back to life spitting fire out their fourteen-inch diameter straight stack exhausts. I heard them throbbing over his calm clear voice on the Air Traffic Control Center's tape. "No visual contact. Going around." But even eight-hundred-fifty horsepower flat out can't lift an airplane with wings spoiled by a thousand pounds of gritty rime ice. Al tried a gentle right 360 to recover the marker, while the plane mushed along at a hundred feet and eighty-five knots, straining at 100% throttle. Until the left wing clipped a construction crane and the classic twin rudder tail wheel Beech-18 nosed into the snow covered foundation of what would eventually become Islip's new Airport Marriott.

I had bailed earlier that evening when Windsor Locks, Connecticut closed for the night due to fog. Woke up in a Motel 6 the next morning, walked across the street to the airport weather station and the meteorologist on duty handed me the Providence Journal.

"See this?"

The front page was dominated by a picture taken from a rescue helicopter lighting up the crash site after the snow flurry cleared out. "Providence Man Dead In Fiery Crash" screamed the headline.

As a few large rocks crumbled from the sagging ceiling of the cave, I came out strongly against a retreat. "We've come up a narrow chute on a steep incline. One boulder or mud slide and we're toast down there. We're inches from daylight. Let's go for it while we can."

"But do you think we can even squeeze through?" Jennifer motioned toward the narrow opening.

The hiss of the rain and the rush of water down the steep, narrow corridor completely masked the sound of an irregularly shaped boulder about the size of a soccer ball falling from the roof of the cave. It splashed into the clay with enough force to splatter us with red mud. Bouncing from a rock protruding out of the floor, the boulder limped a few uneven revolutions downhill towards Katrina. The goat deftly pranced in place to quickstep over it. Behind her the path steepened and we heard over the noise of storm and the water the accelerating projectile careening off the walls.

We were now all facing into the blackness of the steep chute, more fearful than awed by the power a dumb mass has when gravity is allowed to accelerate it. "Like I was saying, we're inches from home. No way do I want to crawl down that incline and risk a boulder or a landslide."

Har'el nodded: "Let's do it. Jennifer, you're the smallest, you'll go first with the rope on one of your feet. Rick and I will push you out. Then I'll tie the rope to Rick's wrist, and you'll pull him while I push from behind. When you guys are out, you can pull me and the pack out."

Jennifer put her two arms through the hole and assumed the pose of a swimmer having just entered the water from a racing dive—hands together above her head and stretched, making her body as long and as narrow as possible. I helped wriggle her through the hole as Har'el clawed impeding rocks and mud away with his fingers. After a minute's struggling, Jennifer was through. Her face looked huge and I felt like a cartoon mouse living inside a hole in the wall as she stooped over to report "I'm fine. It's pouring rain out here. Let's hurry this up and go."

No sooner said than a slide of mud and rocks partially closed the orifice. Har'el and I dug frantically from the inside while Jennifer pulled material off the opening from the outside. Within a few minutes the hole was bigger and rounder than the one Jennifer had to contort herself to get through. "No problem," urged Har'el. "Go for it."

After Har'el tied the rope around my wrist, I followed Jennifer's lead. But because of my wider shoulders I had to back out and drop my left arm to my side while she pulled out my head and right arm first. Once my left shoulder was clear, I popped out pretty quickly, with Har'el pushing from behind.

Now I turned and peeked in at Har'el. "I just have to tie the pack to my end of the rope, then my wrist on the middle. Then you pull me out." As I waited, I surveyed the weather. It was warm, raining hard, and there was thunder in the distance. First time I peeked it was foggy, but now there was a cool wind and the fog was lifting a bit. Cold front, I thought. And with all that rain on the front end, probably the cool wind meant the rain would stop. But then I thought—this time of year? Maybe a hurricane. Cyclonic flow off the ocean would be cool and damp. Whatever it was, the water was pouring out of the sky and within a few seconds I was soaked.

Squatting outside, I untied the rope from my hand as Har'el tied the middle of the rope around his left wrist and his end of the rope to the backpack with our supplies and his brochures. As I staood to balance better while tugging on the rope, I felt the ground shift. I thought I must be dizzy from suddenly standing after the big struggle out of the cave on my stomach. An eerie groan from deep inside the earth startled us. Looking down I saw the tuft of earth over the cave collapse as if the earth gently exhaled a deep breath it had no more use for.

I looked down at the rope. The opening to the cave was gone, covered in rock, mud, torn shards of grass, tree roots, a real mess. My God! Har'el was in there. My heart was beating fast and my legs felt heavy and uncontrollable. I tugged with all my strength on the rope. It wasn't moving. Jennifer and I dropped to our knees and clawed at the dirt around the rope. He was just right there, I thought I said to Jennifer. She didn't even look up, digging and kicking. We loosened some of the mud, but the larger boulders were immobile. Its was futile as trying to dig a hole in a mountainside with bare hands. But we kept at it for five or ten minutes, getting no place but burning off adrenaline...

"Look, he's either safe in there or not. We're getting no place. We need serious help. A team with shovels and picks and machinery. At least he ought to have plenty of air in there. That shaft is miles long. "

"Unless he's crushed inside a small space, closed off from the back."

"Anything could have happened to him. But sitting here clawing at boulders is getting us noplace."

I looked around. The rain had slowed to a light drizzle blown by an occasional gust of wind. The fog was now a ragged cloud deck, maybe two-hundred feet over-head, with grey, furry tendrils reaching down and touching the treetops. I stood up and looked around. "Hey, we must be right about where we left." The remains of police activity were all around. Yellow sticky tape wrapped around trees admonishing people: POLICE LINE: DO NOT CROSS written in blocky black letters. Four wheel drive vehicle tracks crisscross the muddy grass. And the ground in a flat spot a few hundred feet away was littered with paper, probably fast food containers. I took off down the hill to the police line. To my right was an area where they cleared all the bushes and put a big yellow X on the ground. "Helicopter landing pad," I said, and nodded to Jennifer in that direction.

From the helipad we could see the parking lot. It was dusk on Sunday night and there was no one around, not even a parked or abandoned car. This was the lot we drove to last week. OK, can we find the mouth of the tunnel again? Yeah. Just up the hill, helipad on the left, and the rope was lying there—almost twenty feet of it, terminating in what now looked like the weathered side of a hill. Har'el was on the other end of that rope somewhere, possibly crushed or injured, but certainly alive.

Soaked, covered with red mud, sore from getting our ribs scraped exiting the damned cave, and now getting blown by a wet and cool gale, we headed down the dirt road we had driven in on just a few days before, on a warm sunny day for the AeroAstro spelunking fest. That happy little trail now looked sinister and ugly. Water ran down it, cutting steep bruises in the gravel we stepped over—not that we could be any wetter if we stepped in it. The trees were pushed horizontally by the wind and their leaves were partially blown off. The air howled around the branches and the remaining leaves brushed together and hissed at us.

As we walked and jogged, I tried to take stock. We have no money, and no ID. We have no food, and we haven't eaten much all day. We are out in a pretty serious storm without appropriate clothes and shelter. In a few minutes it will be totally dark, and being out in the country and with a thick cloud deck we'll be in virtual total blackness. On the plus side, the rain has subsided at least enough so we can see forward while there's anything to see, and the air is relatively warm. We won't die of hypothermia any time soon.

"Do you remember any farmhouse or store or anything?"

"I wasn't really looking. I think there was a house before we turned onto this road."

As total blackness seemed eminent and as my eyes tried to adjust, I realized that it won't be 100% dark. There is enough random electrical activity in the clouds, that there is a pretty frequent dim flash of what we as kids called heat lightning—cloud to cloud diffuse discharge. We walked, jogging when possible, for over an hour. I remembered that road was several miles long. At the cross street Jennifer thought she remembered that there was a house off to the left so we turned. After about ten minutes we spotted a little house but it looked dark. On closer inspection, it also looked pretty run down, roof and front porch both listing precariously. Is it dark? No, not totally. There's a lantern or a candle burning in one room.

"Any port in a storm?" I asked Jennifer, to make sure she's up for what we might find in there. She nodded. She looked tired and depressed, her eyes sunk deep and trying to catch her breath. We crawled over a large tree downed across the driveway and walked up the steps. My main concern is that they open up for us. At least we could use the phone. I shoved Jennifer in front of the door and stood aside while I rapped sharply on the ragged old door. If they see a woman, maybe they'll be less suspicious. An older woman comes to the tiny window in the door and looked out at Jennifer.

"What do you want?"

"Ma'am, I need to use your telephone. A friend of ours is trapped in a cave out there." Jennifer's voice cracks as she tried to be heard through the door and above the wind. The woman opened the door and for some reason I thought of Bonnie and Clyde. But it worked, and we were in, dripping on her floor, but sheltered at least for a few minutes from the elements. I felt we're like two wild animals allowed inside a human habitat for a few seconds before being shooed out. We were dirty, wet and a bit incoherent. Jennifer asked again about the phone.

"I'm sorry dear", the woman says as she looked me over. Her husband walked in.

"Phone's dead," he says. "See that." He motions out the window. In the flickering lightning I saw that the tree actually took down the power lines on its way. "No electricity. No phones. Real bad storm we've had today. Worst in years." Jennifer explained about the situation with Har'el while I stared around the room. The furniture was fifty years old at least and these people were older, late 60s early 70s. Except for a relatively new TV and VCR in one corner, the farmhouse looked like it could have during the Civil War. The woman looked totally confused by Jennifer's story. What can she really do for us under the circumstances?

"Do you guys have a truck or a car so we could go for help? This could be a life or death emergency. A man is trapped out there. We have to save him!" I figure a little urgency might help.

"We've got a truck, but can't get it out of the garage." The old man led me to another window. Against the spooky black landscape illuminated about every five or ten seconds by a glowing cloud, I saw a stone garage with no door. There's an old pickup handily blocked by a downed tree with about a four-foot diameter. "That tree's stood in the yard since I can remember. No storm coulda blown it over. Lightning musta hit it."

"You got a chain saw?"

"Even with a saw, gonna take you an hour to cut through it."

"How far's the next house?"

"Blackweld farm. Big place. Couple three miles. Gentleman farmers, you know? Probably they ain't there. Probably they's ridin' the storm out in the city."

"Where's the saw?"

The old man takes me out to a shed. He's carrying an ancient Coleman fishing lantern. He turns to me: "Name's Jessup."

"Hi. I'm Rick."

"You all scared my wife nearly t'death. But then ye' looked so sorry out there soakin' wet, she had ta' let ya' in."

"Thanks for taking the chance."

"Not a bit."

Jennifer caught up with us and within a few minutes, I was chomping through the wood with the power saw, Jennifer was chopping away branches with an axe, and Jessup was revving the motor of his pickup truck and shining his headlights on us as we worked. Still, he wasn't far off figuring an hour. A four foot thick tree trunk is a huge amount of wood, soaking wet, covered with small branches. And even cut up, it takes time to separate and move the pieces out of the way.

When we think we've cut a swath big enough for the truck to get out, Jessup ran into the house. "Martha! I'm takin' these two over ta' the Sheriff's."

Pulling out of the driveway Jessup nearly plowed into the other tree trunk we had crawled over to get to their front door. We screeched to a stop, sliding on wet gravel.

"Sorry, but that's not the one you showed me."

"Yeah yeah, forgot about it in all the excitement."

"Now what?"

"Now what's you gotta drive, kid. I can't see nothin' in the dark. We can git around it across the front here."

Jessup and I got out and Jennifer stayed in the middle of the old black pickup truck's torn vinyl bench seat. She and I were both fixated on the time that's passing. What's happening to Har'el? He must think we've abandoned him. That's what we were both thinking as the old man slowly made his way around the truck.

Eventually, and it seems like it was forever, we maneuvered around the tree and Jessup directed us to the Sheriff's office. It too looked dark as we pulled up, but when we got out I could hear a motor running and a light came on.

"Walter? I got me two strangers here. They say they've got an emergency."

Walter looked at us and said nothing.

"Our friend is trapped in a cave. We got out. We need help to dig him out."

Walter looked at Jennifer. "What's your name little lady?" She tells him. "Then who are you?" He looks at me, "Rick?" I nodded. "Shit, we been searchin' for you two for almost a week. Jessup, you never noticed all the police cars runnin' past your house all last weekend?"

"Nope. Can't really see the road from the house, and I don't take much notice."

"Helicopter?"

"Yeah, saw it. Figured you was lookin' for fires."

"Naw, we was lookin' for these two. And a third one. Hard-something."

"Har'el" says Jennifer.

"Yeah, that's it. He foreign or somethin'?"

"No. But he's stuck in the cave we were all lost in. Can you get some help out there ASAP? We marked the spot."

"I can try, but we got ourselves a bit of a hurricane blowin' through right now. People are busy takin' care of their own. Plus I ain't got no telephone. Lines are down all over the county."

"Radio?"

"Yeah, I got the police radio. And I got a civil defense radio. Trouble is, everybody's pretty busy. We got baby's bein' born in station wagons, kid swept away in a river."

"Sounds bad."

"Is bad. Yesterday they said storm's goin' out to sea and today was supposed to be hot and sunny. Instead we got this. Took everybody by surprise it did."

"At least you got shovels, picks and a flashlight? We'll go after him ourselves."

"Heck, I got that back at the house. Walter?" I started to wonder why Jessup always said that name like it was a question "We'll go back to the spot an' git started. You get what help you can."

"Jessup, you oughtn't be out tonight."

"I'm OK. You do your job, and ah'll tend to mahself."

"OK, have it your way. Where's the spot where you'all be diggin'?"

"About two hundred feet from the helipad" I piped up. "If help gets there, we'll see 'em. Please. Do what you can do. It's life or death."

"Everything's life or death tonight. I'll get on the radio right now."

We drove back to Jessup's place. He rounded up digging tools, another ancient lantern and some jackets. Martha figured to come with us rather than spend any more time in the dark house alone, and we all four squeezed back into the pickup and sped down the dirt road to the parking lot. No phones, I think to myself. What a joke. Now I'm in the right world and a damned storm knocks a phone pole over and we might as well be in a different universe—one that never evolved. Total screwup on the weather forecast too, sounds like. Wonder how many people that hurt? I was in an angry mood. I guess I felt like I left Har'el in that hole to die, while I just stood there and somehow I was just not doing my job very well. The four of us in a pickup with some shovels is all we can muster?

Martha brought with her a tin of home made cookies, and Jennifer and I ate nearly all of them before we got to the spot. We set in to digging with picks and shovels while Martha sat in the car with the motor running and headlights on. Jessup tried to be helpful but it was hard for more than two to dig at the same time anyway.

We were at it maybe thirty minutes when the Sheriff's car drove up. Behind him was a fire truck, a rescue squad truck and two private cars. Within a few minutes we had power tools going and ten people standing around. They brought emergency lighting and while they dug, I took a break to dry off in Jessup's truck.

It was almost midnight when we broke through. By now the rain had stopped and a Medevac helicopter and a Channel Four news team joined the excitement. Two men using a sort of portable hoist like you use to pull the engine out of a VW lifted a small boulder, and behind it was Har'el covered in debris and sprawled out in a bizarre geometry of arms, legs, neck and head. But at least all the parts were connected. Broken bones were a given, but was he alive? The two paramedics virtually jumped on him to get vital signs and to set up respiration.

"Weak pulse." says one.

"I think we've got a weak pulse" repeats the other into his radio, a microphone on a gooseneck planted near his face. "Out of our way." They brought in a bright yellow backboard with red straps, slid Har'el onto it and whisked him over to the helicopter where there was a waiting IV and a portable monitor all warmed up and ready to go.

"Where are you taking him?" I yell over the rotor blades spinning up.

"County Hospital. Sheriff'll take you there." And with that, they were gone. The other rescue workers started clearing out. One of them came up to me.

"What happened to you all?"

"We got lost in a cave. We finally found a way out, but Har'el didn't get out in time."

"You been in a cave four days and nights?"

I thought about the question. "Yeah. Plenty of water in there. We had food in a pack." The pack! I run back to the cave with Jessup's lantern and the rescue volunteer following me. "Where's the rope? Right there—but where's the other end, with the pack attached?"

Jennifer pulled the rope into her hand. The other end was still buried in rubble.

"We'll get to that in the mornin'" the volunteer said to me.

"Yeah. We need to get to the hospital."

"I'll drive you." Jennifer and I said goodbye and thanks to Jessup and Martha. We told them we think they saved Har'el's life. Jessup just smiled a little and says "You get along to the hospital now. We'll take care here."

We rode to the County hospital virtually in silence. All I can think is Har'el has been injured before—skateboarding, hotdogging on skis. He can take it. I feel sorry. He looked so bad. I imagine him in a body cast and how long and uncomfortable that kind of recovery is bound to be. Poor guy. Luckily he is only twenty-three years old. He'll be fine, maybe in six months or so.

The hospital was an island of light on a dark, drizzly night in a rural county with no electricity and no phones. We pulled up to the front door. Locked. Of course. We drove around to the brightly lit ER entrance and our volunteer escort says he's gotta get home. I thanked him for the ride, and he disappeared through the parking lot. Jennifer and I look like we could be patients in the ER, but nobody pays much attention to us. They are swamped. We go from nurse to doctor to nurse, and finally someone points us to admitting. There's a nurse staffing the night shift and we give her Har'el's name.

"Not on my list."

"How current's your list? He just came in the last hour or so. Via helicopter" I say, wondering if maybe the helicopter didn't make it through the lousy weather. Maybe it diverted?

"I don't have him. I think the helicopter crew is around someplace. Ask around, probably they're in the cafeteria." She pointed us around the back of her desk.

Sure enough, the helicopter driver and the paramedic were drinking coffee and sharing a sandwich. Their two VHF walkie-talkies were sitting on the table in front of them squawking nonsense sounds.

I noticed the wall clock in the tiny basement cafeteria. 2:40 a.m. I didn't feel especially tired despite the two or three hour hilly trek through the cave and the all night struggle to get help for Har'el. Adrenaline I guess. After all that, I felt pretty good. We made it back and the rain soaked ordeal is about over except for finding Har'el among all the confusion. This small rural hospital is overrun with admissions injured in the hurricane. But according to the AM radio in the rescue volunteer's Subaru the storm is now moving out to sea. The helicopter pilot and paramedic don't seem to recognize Jennifer or me. Uncharacteristically Jennifer interrupted their conversation. "Where is Har'el?"

The two big men look up at her. "Huh?"

"The guy you pulled out of the cave."

"Who are you?"

Jennifer points alternately at me and herself. "We're the other two who were lost with him."

The paramedic hesitates but the pilot explains "'T's ok Guy. I recognize 'em from the pictures. He's Rick and she's Jennifer."

"Oh yeah." Guy nods to the adjacent two chairs. "Nice to meet you after three, four days searching the woods for y'all."

Guy's accent was that combination southern friendly plus northern precision you hear at colleges like Virginia Tech and Duke. "Every rescue company in the valley's got your picture. You know'f they called off the search yet? Officially I mean. You people been checked out yet?"

"They're a little busy upstairs. Anyway, we're fine."

"Glad uh that. Y'all looked a little finer in your pictures though. Coffee?"

Jennifer sits down and stares at Guy. "Where's Har'el?"

Guy looks at Jennifer, then the pilot who is flexing his wrist in circles swirling his coffee, then at me, then back at Jennifer. "Ah'm not really the one to be saying this, but ah'm afraid your friend didn't make it."

"Didn't make it?

"Time we got to him, he was purdy shocky. Erratic weak pulse. Ah couldn't git a BP on him at all. We trah'd everything. He didn't live to make it back here." Guy's voice recovers a matter of fact tone "No pathologist on duty here nights. ER Doc signed off on the case. Trauma I think he wrote. We took him down stairs. Ah'm really sorry. It happens sometahms."

Chapter 12

The realization of tragedy comes only slowly to me. I can't feel it 'till I'm alone with it in my bed at night or biking to work in a bright sunrise. My eyes tear and I blink behind my Oakleys and it's just the cool autumn air. Despite ninety-five decibels of DC 101 plugged into my ears by my banana yellow digital Sports Walkman, suddenly I hear nothing but the drilling of the rain on our cave, the boulder smashing against the black walls of the narrow twisting passage and my own voice urging us to press on. The next seven hours replay like a nightmare that in reality is over in three minutes.

How humane and heroic we all are. We dispatch police and firemen and paramedics into the teeth of a storm. Even a rural Blue Ridge county owns $ million in rescue gear. Heroic astronauts climb atop billion dollar rockets. They risk their lives for the sake of the final frontier. Supreme ego leveraged on the backs of the rest of us. One percent of the Shuttle's budget would buy the satellite weather forecast system taken for granted by Gib and Hilary. A hurricane is then no more a surprise than tomorrow's sunrise. How noble is a society which has spent $500 billion on space but whose elderly can be cut off from heat, light, communication and transportation by the felling of one tree? Our leaders indulge a religious worship of our crews orbiting uselessly among the stars. Actually they are only two hundred miles away from us, still light years away from those distant stars, same as all three billion of us down here.

As they arch overhead angelic in white suits with red and blue NASA emblems we dig frantically, futilely, against the pouring rain and oozing mud, with just our fingernails, trying to rescue a friend. A heartbeat falters as we slog undetected and unassisted down a long muddy park service road, carve a mammoth tree trunk into firewood and rouse a policeman who also is without even a telephone. The seconds fly by. We crawl like bugs lured into the sweet glycerine of our inappropriate technologies.

We had to go back to the Blue Ridge later in the day Har'el died to officially identify his body and to officially end the search. And, at the end of a long bureaucratic conclusion to our adventure, unofficially to bring shovels to the spot where Har'el was cut from our rope and carried into the Medivac helicopter.

The severed rope led into a mound of mud blocking the cave. Rescuers had moved the heavy boulders and after a few minutes' digging and poking we could peer into the dark, rocky corridor. I flicked the rope and gently pulled it free of a few rocks and some congealed clay. Jennifer reeled it back to us easily. Too easily. The other end, which Har'el had tied to his backpack filled with souvenirs, was a soggy knot of frayed nylon fibers. Jennifer and I looked at each other. Had the goat gnawed it off? Bounded by a wall of tightly compressed, irregular rocks, the opening we had dug, actually punched through with the handle of the shovel, was only about the diameter of my shoulder. I pressed my right eye and a flashlight against the hole. Looking fifty or so feet farther in I saw nothing. Maybe the water carried it back down the chute? But how did the rope come unraveled? We couldn't get through the hole to go search for it, and anyway we didn't have much appetite for another underground adventure, particularly in an unstable cave.

God no longer lives in the simplistic naivete of the surging spring, the rock and the tree. We have promoted the concept to a height few of us can reach and God's relevance to daily life doesn't even rival the nightly news let alone the price of a gallon of gasoline. Space is accessible to even fewer of us—those who have the clout to control hundreds of millions of dollars and who can devote a working lifetime to its pursuit.

Yet all those miracles space gave to the world of Gib and Hilary, global mobile voice and data links, perimeter monitoring and asset tracking affordable to the simplest farmers, accurate medium term weather forecasting, instantaneous emergency notification and location, traffic reporting and feedback on our environmental impact, we have had the technology to do it all for over decades. And at a price to the public, for all of it, dwarfed by just one year of astronaut flights on Space Shuttle. When we want those blessings, they are ours to enjoy. Meanwhile they stand at our side, the ghosts of decisions not taken. The outdoors people who needlessly die, the crops we never sow nor harvest, the fossil materials we carelessly burn and then inhale, the time we spend in traffic instead of in our living rooms with our families. They are the casualties in our holy war waged in the name of the exalted and deified mental model of space we lug around with us

No pain, no gain? Only get what you pay for? American physics, without its $15 billion super collider, is banished to second class status? Does exercise have to hurt to help? Or can a casual walk with friends, outside among trees, birds and sky be admitted to having real benefit?

Is a space program without five or ten shuttle flights a year necessarily uninteresting? Not to farmers whose crops are saved or to families whose loved ones are rescued, or to people everywhere who can live in a cleaner, safer environment.

The telephone is a vital, integral part of the world society. A telephone is as mundane as a drawer of socks. Space as spectacle means space as a realm limited to techies and heroes. Know any utilitarian spectacles? It's ok to ask: "What has my space dollar done for me lately?" same as we ask of every dollar spent on medicine, groceries or tuition. When we hold space up to a common standard of everyday utility and relevance instead of its potential for religious inspiration and the pursuit of some hypothetical human destiny among the stars, only then we will trigger its renaissance.

Popping open the front door, I coast through the AeroAstro office in Herndon. I swing the Trek 930 into its parking place behind my desk and turn to sit down in front of a screenful of e-mail. Pausing just a second I shed the iridium Oakleys, wipe some road grit out of my eyes and look up at a red-mud stained coil of climbing rope hanging over my whiteboard, one end cleanly cut, the other ragged and unraveled.

Index

O

P

Q

R

W

X

Y

Z